# Lecture Notes in Electrical Engineering

Volume 622

The book series *Lecture Notes in Electrical Engineering* (LNEE) publishes the latest developments in Electrical Engineering—quickly, informally and in high quality. While original research reported in proceedings and monographs has traditionally formed the core of LNEE, we also encourage authors to submit books devoted to supporting student education and professional training in the various fields and applications areas of electrical engineering. The series cover classical and emerging topics concerning:

- Communication Engineering, Information Theory and Networks
- Electronics Engineering and Microelectronics
- Signal, Image and Speech Processing
- Wireless and Mobile Communication
- Circuits and Systems
- Energy Systems, Power Electronics and Electrical Machines
- Electro-optical Engineering
- Instrumentation Engineering
- Avionics Engineering
- Control Systems
- Internet-of-Things and Cybersecurity
- Biomedical Devices, MEMS and NEMS

For general information about this book series, comments or suggestions, please contact leontina.dicecco@springer.com.

To submit a proposal or request further information, please contact the Publishing Editor in your country:

**China**

Jasmine Dou, Associate Editor (jasmine.dou@springer.com)

**India, Japan, Rest of Asia**

Swati Meherishi, Executive Editor (Swati.Meherishi@springer.com)

**Southeast Asia, Australia, New Zealand**

Ramesh Nath Premnath, Editor (ramesh.premnath@springernature.com)

**USA, Canada:**

Michael Luby, Senior Editor (michael.luby@springer.com)

**All other Countries:**

Leontina Di Cecco, Senior Editor (leontina.dicecco@springer.com)

**** Indexing: The books of this series are submitted to ISI Proceedings, EI-Compendex, SCOPUS, MetaPress, Web of Science and Springerlink ****

More information about this series at http://www.springer.com/series/7818

Zhongliang Jing
Editor

# Proceedings of the International Conference on Aerospace System Science and Engineering 2019

Springer

*Editor*
Zhongliang Jing
Shanghai Jiao Tong University
Shanghai, China

ISSN 1876-1100 ISSN 1876-1119 (electronic)
Lecture Notes in Electrical Engineering
ISBN 978-981-15-1772-3 ISBN 978-981-15-1773-0 (eBook)
https://doi.org/10.1007/978-981-15-1773-0



This Springer imprint is published by the registered company Springer Nature Singapore Pte Ltd.
The registered company address is: 152 Beach Road, #21-01/04 Gateway East, Singapore 189721, Singapore

# Contents

# Spacecraft Attitude Maneuver Using Fast Terminal Sliding Mode Control Based on Variable Exponential Reaching Law

**Yue-Dong Wu, Shu-Fan Wu, De-Ren Gong, Ze-Yu Kang and Xiao-Liang Wang**

**Abstract** As a typical multi-input and multi-output nonlinear system, spacecraft faces the challenges of rapid maneuvering at a large angle and can control rapidly as well as stably. However, there are several problems in the sliding mode control method commonly used in the aerospace field, such as slow convergence, singularity, and chattering. Taking the consideration of aforementioned conditions, this paper proposes a non-singular fast terminal sliding mode control based on variable exponential reaching law (NFTSMC-VERL). Firstly, using a continuous hyperbolic tangent function instead of symbolic function, variable exponential reaching law is designed to solve the problem of chattering in exponential reaching law. The reachability and finite time convergence of this method are proved by theory. Secondly, a non-singular fast terminal sliding mode (NFTSM) surface based on satellite attitude dynamics is designed as a sliding mode function. Thirdly, the controller is designed by combining the VERL with the NFTSM function. Finally, not only has the simulation results showed the good tracking performance of the spacecraft attitude control system, but also the chattering and singularity phenomenon of the traditional sliding mode controller is avoided by the designed controller. Besides, the robustness of designed controller is improved, compared with the traditional sliding mode controller.

**Keywords** Rapid attitude maneuver · Variable exponential reaching law · Fast convergence rate · Non-singularity

## 1 Introduction

In recent years, with the continuous development of global space science and technology, the development of spacecraft has reached an unprecedented height. The

Y.-D. Wu · S.-F. Wu (✉) · D.-R. Gong · Z.-Y. Kang · X.-L. Wang
School of Aeronautics and Astronautics, Small Satellite Technology Center, Shanghai Jiao Tong University, Shanghai, China
e-mail: shufan.wu@sjtu.edu.cn



Z. Jing (ed.), *Proceedings of the International Conference on Aerospace System Science and Engineering 2019*, Lecture Notes in Electrical Engineering 622,
https://doi.org/10.1007/978-981-15-1773-0_1

rapid attitude maneuvering and stable control of payload as one of the key technologies of spacecraft are increasingly concerned by scholars [1]. However, the attitude dynamics of the spacecraft is nonlinear. Therefore, the attitude control problem of the spacecraft becomes a nonlinear control problem [2], which has a serious impact on large-angle rapid maneuvering and rapid stability [3, 4].

Sliding mode control (SMC) has been used by many scholars for precise spacecraft attitude control [5–8] because of its strong robustness. However, the traditional sliding mode control has problems such as infinite convergence of tracking error, so the terminal sliding mode (TSM) which can converge in finite time was proposed and gained extensive attention [9]. On the basis of TSM, combining with reaching law, the quality of controller reaching motion and sliding mode motion can be improved as a whole, which has attracted the attention of scholars.

In reference [10], exponential reaching law was combined with non-singular terminal sliding mode (NTSM) for the first time, which improved the convergence speed of the system and shortened the adjustment time on the basis of overcoming singular problems. However, as the sign function of exponential reaching law will cause chattering when approaching the sliding surface, it is particularly important to improve the exponential reaching law and weaken chattering. In reference [11], the improved saturation function was used to replace the sign function, which weakened the chattering and then combined with the sliding surface. In reference [12], based on the exponential reaching law, an adaptive variable speed exponential reaching law was designed, and then combined with the NTSM surface. In reference [13], the chattering was suppressed by changing the constant velocity reaching term into the variable speed reaching term, which was combined with the sliding surface.

However, in the design process of the controller mentioned above, the methods to solve exponential reaching law chattering were too complex, and the selection of sliding surface did not consider the fast convergence. Therefore, when designing the controller, this paper has proposed to use the continuous hyperbolic tangent function instead of the symbol function to suppress the chattering of exponential reaching law. Secondly, when choosing the sliding surface, the NFTSM surface is selected based on the satellite attitude described by the modified Rogers parameter (MRP) [14]. Then, the controller is designed by combining the VERL with the NFTSM. Finally, under the same conditions, not only has the simulation results showed the good tracking performance of the spacecraft attitude control system, but also the chattering and singularity phenomenon of the traditional sliding mode controller is avoided by the designed controller.

This paper is organized as follows: Spacecraft attitude kinematics and dynamics is introduced in Sect. 2. Variable exponential law is presented in Sect. 3. A NFTSM-VERL controller for attitude fast maneuver is designed in Sect. 4. The simulation results are depicted in Sect. 5. Finally, some conclusions are given in Sect. 6.

## 2 Spacecraft Attitude Kinematics and Dynamics

The spacecraft attitude dynamics can be composed of kinematic equations and dynamic equations [15]. In the body coordinates, based on the MRP attitude description, the nonlinear attitude kinematics and dynamics equations of a rigid spacecraft external disturbances can be summarized as [16]

$$\dot{\sigma} = G(\sigma)\omega \tag{1}$$

$$\dot{\omega} = J^{-1}(-\omega^{\times} J\omega + u + d) \tag{2}$$

where $J \in R^{3\times 3}$ is the inertia matrix of spacecraft. $\sigma = [\sigma_1, \sigma_2, \sigma_3]^T \in R^3$ is the MRP vector, $\omega = [\omega_1, \omega_2, \omega_3]^T = [\omega_x, \omega_y, \omega_z]^T \in R^3$ is the angular velocity vector.$u = [u_1, u_2, u_3]^T = [u_x, u_y, u_z]^T \in R^3$ denotes the three-axis control torque of the spacecraft, which can be generated by the flywheel, thruster, magnetic torquer, and other actuators of the satellite. $d = [d_1, d_2, d_3]^{\mathrm{T}} = [d_x, d_y, d_z]^{\mathrm{T}} \in R^3$ denotes bounded disturbance vector acting on rigid body spacecraft.

The notation $\omega^{\times}$ is used to denote a skew-symmetric matrix generated by $\omega$, having the following structure

$$\omega^{\times} = \begin{bmatrix} 0 & -\omega_z & \omega_y \\ \omega_z & 0 & -\omega_x \\ -\omega_y & \omega_x & 0 \end{bmatrix} \tag{3}$$

The matrix function $G(\sigma)$ is given by

$$G(\sigma) = \frac{1}{4}[(1 - \sigma^{\mathrm{T}}\sigma)I + 2\sigma\sigma^{\mathrm{T}} + 2\sigma^{\times}] \tag{4}$$

$I \in R^{3\times 3}$ denotes a $3 \times 3$-dimensional identity matrix.

$d$ is a bounded function satisfying

$$||d|| \le \rho \tag{5}$$

where $\rho$ denotes the upper bound of disturbances.

# 3 Variable Exponential Reaching Law Design and Analysis

## 3.1 Variable Exponential Reaching Law Design

The motion of sliding mode variable structure control consists of two stages: The first stage is that the moving point tends to switch from an arbitrary initial state to a switching surface, which we called this phase as approaching motion. The second stage is the stage in which the moving point moves along the switching surface toward the stable point, which we called this phase as sliding mode motion. Since there is no limit to the trajectory in the approaching process, the approach of the reaching law can be used to improve the dynamic quality of the approaching motion. In the 1990s, Academician Gao Wei-bing first proposed the concept of exponential reaching law [17]. Exponential reaching law is as follows:

$$\dot{s} = -ks - \varepsilon\,\mathrm{sgn}(s) \quad k > 0, \varepsilon > 0 \tag{6}$$

where $\dot{s} = -ks$ is the exponential reaching term, and $\dot{s} = -\varepsilon \mathrm{sgn}(s)$ is the constant velocity reaching term. Using the exponential reaching law can not only make the moving point reach the sliding surface quickly at the constant velocity and exponential speeds, but also make the speed of the moving point reach the sliding surface very small and improve the dynamic quality of the system.

However, since the moving point reaches the sliding surface, the constant velocity approaching term $\dot{s} = -\varepsilon \mathrm{sgn}(s)$ is not equal to zero, so a high-frequency buffeting band with bandwidth $2\varepsilon$ is formed near the origin, which increases the burden on the controller. Therefore, the authors propose a method of using a hyperbolic tangent function $\tanh(\mu s)$ to replace the sign function $\varepsilon \mathrm{sgn}(s)$ . The variable exponential reaching law is as follows:

$$\dot{s} = -ks - \varepsilon \tanh(\mu s) \tag{7}$$

where $k > 0, \varepsilon > 0,\ \mu$ is a positive constant. Since $\tanh(\mu s)$ is close to $\mathrm{sgn}(s)$ as $\mu$ increases, $\mu$ is limited to a range of intervals.

In terms of the variable exponential reaching law, the system motion point can reach the sliding surface quickly at both exponential and variable speeds, ensuring that the sliding mode motion moves along the designed sliding surface. At the same time, when the system motion point reaches the sliding surface, the exponential term reaching term $\dot{s} = -ks$ is close to 0, and the shifting approach term $\dot{s} = -\varepsilon \tanh(\mu s)$ is getting smaller, so that the bandwidth $2\varepsilon$ of the switching band is continuously reduced, and finally the system motion mode is stabilized at the origin, thereby suppressing chattering caused by sliding mode variable structure control.

### 3.2 Reachability Analysis

The reachability condition of the sliding mode is

$$s \cdot \dot{s} < 0 \tag{8}$$

Substituting (7) into (8), it follows that

$$\begin{aligned} s \cdot \dot{s} &= s(-ks - \varepsilon \tanh(\mu s)) \\ &= -ks^2 - \varepsilon s \tanh(\mu s) \\ &= -ks^2 - \varepsilon |s \tanh(\mu s)| < 0 \end{aligned} \tag{9}$$

Therefore, this VERL satisfies the reachability condition.

### 3.3 Finite Time to Reach the Sliding Surface Analysis

The system is divided into two stages from the initial state to the sliding surface $s = 0$: The first stage is from the initial position to the reaching switching surface. Since s is large, the exponential reaching term $\dot{s} = -ks$ plays a leading role in this process, ignoring the influence of the second term and calculating the time $t_1$ required for the first stage.

$$\dot{s} = -ks \tag{10}$$

Integrating (10), it follows that

$$s = s(0)e^{-kt} \tag{11}$$

Thus,

$$t_1 = -\frac{1}{k} \ln \frac{s}{s(0)} \tag{12}$$

The second stage is from reaching switching surface to the switching surface. Since $s$ is small, $\dot{s} = -\varepsilon \tanh(\mu s)$ plays a leading role in the process, ignoring the influence of the first item and calculating the time $t_2$ needed for the second stage.

$$\dot{s} = -\varepsilon \tanh(\mu s) \tag{13}$$

Since the second stage s is small, if $\tanh(\mu s)$ is approximated to $\mu s$, then (13) becomes

$$\dot{s} = -\varepsilon\mu s \tag{14}$$

Integrating (14), it follows that

$$s = s(0)e^{-\varepsilon\mu t} \tag{15}$$

Thus,

$$t_2 = -\frac{1}{\varepsilon\mu}\ln\frac{s}{s(0)} \tag{16}$$

Considering the time obtained in Eqs. (12) and (16) was obtained by ignoring one of the expressions of the reaching law, the total time required to reach the switching surface from the initial position should be

$$\begin{aligned} t < t_1 + t_2 &= -\frac{1}{k}\ln\frac{s}{s(0)} - \frac{1}{\varepsilon\mu}\ln\frac{s}{s(0)} \\ &= \ln\left(\frac{s}{s_0}\right)^{-\frac{1}{k}-\frac{1}{\varepsilon\mu}} \end{aligned} \tag{17}$$

Since $k$, $\varepsilon$, $\mu$ are fixed constants, the VERL can reach the sliding surface in a finite time.

## 4 NFTSMC-IRL for Attitude Dynamics

### 4.1 Sliding Mode Surface Function Design

In order to avoid the singular problem of the traditional sliding mode and improve the convergence speed of the sliding mode system, Yang [18] proposed the NFTSM method and proved its characteristics of fast convergence in a limited time and avoiding singularity problem. The structure of NFTSM is given as

$$s(t) = x + k_1'\mathrm{sgn}^{\alpha_1'}x + k_2'\mathrm{sgn}^{\alpha_2'}\dot{x} = 0 \tag{18}$$

in which $k_1' > 0, k_2' > 0, \alpha_1' > \alpha_2'$, and $1 < \alpha_2' < 2$.

Based on (18), the structure of NFTSM with MRP and angular velocity can be defined as

$$s = \sigma + \Lambda_1\mathrm{sgn}^{\Gamma_1}\sigma + \Lambda_2\mathrm{sgn}^{\Gamma_2}\dot{\sigma} \tag{19}$$

where $s$ denotes three-dimensional sliding surface, and the involved matrices are represented by

$$\Lambda_1 = \mathrm{diag}(\lambda_{11}, \lambda_{12}, \lambda_{13}) \quad \Gamma_1 = \mathrm{diag}(\gamma_{11}, \gamma_{12}, \gamma_{13})$$
$$\Lambda_2 = \mathrm{diag}(\lambda_{21}, \lambda_{22}, \lambda_{23}) \quad \Gamma_2 = \mathrm{diag}(\gamma_{21}, \gamma_{22}, \gamma_{23}) \tag{20}$$

with $\lambda_{1i} > 0$, $\lambda_{2i} > 0$, $\gamma_{1i} > \gamma_{2i}$, $1 < \gamma_{2i} < 2$, for every $i = 1, 2, 3$.

$\mathrm{sgn}^{\Gamma_1}\sigma$ is a vector defined as

$$\begin{aligned}\mathrm{sgn}^{\Gamma_1}\sigma &= [\mathrm{sgn}^{\gamma_{11}}\sigma_1, \mathrm{sgn}^{\gamma_{12}}\sigma_2, \mathrm{sgn}^{\gamma_{13}}\sigma_3]^{\mathrm{T}} \\ &= [|\sigma_1|^{\gamma_{11}}\mathrm{sgn}\sigma_1, |\sigma_2|^{\gamma_{12}}\mathrm{sgn}\sigma_2, |\sigma_3|^{\gamma_{13}}\mathrm{sgn}\sigma_3]^{\mathrm{T}}\end{aligned} \tag{21}$$

$\mathrm{sgn}^{\Gamma_2}\dot{\sigma}$ is a vector defined as

$$\begin{aligned}\mathrm{sgn}^{\Gamma_2}\dot{\sigma} &= [\mathrm{sgn}^{\gamma_{21}}\dot{\sigma}_1, \mathrm{sgn}^{\gamma_{22}}\dot{\sigma}_2, \mathrm{sgn}^{\gamma_{23}}\dot{\sigma}_3]^{\mathrm{T}} \\ &= [|\dot{\sigma}_1|^{\gamma_{21}}\mathrm{sgn}\dot{\sigma}_1, |\dot{\sigma}_2|^{\gamma_{22}}\mathrm{sgn}\dot{\sigma}_2, |\dot{\sigma}_3|^{\gamma_{23}}\mathrm{sgn}\dot{\sigma}_3]^{\mathrm{T}}\end{aligned} \tag{22}$$

The time derivatives of $\mathrm{sgn}^{\Gamma_1}\boldsymbol{\sigma}$ and $\mathrm{sgn}^{\Gamma_2}\dot{\boldsymbol{\sigma}}$ are given by

$$\begin{aligned}\frac{d}{dt}(\mathrm{sgn}^{\Gamma_1}\sigma) &= \Gamma_1\mathrm{diag}\big(|\sigma|^{\Gamma_1 - I}\big)\dot{\sigma} \\ \frac{d}{dt}(\mathrm{sgn}^{\Gamma_2}\dot{\sigma}) &= \Gamma_2\mathrm{diag}\big(|\dot{\sigma}|^{\Gamma_2 - I}\big)\ddot{\sigma}\end{aligned} \tag{23}$$

## 4.2 Control Law Design

In the design of the control law, the system will follow the variable exponential reaching law to reach the non-singular fast terminal sliding surface.

From (19), its time derivative can be given by

$$\dot{s} = \dot{\sigma} + \Lambda_1\Gamma_1\mathrm{diag}\big(|\sigma|^{\Gamma_1 - \mathrm{I}}\big)\dot{\sigma} + \Lambda_2\Gamma_2\mathrm{diag}\big(|\dot{\sigma}|^{\Gamma_2 - \mathrm{I}}\big)\ddot{\sigma} \tag{24}$$

Jointing (7) and (24), it follows that

$$\dot{\sigma} + \Lambda_1\Gamma_1\mathrm{diag}\big(|\sigma|^{\Gamma_1 - \mathrm{I}}\big)\dot{\sigma} + \Lambda_2\Gamma_2\mathrm{diag}\big(|\dot{\sigma}|^{\Gamma_2 - \mathrm{I}}\big)\ddot{\sigma} = -ks - \varepsilon\tanh(\mu s) \tag{25}$$

From (1), its time derivative can be given by

$$\ddot{\sigma} = \frac{d}{dt}(G(\sigma)\omega) = \frac{1}{2}\big[(\sigma^{\mathrm{T}}\omega)I + \sigma\omega^{\mathrm{T}} - \omega\sigma^{\mathrm{T}} - \omega^{\times}\big]\dot{\sigma} + G(\sigma)\dot{\omega} \tag{26}$$

In order to simplify (26), define

$$H(\sigma, \omega) = \frac{\partial\dot{\sigma}}{\partial\sigma} = \frac{1}{2}\big[(\sigma^{\mathrm{T}}\omega)I + \sigma\omega^{\mathrm{T}} - \omega\sigma^{\mathrm{T}} - \omega^{\times}\big] \tag{27}$$

Substituting (27) into (26), it follows that

$$\ddot{\sigma} = H(\sigma, \omega)\dot{\sigma} + G(\sigma)\dot{\omega} \tag{28}$$

Jointing (1), (2), (25), and (28), it follows that

$$\begin{aligned} &\left[\mathrm{I} + \Lambda_1 \cdot \Gamma_1 \cdot \mathrm{diag}(|\sigma|^{\Gamma_1 - \mathrm{I}}) + \Lambda_2 \cdot \Gamma_2 \cdot \mathrm{diag}(|\dot{\sigma}|^{\Gamma_2 - \mathrm{I}}) \cdot H(\sigma, \omega)\right]\dot{\sigma} \\ &\quad + \Lambda_2 \cdot \Gamma_2 \cdot \mathrm{diag}(|\dot{\sigma}|^{\Gamma_2 - \mathrm{I}}) \cdot G(\sigma) \cdot J^{-1} \cdot (-\omega^{\times} \cdot J \cdot \omega + d) + ks + \varepsilon \cdot \tanh(\mu s) \\ &= -\Lambda_2 \cdot \Gamma_2 \cdot \mathrm{diag}(|\dot{\sigma}|^{\Gamma_2 - \mathrm{I}}) \cdot G(\sigma) \cdot J^{-1} u \end{aligned} \tag{29}$$

In order to simplify (29), define

$$M = \Lambda_2 \cdot \Gamma_2 \cdot \mathrm{diag}(|\dot{\sigma}|^{\Gamma_2 - \mathrm{I}}) \tag{30}$$

Substituting (5) and (30) into (29), we could get the following control law

$$u = -JG^{-1}M^{-1}\left\{ \begin{array}{l} \left[\mathrm{I} + \Lambda_1 \cdot \Gamma_1 \cdot \mathrm{diag}(|\sigma|^{\Gamma_1 - \mathrm{I}}) + M \cdot H(\sigma, \omega)\right]\dot{\sigma} \\ + M \cdot G(\sigma) \cdot J^{-1} \cdot (-\omega^{\times} J\omega + \rho \mathrm{I}) + ks + \varepsilon \cdot \tanh(\mu s) \end{array} \right\} \tag{31}$$

## 5 Numerical Simulation

In order to verify the effectiveness of the algorithm proposed in this paper, both the traditional sliding mode control method and the algorithm proposed in this paper are simulated under the same physical conditions. The specific physical parameters are as follows [16]:

The inertia matrix of a rigid spacecraft

$$J = \begin{bmatrix} 420 & 18 & -15 \\ 18 & 256 & -12 \\ -15 & -12 & 618 \end{bmatrix} (\mathrm{kg\ m^2}) \tag{32}$$

The initial angular velocity and MRP orientation are, respectively, set to

$$\sigma(0) = [0.3\, 0.2\, -0.3]^{\mathrm{T}},\ \omega(0) = [0\, 0\, 0]^{\mathrm{T}} \tag{33}$$

The external disturbance torque is assumed to be

$$d = \begin{bmatrix} 2 + 2\sin(0.042t) \\ 1 + 3\sin(0.042t) \\ 3 + 2\sin(0.042t) \end{bmatrix} \times 10^{-3}(\mathrm{N\ m}) \tag{34}$$

The coefficients in control law (31) are set to

$$\begin{aligned} &\Lambda_1 = 0.9\mathrm{I}\ \Lambda_2 = 8\mathrm{I} \\ &\Gamma_1 = 0.7\mathrm{I}\ \Gamma_2 = \mathrm{I} \\ &K_1 = 0.8\mathrm{I}\ \mu = 1000 \\ &\rho = 0.005\ \varepsilon = 0.001 \end{aligned} \tag{35}$$

The simulation results of the proposed method are shown in Fig. 1.

By analyzing Fig. 1, we can get the algorithm proposed in this paper which solves the problem of chattering. In addition, under the same external conditions, the convergence time of the proposed algorithm is much less than that proposed in the reference [15], which indicates that the proposed algorithm in this paper has better fast convergence.

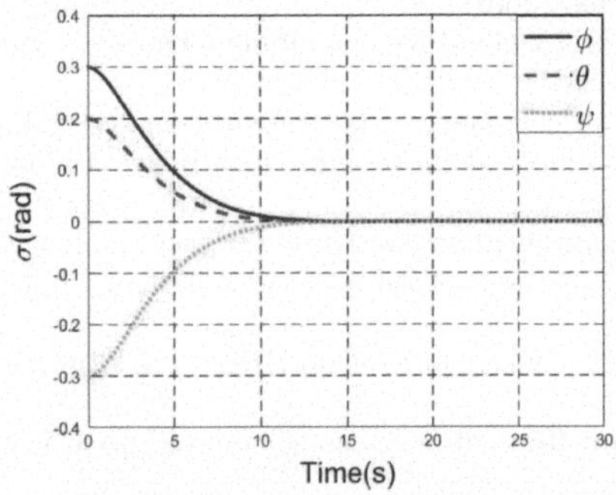


(a) Time history of MRP vector

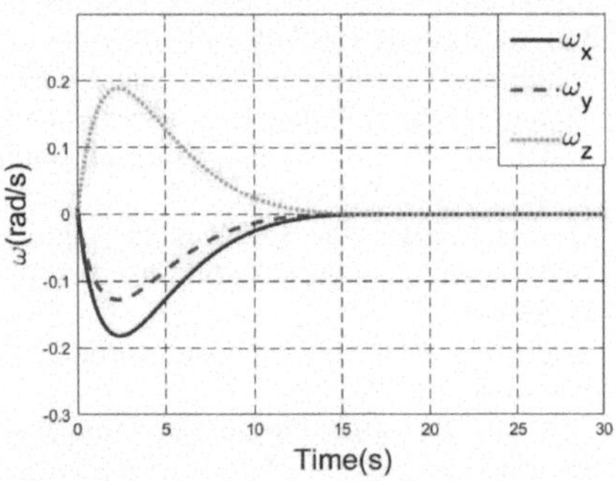


(b) Time history of angular velocity

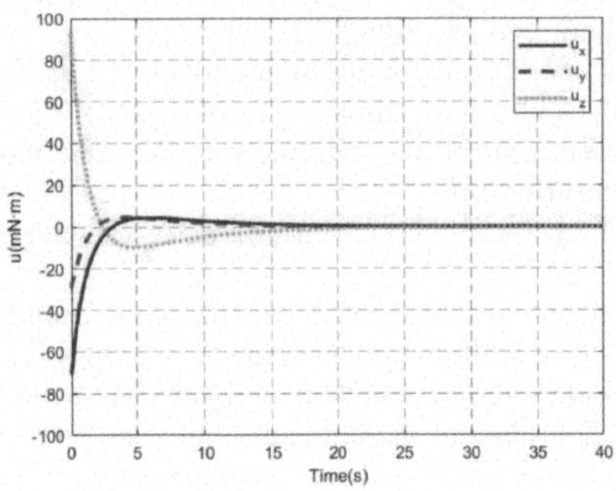


(c) Time history of control input

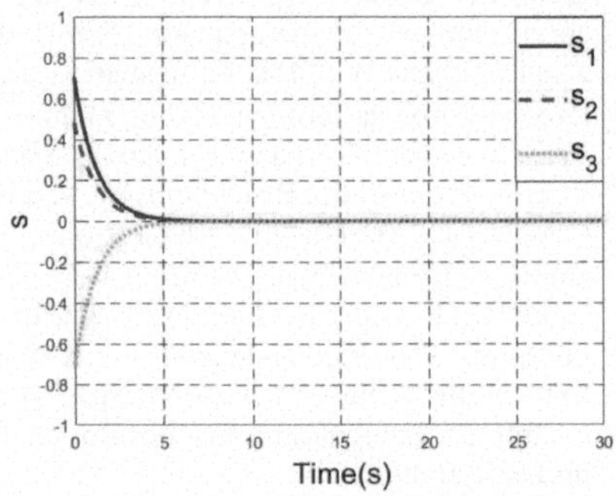


(d) Time history of sliding surface

**Fig. 1** Time history of spacecraft attitude

## 6 Conclusion

In view of the challenges of spacecraft rapid maneuver and rapid stable control in space flight, combined with the idea of reaching law, this paper proposes a design method of NFTSMC-VERL. On the basis of overcoming the chattering and singularity problems, the speed of approaching the sliding surface is improved, and the adjustment time is shortened. The simulation results show that the designed algorithm has the characteristics of fast convergence and strong anti-interference ability, which is more suitable for spacecraft attitude control.

## References


1. Wu HX (2014) Research and prospect on the control theory and method in the engineering. Control Theory Appl (12)
2. Scrivener SL, Thompson RC (1994) Survey of time-optimal attitude maneuvers. J Guid Control Dyn 17(2):225–233
3. Sun DQ, Wu HX (2005) Characteristic modelling and adaptive fuzzy control method of MIMO higher-order linear time-varying systems. J Astronaut 26(6)
4. Lei YJ, Tan SP, Liu YW (2010) Spacecraft control method for fast attitude maneuver and stabilization. Chin Space Sci Technol V30(5):48–53
5. Slotine JJ (1984) Sliding controller design for non-linear systems. Int J Control 40:421–434
6. Vadali SR (1986) Variable-structure control of spacecraft large-angle maneuvers. J Guid Control Dyn 9(2):235–239
7. Ding H, Li SH (2006) Sliding mode control of spacecraft attitude with finite-time convergence. In: Proceedings of the 6th world congress on intelligent control and automation, Dalian, China, pp 830–834
8. Yildiz Y (2007) Sliding mode neuron controller for uncertain systems. IEEE Trans Industr Electron 54:1676–1684
9. Ji X, Hu DB (2011) Progress in terminal sliding mode control research. Control Instr Chem Ind 38(9)
10. Zhang WW, Wang J (2012) Nonsingular terminal sliding model control based on exponential reaching law. Control Decis 6:909–913
11. Zhang JH, Liu B, Jiang Z (2018) The application of sliding mode control based on improved approaching law in manipulator. Chin High Technol Lett 28(06):62–74
12. Zhu LF, Zhang W (2018) Permanent magnet low-speed synchronous motor adaptive variable-rated nonsingular terminal sliding model control. Control Eng China 25(1):112–116
13. Yang Y, Huang F, Zhang JM (2018) Sliding mode control of electronic throttle value with variable Exponential Reaching Law. Veh Eng 239(06):40–47
14. Crassidis JL, Markley FL (1996) Sliding mode control using modified rodrigues parameters. J Gudiance Control Dyn 19:1381–1383
15. Gong DR, Li W, Chai YL et al (2011) Little satellite rapid attitude maneuver using nonsingular fast terminal sliding mode control. In: Small Satellite Technology Seminar
16. Li C, Wang Y, Xu L et al (2010) Spacecraft attitude stabilization using optimal sliding mode control. In: International Symposium on Systems & Control in Aeronautics & Astronautics, pp 1085–1089
17. Gao WB (1996) Theory and design method for variable sliding mode control. Science Press, Beijing
18. Yang L, Yang J (2011) Nonsingular fast terminal sliding-mode control for nonlinear dynamical systems. Int J Robust Nonlinear Control 21(16):1865–1879

# Ultra-Rapid Direct Satellite Selection Algorithm for Multi-GNSS

**Cheng Chi, Xingqun Zhan, Tong Wu and Xin Zhang**

**Abstract** Global navigation satellite systems (GNSS) provide many more satellites than ever before. However for applications extremely sensitive to power consumption, not all satellites can be incorporated into the measurement vector, either because of the sheer computation overload or for purpose of power saving. These applications include but are not limited to unmanned aerial system (UAS), flying cars, and asset tracking. Thus, satellite selection methodology should be used to obtain subset satellites with good geometry. Recently, a downdate method proposed in receiver autonomous integrity monitoring (RAIM) can be used for reference in satellite selection, although RAIM and GNSS positioning are quite different. In this paper, a DOP-based ultra-rapid satellite selection methodology, the direct satellite selection (DS) method, is proposed according to the downdate method. Furthermore, to compensate for the shortcomings of the DS method, a constrained direct satellite selection (CDS) method is then proposed by adding error monitoring and restrictive conditions. The two algorithms are examined for precision performance and computational performance. Simulations show the DS method performs about three orders of magnitude faster than the recursive method, which is the existing fastest DOP based algorithm, with 0.25 increase in DOP on average relevantly when excluding the number of satellites from 42 to 8. And the CDS method performs about two orders of magnitude faster than the recursive method with only 0.15 increase in DOP even when excluding satellites from 42 to 6. Consequently, both the two methods have much lower computation time than all the existing DOP-based algorithms, with very little reduction in precision. Comparatively, the DS method has lower computational load and the CDS method has higher precision. Thereby, the algorithms proposed in this paper successfully address the satellite selection problem in two scenarios; the CDS method fits in fast satellite selection and high precision situation; the DS method can be employed in some extremely speed demanding circumstances.

C. Chi · X. Zhan (✉) · T. Wu · X. Zhang
School of Aeronautics and Astronautics, Shanghai Jiao Tong University, No. 800, Dongchuan Road, Shanghai 200240, China
e-mail: xqzhan@sjtu.edu.cn



Z. Jing (ed.), *Proceedings of the International Conference on Aerospace System Science and Engineering 2019*, Lecture Notes in Electrical Engineering 622,
https://doi.org/10.1007/978-981-15-1773-0_2

**Keywords** Direct satellite selection · Ultra-rapid · Multi-constellation · GDOP · Recursive method

## 1 Introduction

The development of multiple constellations provides a higher positioning performance by increasing the visible probability of satellite. It is common for a GNSS receiver observing a huge number of satellites in some ideal areas. Generally, more observed satellites lead to higher positioning accuracy. But satellite tracking and position calculations are time and resource consuming for GNSS receivers. As the number of observed satellites increasing, more tracking channels are needed, which means better processor and higher cost. Meanwhile, GNSS positioning precision depends more on satellite geometry, rather than number of satellites. Thus, it is impossible for a civilian grade GNSS receiver to take all-in-view satellites into account. Satellite selection method should be taken to address this issue.

There are many methods for selecting a set of satellites to use for GNSS positioning solution. The basic principle is to minimize the dilution of precision (DOP). Some authors employ satellite measurement error into account, such as using protection level (VPL and HPL) as the judgment condition. While, no matter which principle is used, the ideas are similar and universal. Hence, we just apply DOP as the judgment condition in this paper.

Most of the researchers who use DOP as the judgment condition often use the geometric dilution of precision (GDOP). A simple method is the brute force, which enumerates all the permutations to find the optimal one, namely brute force. Obviously, this method is the most time-consuming. There are several methods to optimize the GDOP-based satellite selection. Some authors take efforts to optimize the global search process using heuristic algorithms. One such method is to reduce global search time by using genetic algorithms (GAs) [1]. Similarly, [2] uses a chaotic particle swarm optimization (CPSO) method to reduce global search. These methods can be more optimal than other optimization algorithms, but they are both more time-consuming than others. As to more efficient methods, inverse lemma is used to simplify the GDOP calculation [3, 4]. It uses recursive thinking to generate each n-subset from its immediate predecessor by deleting a single satellite for one time. These methods can be induced as the recursive method and are much time reducing.

The following shows the computational load of existing DOP-based algorithms. Firstly is the brute force, which is the most time-consuming. Supposing that the observed satellite number is $n$, and the desired number is $k$, the brute force method should calculate the GDOP, which contains a huge matrix calculation, for $C_n^k$ times. If set $n$ as 30 (It is a general situation in GPS-GLONASS-BDS system), and set $k$ as 10, the calculation complex will be 30045015 times. It is hard for a civilian grade GNSS receiver operating this algorithm in a short time. The CPSO method given by [2] reduced the calculation time to about 37.5% of the brute force method, but it is still time-consuming. The recursive method proposed by [4] operated a time-saving performance. It reduced the calculation times to $\sum_{i=k}^{n} i$. For $n = 30$ and $k = 10$, the calculation time is 410, which is much smaller than the above method. But the computational time will grow quadratically when the number of observation increases.

Besides, a number of authors employ alternative performance measures beyond DOP [5]. A kind of them is the volume of the polyhedron method [6, 7], which calculate the volume of the polyhedron formed by satellites and the user, or employ elevation angle and azimuth angle to divide satellites into several block and allocate with some logic [8, 9]. While, the polyhedron-based algorithm can reach sub-optimal but time-consuming, and the latter performs faster than the recursive method but failed in precision.

Consequently, existing satellite selection algorithms either have unsatisfactory time performance, or failed in precision. Currently, [10] provided a downdate method in autonomous integrity monitoring (RAIM) for satellite-based augmentation systems (SBAS), with protection level. It can allow us to directly sort the satellite from the all-in-view matrix, rather than calculating each subset matrix. To address the satellite selection problem in low computational load demanding situation, we apply the downdate method to the geometric condition, namely direct satellite selection method, using GDOP as the basis for selecting. Then we propose a constrained direct satellite selection method to increase accuracy and enlarge usage scope. Comparisons with other algorithms are also given in this paper. Besides, this paper will discuss how many satellites should be deleted to get appropriate GDOP value in a global sight.

## 2 Direct Satellite Selection Method

Since the downdate method proposed by [10] was proposed to select subset satellites with protection level, it is not suitable for the multi-GNSS DOP-based satellite selection condition. Formulas need to be redefined.

For a single constellation navigation algorithm, the geometry matrix $\boldsymbol{G}$ contains four columns. But in multi-constellation system, the most existing constellations are not synchronized with each other, the geometry matrix is defined as follows [11] (supposing there are $L$ unsynchronized constellations).

$$\boldsymbol{G} = \begin{pmatrix} -e_{rx,1}^{1} & -e_{ry,1}^{1} & -e_{rz,1}^{1} & 1 & 0 & 0 & \cdots & 0 \\ \vdots & \vdots & \vdots & \vdots & \vdots & \vdots & & \vdots \\ -e_{rx,1}^{m_1} & -e_{ry,1}^{m_1} & -e_{rz,1}^{m_1} & 1 & 0 & 0 & \cdots & 0 \\ -e_{rx,2}^{1} & -e_{ry,2}^{1} & -e_{rz,2}^{1} & 0 & 1 & 0 & \cdots & 0 \\ -e_{rx,2}^{2} & -e_{ry,2}^{2} & -e_{rz,2}^{2} & 0 & 1 & 0 & \cdots & 0 \\ \vdots & \vdots & \vdots & \vdots & \vdots & \vdots & & \vdots \\ -e_{rx,L}^{m_L} & -e_{ry,L}^{m_L} & -e_{rz,L}^{m_L} & 0 & 0 & 0 & \cdots & 1 \end{pmatrix} \quad (1)$$

where the second of the two subscripts is the constellation number, from 1 to $L$; and the superscript is the satellite number within the $k$th constellation, from 1 to $m_k$. And the *GDOP* is formed by the $\boldsymbol{G}$ matrix.

$$GDOP = \sqrt{\mathrm{trace}\left[\left(\boldsymbol{G}^T \cdot \boldsymbol{G}\right)^{-1}\right]} \tag{2}$$

Readers should notice that the GDOP cannot directly represent the positioning error. The GDOP is just a conversion from satellite measurement error to positioning error. It is more suitable to employ protection level to quantify the positioning error [12]. While, in this paper, we only use GDOP to get the optimal geometry.

The weight coefficient matrix H is given below

$$\boldsymbol{H} = \left(\boldsymbol{G}^T \cdot \boldsymbol{G}\right)^{-1} \tag{3}$$

Rather than computing the inverse matrix, we can use the inverse lemma to get a recursive formula [10]

$$\boldsymbol{H}^{(i)} = \boldsymbol{H} + \frac{\boldsymbol{S}_i \cdot \boldsymbol{S}_i^T}{p_{i,i}} \tag{4}$$

$$\boldsymbol{S} = \left(\boldsymbol{G}^T \cdot \boldsymbol{G}\right)^{-1} \cdot \boldsymbol{G}^T \tag{5}$$

$$\boldsymbol{P} = \boldsymbol{I} - \boldsymbol{G} \cdot \left(\boldsymbol{G}^T \cdot \boldsymbol{G}\right)^{-1} \cdot \boldsymbol{G}^T \tag{6}$$

where $\boldsymbol{H}^{(i)}$ is the weight coefficient matrix with the $i$th satellite removed, $\boldsymbol{S}_i$ is the $i$th column of the $/bmS$ matrix, $p_{i,i}$ is the element of line $i$, column $i$ of the $\boldsymbol{P}$ matrix. Observe the formula (4), we can see that

$$h_{j,j}^{(i)} = h_{j,j} + \frac{s_{j,i}^2}{p_{i,i}} \tag{7}$$

It indicates that when deleting a single satellite, the increase in *GDOP* can be obtained by calculating the term $s_{j,i}^2/p_{i,i}$. From formula (2), when deleting the $i$th satellite, we can use the following formula to express the increase of $GDOP^2$

$$C_{GDOP^2}^{(i)} = \left(\sum_{j=1}^{L+3} s_{j,i}^2\right)/p_{i,i} \tag{8}$$

We can calculate each $C_{GDOP^2}^{(i)}$ of all observed satellite, and select the satellites with k largest $C_{GDOP^2}^{(i)}$ values.

We name this method as the direct satellite selection (DS) method. The process is explained as follows.

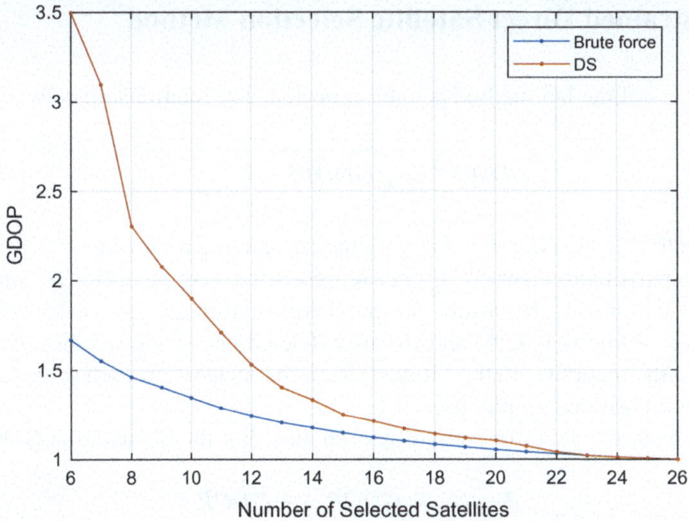


**Fig. 1** Precision performance of the direct satellite selection method, compared with the brute force method

(1) Use formula (1) to compute the $\boldsymbol{G}$ matrix with all-in-view satellites, and get $\boldsymbol{H}$, $\boldsymbol{S}$ and $\boldsymbol{P}$ matrix and $C_{GDOP^2}^{(i)}$ from formula (3), (5), (6) and (8), respectively.
(2) Find $k$ maximum $C_{GDOP^2}^{(i)}$ and take the corresponding $k$ satellites as the desired subset satellites.

We use a group of data collected in Shanghai to examine this algorithm. The data contains three constellations, including GPS, GLONASS and BDS, and the observed satellite number is 27. We use both the DS method and brute force to operate the data. Figure 1 shows the curve of *GDOP* changing with selected satellite number $k$.

According to Fig. 1, the DS method is well agreed with the brute force method if $k$ has a large value. But when $k$ is getting smaller, the error between the DS method and the brute force method is gradually increasing, and the error dramatically increases when $k$ has down to a small value. It is impermissible to use the subset satellites as working satellites for positioning when the $k$ falls to small. In short, the experimental result indicates that the DS method does not fit in long step judgment. This method should be corrected by some restrictive conditions.

Generally, the DS method is a preferred algorithm, because it has a low calculation burden and high precision in a short step. If users are demanding for a fast satellite selection methodology and do not need to exclude too many satellites, this method is the most appropriate.

## 3 Constrained Direct Satellite Selection Method

The weakness of the DS method is to be expected. The relationship between $GDOP$ and $C^{(i)}_{GDOP^2}$ is

$$GDOP^{(i)} = \sqrt{GDOP^2 + C^{(i)}_{GDOP^2}} \quad (9)$$

where $GDOP^{(i)}$ is the $GDOP$ value with the $i$th satellite removed.

The standard of judgment $C^{(i)}_{GDOP^2}$ can represent the increase of $GDOP^2$ unbiasedly only when $k = n - 1$. To ensure the precision of formula (9), matrices must be refreshed according to (4), (5) and (6) after deleted one single satellite. Actually, it is similar to the recursive method if we refresh the matrixes in each step. And it will add a lot of algorithm complexity.

We can examine the error caused by each step. For the $l$th step, the $GDOP$ error is

$$E_{GDOP_l} = GDOP_l^{(i)} - GDOP_l \quad (10)$$

where the $GDOP_l^{(i)}$ is the biased $GDOP$ value calculated by equation (9).$GDOP_l$ is the $GDOP$ value with $l$ selected satellites deleted, it is given by

$$GDOP_l = \sqrt{\text{trace}\left[\left({G_l}^T \cdot G_l\right)^{-1}\right]} \quad (11)$$

The absolute value of $E_{GDOP_l}$ will continuously increase as steps go on and can get a large value. For the data used before, $E_{GDOP_l}$ rises to $-0.545$ when $k = 10$, and the value is about 31% of its $GDOP_l$. It obviously performs an enormous influence to the DS step for tracking the optimal geometry. Thus, the error $E_{GDOP_l}$ should be limited in an appropriate range. Consider that the gradient of $GDOP$ descent rises with the increase of $l$, it is not suitable to set a constant threshold to judge $E_{GDOP_l}$. Notice that large $E_{GDOP_l}$ will influence the minimal $C^{(i)}_{GDOP^2}$ judgment, we employ $C^{(i)}_{GDOP^2}$ as the threshold. The constraint condition is given as follows:

$$E_{GDOP_l} > \min_{1 \le i \le n-l} C^{(i)}_{GDOP^2} \quad (12)$$

The inequality searches $C^{(i)}_{GDOP^2}$ of the remaining satellites for each step. If the constraint condition is satisfied, matrixes should be refreshed according to (4), (5) and (6) to reset $E_{GDOP_l}$ and enter into a new epoch, this process continues until $l = k$. By the way, operators can also set the threshold as the above elements divided by $d$ for higher precision:

$${E_{GDOP_l}}^2 > \min_{1 \le i \le n-l} C^{(i)}_{GDOP^2}/d(d \ge 1) \quad (13)$$

The value of $d$ is related to the trade-off between precision and computational efficiency. In the following sections, our examinations will mainly focus on inequality (13), since it already has a good performance.

In order to examine the error value $E_{GDOP_l}$, we need to obtain the $GDOP_l$ in each step. Instead of computing the inverse matrix, which is much time-consuming, inversion lemma can be used to optimize the calculation [4]. As a deformation of formula (4), the recursive formula is as follows:

$$\boldsymbol{H}^{(i)} = \boldsymbol{H} + \frac{\boldsymbol{H} \cdot \boldsymbol{g}_l \cdot \boldsymbol{g}_l^T \cdot \boldsymbol{H}}{1 - \boldsymbol{g}_l^T \cdot \boldsymbol{H} \cdot \boldsymbol{g}_l} \tag{14}$$

where $\boldsymbol{g}_l$ is the row vector deleted from the $\boldsymbol{G}$ matrix, and $GDOP_{l-1}$ is the $GDOP$ value in the last step. This formula can recursively compute the new $GDOP$ when one single satellite is deleted by the DS algorithm. We name this method as constrained direct satellite selection (CDS). The core structure of the CDS method is represented in the scheme in Fig. 2, and the methodology is explained step by step in the following:

(1) Use formula (1) to compute the $\boldsymbol{G}$ matrix with all-in-view satellites, and get $GDOP$, $\boldsymbol{H}$, $\boldsymbol{S}$ and $\boldsymbol{P}$ matrix and $C^{(i)}_{GDOP^2}$ from formula (2), (3), (5), (6) and (8) respectively.
(2) Find minimum $C^{(i)}_{GDOP^2}$ from all remaining satellites and exclude the corresponding satellite in this step.
(3) Compute the new $GDOP$ of this step by formula (14) and (2).

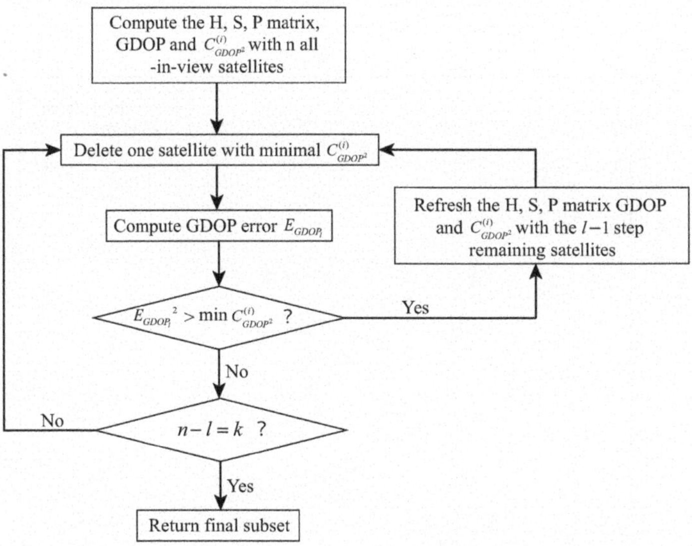


**Fig. 2** Schematic overview of the constrained direct satellite selection method

(4) Use formula (9) to calculate the biased $GDOP_l^{(i)}$. Clear that this term is not the true *GDOP*. It is just a roughly estimated value computed by corresponding elements of the ***S*** and ***P*** matrix.
(5) Compute the GDOP error $E_{GDOP_l}$ of the $l$th step by formula (10).
(6) Judge the constraint condition given by (12), or (13).

    (a) If the constraint condition is satisfied, refresh the elements included in step 1 with the remaining satellites of last selection step.
    (b) If the constraint condition is not satisfied, it indicates that the error is not large enough to enormously influence the selection process. So the process can be continued to step 7.

(7) Examine whether the process reaches the desired satellite number. If not, jump to step 2 to delete more satellites. Or return the final subset satellites.

Using the same data as before, we can get a performance comparison between the DS method and the CDS method. The result is shown in Fig. 3. It indicates that the CDS method addressed the problem of the DS method and keep closer to the brute force. While the result discussed before is just based on a single group of data. The detailed test will be given in the later section.

Inevitably, the CDS method invested in more computing time. If the algorithm refreshed elements for $\xi$ times, the big matrix computation (which is the main part of computing time) contains two parts. On the one hand, the recursive formula ($6\times6$ matrix for 3 constellations) is calculated for $n-k+\xi$ times. On the other hand, the ***H***, ***S*** and ***P*** matrix ($n-l$ dimensions) is calculated for $\xi$ times. Normally, the refreshing times $\xi$ is a small number, which is just 4 when selecting 6 satellites out

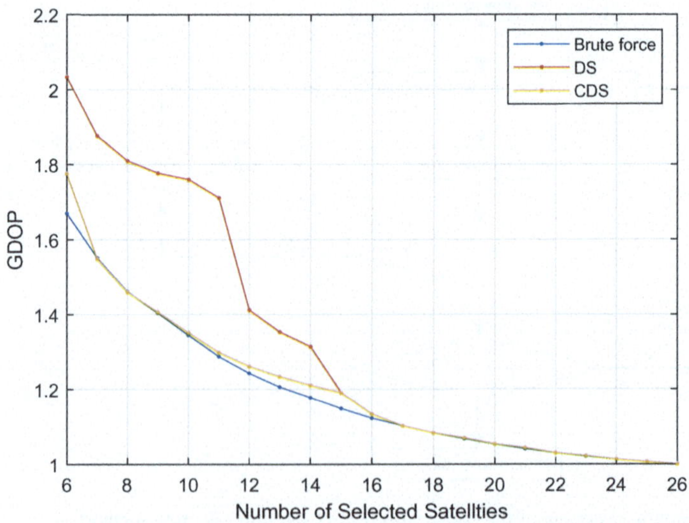


**Fig. 3** Performance comparison between the DS method and the CDS method

of 27 observations (Using the data above). Though its computation is larger than the DS method, it is much smaller than recursive method. The time performance will be examined in detail later.

## 4 Experiments and Discussion

In this section, an assessment of the proposed methodology will be given by comparing the results with the other methods, including the brute force and the recursive method.

Firstly, to obtain the performance of the proposed methods and the others when setting different selected number of satellites, we observed the satellites, including GPS, GLONASS and BDS, at Shanghai for 24 h, with few signal occlusion. We operated four satellite selection algorithms every 10s, setting the desired number to 16, 12 and 10 separately, and compared them in each measurement time. Results are shown in Fig. 4.

According to Fig. 4a, when set $k = 16$ the DS method and the CDS method both have good performance compared with the recursive method. The DS method seldom has protruding values. While, according to Fig. 4b, there are more different values between the two proposed methods. According to Fig. 4c, there are a lot of dramatic increases when using the DS method. Meanwhile, the CDS method is far more close to the recursive method, and there is no sharply increasing point during the whole testing time, though it performed worse than $k = 16$.

To examine the relationship between the desired number of satellites and the performance of two proposed methods, we compared them to the recursive method separately in each desired number. The result is shown in Fig. 5. It is clear that the performance of the two is similar when the number of selected satellites are above 14. The performance of the DS method get worse when the number of selected satellite falls down below 14, and it sharply increases when the number is below 8. Though the performance of CDS method also get worse when the number of satellites decreasing, it is still much better than the DS method.

In order to verify the universality of the algorithms, we simulated satellites situation by MATLAB in the global range, also included the three constellations. We examined the DS method and CDS method globally in 24 h and compared them to the recursive method. Both desired number of satellites $k$ are set as 10. Results are shown in Fig. 6. We can see that the $\Delta GDOP$ of CDS method is far smaller than that of DS method. Even the maximum value of CDS method approximately equals to the minimum value of the DS method.

We can conclude that the two proposed methods perform similar precision with each other when setting $k$ as a big number. While, the CDS performs much better precision than the DS method when $k$ is small.

Then we tested the computational load. It was discussed in the previous sections. In the simulation, we will give the computing time through specific tests. Note that

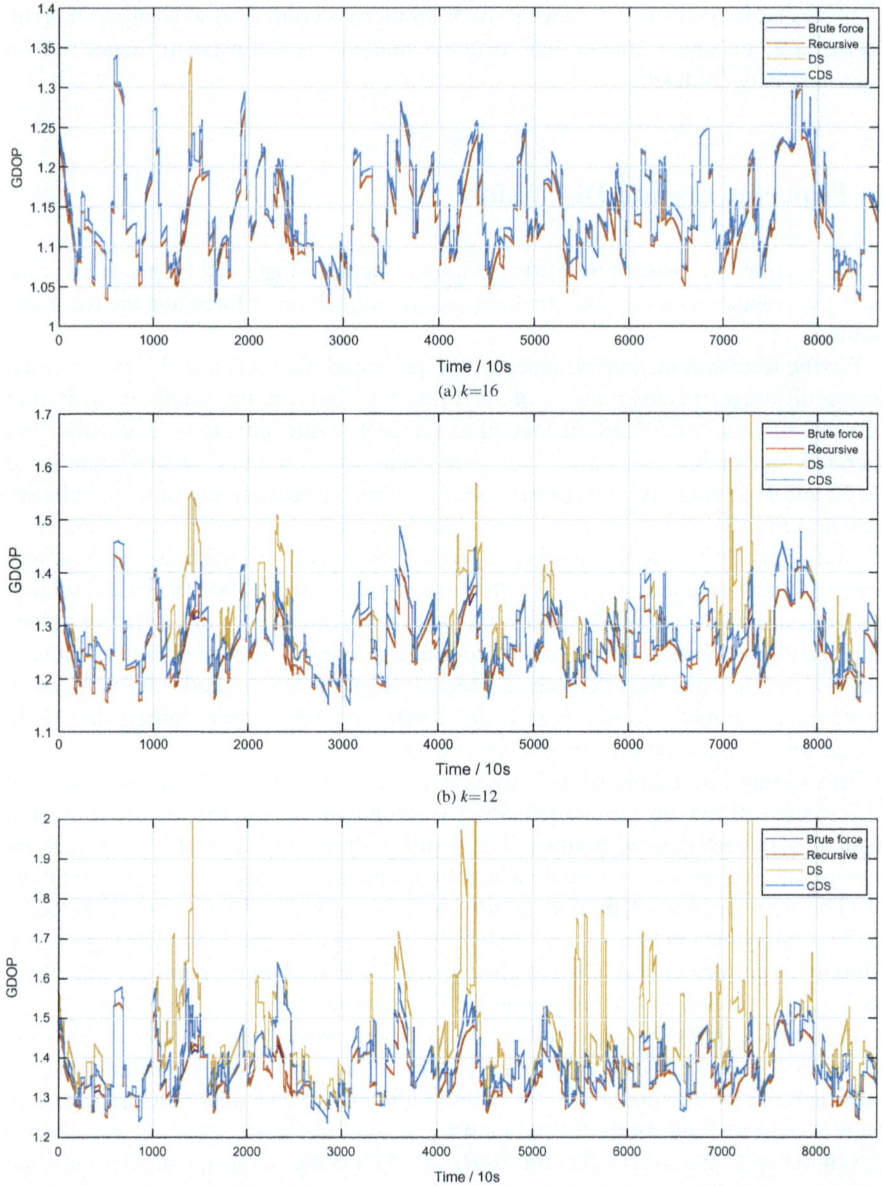


(a) $k$=16

(b) $k$=12

(c) $k$=10

**Fig. 4** Comparison of *GDOP* using the brute force, recursive method, proposed DS method and CDS method for 24 h. The recursive method and the brute force are almost overlapped, so only compare the proposed method with the recursive method is sufficient

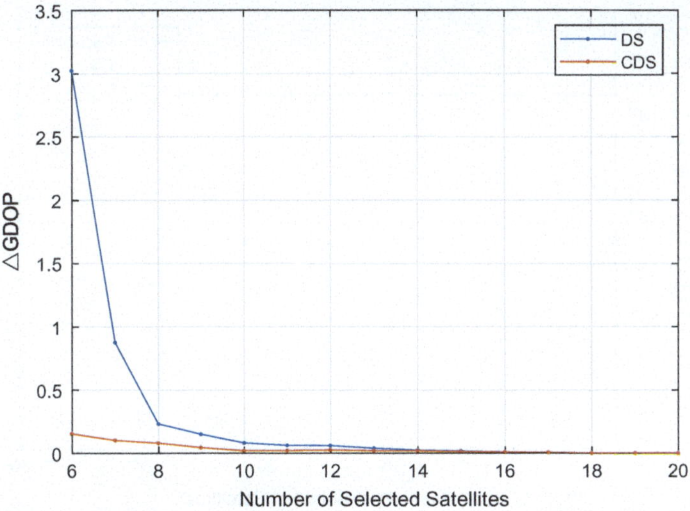


**Fig. 5** Mean error of 24 h relative to the recursive method

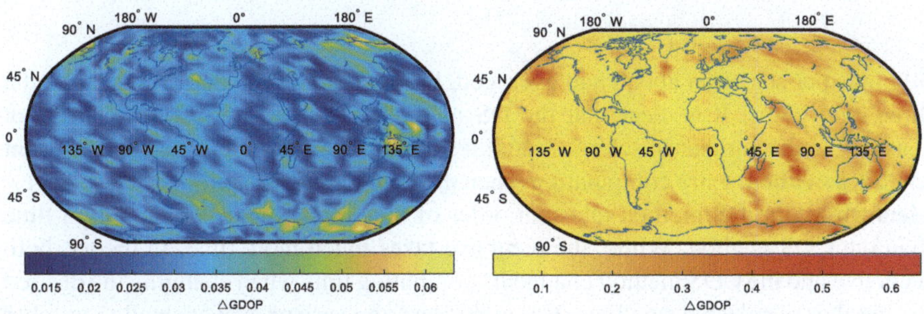


(a) The GDS method

(b) The DS method

**Fig. 6** Global error relative to the recursive method ($k = 10$). Note that the error of the GDS method is about one order of magnitude lower than that of the DS method

all tests in this section are operated on the MATLAB environment, and the hardware setup includes an Intel Core i7-8700 CPU @ 3.20 GHz and a 16 G RAM @ 2666 MHz.

Following the discussion in the previous sections, we can get the time complexity of the brute force method is $O(C_n^k)$. And the time complexity of the DS method and the CDS method is constant order $O(1)$ and linear order $O(n)$. While the recursive is square order $O(n^2)$. Apparently, the computational load of brute force is several orders of magnitude higher than the recursive method and the two proposed methods in almost all cases. It is no need to compare the brute force method with others. Thus, we will not invest the brute force method into comparison.

Firstly, we simulated 42 observed satellites and used the recursive method, DS method and CDS method separately to select 6 to 30 satellites. All data are computed

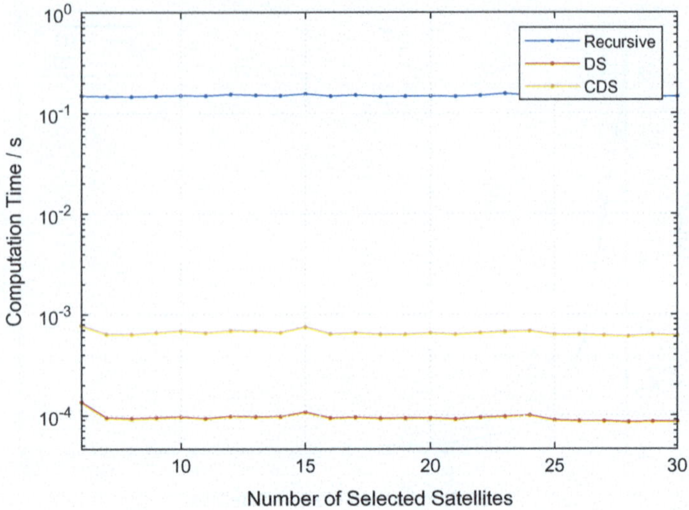


**Fig. 7** Average computation time per selection when selecting 6–30 satellites out of 42 observed satellites

30 times per selection epoch and averaged. Figure 7 shows the performances of the three algorithms. It is clear that the DS method takes the shortest time. It is not sensitive to the number of selected satellites, because it can directly select the subset satellites, without extra computation when the number of selected satellites changes. Meanwhile, the DS method takes an order of magnitude higher computation time than the CDS method. While, the recursive takes much more time. In theory, both recursive and the CDS method change its computing time when changing the desired the number of selected satellites. But in this test, the variances over satellite number are so small that it is drowned in computer time error due to unstable frequency. Anyway, it is highly discernable to judge the computing time between the three algorithms.

In Fig. 8, the computation time versus the number of observed satellites is given. We set the desired satellite number as 10, and generated 18–42 observed satellites in each epoch. It is shown in the figure that sequence of the three methods does not change in this case, and both the DS method and the CDS method are also not sensitive to the number of observed satellites. While, as for the recursive method, computation time increase as the observed satellites increasing. It is reasonable, because it must compute bigger matrix for several times when all-in-view number rising.

Consequently, the DS method performed the best in the computation load case, and the CDS method did not cause so much extra computation time compared to DS. However, the recursive method performed a much longer computation time, especially when the number of observed satellites is great. Hence, the best choice is the DS method in terms of efficiency, followed by the CDS method.

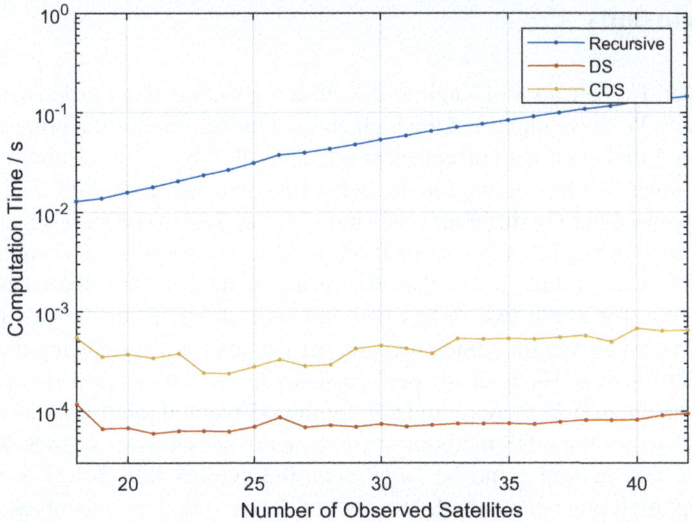


**Fig. 8** Average computation time per selection when selecting 10 satellites out of 18–42 observed satellites

According to the performance tests shown above, it is safe to say that either the DS method or the CDS method is very fast satellite selection algorithm, with decent performance. To comparison, the DS method has the most efficient computational capacity but has unsatisfactory precision when the number of subset satellites is low. The CDS method has excellent performance covering all satellite quantity circumstances, with a larger computational time than the DS method. We suggest users use the CDS method when operating satellite selection because of its high adaptability and fast computational performance, unless very short computational time is demanded.

The methods can also be used to automatically optimize GDOP value. When considering on the desired precision, it is more suitable to focus on maximal (or alert) GDOP, rather than the number of desired satellites, since the number of satellites and the value of GDOP are not absolutely corresponding and for fixed satellite numbers, satellites with better geometry can have a smaller GDOP value. Just as what Fig. 4a–c indicate, take the most optimal one, recursive method, as an example, the GDOP changes with time changing when setting the stable number of subset satellites. Hence, in the actual operation, manufacturers can take the discussion above for guidance to decide how the number of subset satellites fits the GDOP value.

As we mentioned before, satellite measurement error is not included in the geometric selection method. Thus, users should take care of the multipath, ionosphere error, and others for some satellite. The DS and the CDS method both prefer to select satellites with low elevation angle, which may have huge measurement error, such as multipath. The kind of satellites with low performance should be excluded before operating the satellite selection algorithms.

## 5 Conclusions

In this paper, two ultra-rapid satellite selection algorithms were proposed, namely DS and CDS. We have identified the high computational time of the brute force and demonstrated that even the current most efficient DOP-based algorithm, recursive, could cost several orders of magnitude higher time to achieve satellite selection. We have also pointed out the different precision range of the two proposed methods.

Simulations showed that the DS method could save computational time for about three orders of magnitude lower than the recursive method, and the CDS method saved the time for about two orders of magnitude lower. Both the two proposed methods showed almost the same precision performance when reducing the number of satellites from 42 to 16. Precision performances decrease when further reducing the number, with about 0.25 increase in DOP for the DS method relative to the recursive and about 0.10 for the GDS method, when reducing subset from 42 to 8. While the error of the DS method could be large when the number falls below 8, the CDS method kept high precision, which was about 0.2 error relative to recursive method. Thus, both of the two proposed methods have much lower computational load than existing DOP based satellite selection algorithms and have very little decrease in precision.

Operators are suggested to use the CDS method for satellite selection for its high adaptability and fast computational performance. In very short computational time desired situation, special embedded system or other similar conditions, the DS method can be employed.

## References


1. Mosavi M, Divband M (2010) Calculation of geometric dilution of precision using adaptive filtering technique based on evolutionary algorithms. In: 2010 international conference on electrical and control engineering, IEEE. https://doi.org/10.1109/icece.2010.1171
2. Wang E, Jia C, Feng S, Tong G, He H, Qu P, Bie Y, Wang C, Jiang Y (2018) A new satellite selection algorithm for a multi-constellation GNSS receiver. In: Proceedings of the 31st International Technical Meeting of The Satellite Division of the Institute of Navigation (ION GNSS+ 2018), Institute of Navigation. https://doi.org/10.33012/2018.15993
3. Phatak M (2001) Recursive method for optimum GPS satellite selection. IEEE Trans Aerosp Electron Syst 37(2):751–754. https://doi.org/10.1109/7.937488
4. Liu M et al (2009) A recursive quasi-optimal fast satellite selection method for GNSS receivers. In: Proceedings of the 22nd International Technical Meeting of the Satellite Division of The Institute of Navigation (ION GNSS 2009), Savannah, GA, September 2009, pp 2061–2071
5. Swaszek PF, Hartnett RJ, Seals KC (2016) Multi-constellation GNSS: new bounds on DOP and a related satellite selection process. In: Proceedings of the 29th International Technical Meeting of The Satellite Division of the Institute of Navigation (ION GNSS+ 2016), Institute of Navigation. https://doi.org/10.33012/2016.14748
6. Kihara M, Okada T (1984) A satellite selection method and accuracy for the global positioning system. Navigation 31(1):8–20. https://doi.org/10.1002/j.2161-4296.1984.tb00856.x

7. Blanco-Delgado N, Nunes FD (2010) Satellite selection method for multi-constellation GNSS using convex geometry. IEEE Trans Veh Technol 59(9):4289–4297. https://doi.org/10.1109/tvt.2010.2072939
8. Zhang M, Zhang J (2009) A fast satellite selection algorithm: Beyond four satellites. IEEE J Sel Top Signal Process 3(5):740–747. https://doi.org/10.1109/jstsp.2009.2028381
9. Ward L et al (1999) GPS receiver satellite/antenna selection algorithm for the Stanford gravity probe b relativity mission. In: Proceedings of the 1999 National Technical Meeting of The Institute of Navigation, San Diego, CA, January 1999, pp 541–550
10. Walter T, Blanch J, Kropp V (2016) Satellite selection for multi-constellation SBAS. In: Proceedings of the 29th International Technical Meeting of The Satellite Division of the Institute of Navigation (ION GNSS+ 2016), Institute of Navigation. https://doi.org/10.33012/2016.14608
11. Swaszek PF, Hartnett RJ, Seals KC, Swaszek RMA (2017) A temporal algorithm for satellite subset selection in multi-constellation GNSS. In: Proceedings of the 2017 International Technical Meeting of The Institute of Navigation, Institute of Navigation. https://doi.org/10.33012/2017.14904
12. Blanch J, Walter T, Enge P, Wallner S, Fernandez FA, Dellago R, Ioannides R, Hernandez IF, Belabbas B, Spletter A, Rippl M (2013) Critical elements for a multi-constellation advanced RAIM. Navigation 60(1):53–69. https://doi.org/10.1002/navi.29

# On Ground Test of an IMU/GNSS Receiver for Atmospheric Re-Entry Vehicle Applications

**Xiao-Liang Wang, Shu-Fan Wu, De-Ren Gong and Zhe Su**

**Abstract** This article introduced the on ground test of a novel IMU/GNSS integrated navigation system foratmospheric re-entry applications. Dynamic and kinetic model of re-entry vehicle are introduced with proper coordinates, and dynamic and kinematic equations for angular motion are also provided by using Euler's Law and quaternions. The sensor bias, scale factor and misalignment in inertial system are provided in detail, and the principle of SINS mechanization is given, which can easily be inserted to GNC closed loop. On ground and underground vehicle test results are provided that demonstrated the promising performance of this IMU/GNSS receiver for atmospheric re-entry applications.

**Keywords** IMU/GNSS · HiSGR · SINS · Atmospheric re-entry · Acquisition and tracking

This paper is supported by Nature Science Fund of Shanghai (No. 19ZR1426800), National Nature Science Fund of China (No. 91438107) and SJTU Global Strategic Partnership Fund (2019 SJTU-UoT No. WF610561702).

X.-L. Wang · S.-F. Wu (✉) · D.-R. Gong
The School of Aeronautics and Astronautics, Shanghai Jiaotong University,
Dongchuan Road No. 800, Shanghai 200240, China
e-mail: shufan.wu@sjtu.edu.cn

X.-L. Wang
e-mail: xlwang12321@sjtu.edu.cn

D.-R. Gong
e-mail: drgong@sjtu.edu.cn

Z. Su
Institute of Satellite Navigation and Intra-Satellite-Link Technology, Academy of Space Electronic Information Technology, East Chang'an Street No. 504, Xi'an 710100, China
e-mail: suzhe504@163.com

 
Z. Jing (ed.), *Proceedings of the International Conference on Aerospace System Science and Engineering 2019*, Lecture Notes in Electrical Engineering 622,
https://doi.org/10.1007/978-981-15-1773-0_3

## 1 Introduction

The atmospheric re-entry vehicle as X-38 Crew Return Vehicle (CRV) is a key component of the international space station (ISS). It is a spacecraft connected to the ISS, which will serve as a "lifeboat" for astronauts to return to earth when they are illness or in other emergencies. It will leave the docking port, glide unpowered from orbit, re-entry the earth's atmosphere and use a maneuvrable parachute to reach a specific landing site [1–3].

For re-entry vehicle navigation application in this paper, conventional techniques of IMU/GNSS integration may not be used. During the phase of blackout that covers most of the re-entry time, no GNSS signals would be available for receiver use. Therefore, the IMU will be the only navigation sensor on board, and the GNSS data can be used for sensor calibration only during the outer-atmospheric and de-orbiting phases.

With the supporting of China Nature Science Fund, a novel compact spaceborne IMU/GNSS integrated navigation receiver "HiSGR" is developed with the cooperation of Shanghai Jiaotong University and Academy of Space Electronic Information Technology (ASEIT), aiming for future space applications, as shown in Fig. 1. HiSGR

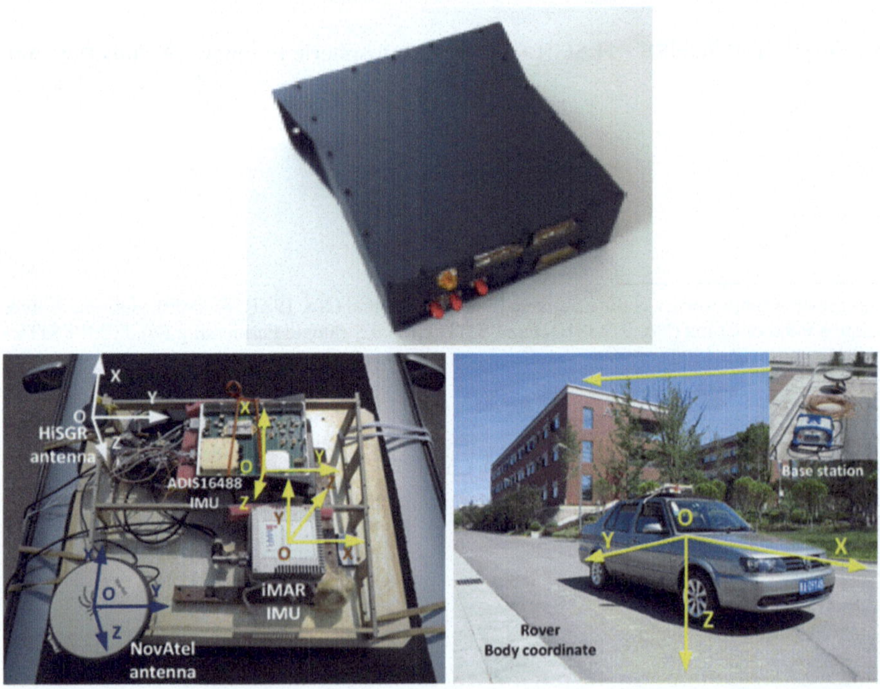


**Fig. 1** HiSGR receiver and on ground test

is a multifunctional receiver that can be applied in many platforms, depending on the different version burned software, which has been strictly tested both indoor and outdoor [4].

This article concerns the on ground test of such IMU/GNSS integrated navigation system for atmospheric re-entry applications. Dynamic and kinetic model of re-entry vehicle are introduced with proper coordinates, and dynamic and kinematic equations for angular motion are also provided by using Euler's Law and quaternions. The sensor bias, scale factor and misalignment are introduced in detail, and the principle of SINS mechanization is given, that can easily be inserted to GNC closed loop. On ground and underground vehicle test results are provided which demonstrated the promising performance of this IMU/GNSS receiver for atmospheric re-entry applications.

## 2 The Dynamic and Kinetic Model of Re-Entry Vehicle

The re-entry dynamic motion of a spacecraft vehicle is of six DOF, consisting of three DOF translational dynamics representing the point-mass trajectory movements and three DOF angular motion dynamics representing the rigid body attitude maneuvers. Both dynamic and kinematic movements should be modeled by equations to get a full dynamic description.

### 2.1 *Translational Motion*

The vehicle can be considered as a mass point in the space. Thus, the Newton's Second Law could be used for the modeling. With reference to the rotating R-frame (see Appendix), the dynamic and kinematic equations have following forms [5].

$$\begin{aligned}\vec{F}_R = m\frac{\mathrm{d}^2\vec{r}_{cm_R}}{\mathrm{d}t^2} + 2m\vec{\omega}_{R/I} \times \frac{\mathrm{d}\vec{r}_{cm_R}}{\mathrm{d}t} \\ + m\vec{\omega}_{R/I} \times (\vec{\omega}_{R/I} \times \vec{r}_{cm_R})\end{aligned} \tag{1}$$

$$\frac{d\vec{r}_{cm_R}}{\mathrm{d}t} = \vec{V}_R \tag{2}$$

where $F_R$: summation of all external forces acting on the vehicle, expressed in the rotating R-frame; $r_{cm_R}$: the position vector of the vehicle in the rotating frame; $d^2r_{cm_R}/\mathrm{d}t^2$: apparent acceleration of the vehicle in the rotating frame; $\omega_{R/I}$: the rotational rate of the R-frame with reference to the inertial frame, which is equal to the rotational rate of the earth; $V_R$: the velocity vector of the vehicle in the rotating frame; $2\omega_{R/I} \times V_R$: Coriolis acceleration due to the motion of the vehicle in the rotating frame; $\omega_{R/I} \times (\omega_{R/I} \times r_{cm_R})$: apparent (or transport) acceleration of the vehicle due to angular rate of the rotating frame.

The position and velocity vectors can be expressed in both Cartesian and spherical coordinates, or their mixtures, resulting in different equations of motion. For convenience in the navigation computation, a mixture of the spherical position variables (R, τ, δ) and the Cartesian velocity variables $(v_\delta, v_\tau, v_r)$ (ground speed) defined in the vertical frame (see Appendix) are used to describe the dynamic equations [5, 6].

With spherical coordinates, the position vector in the vertical frame is

$$r_{cm_V} = \begin{bmatrix} 0 \; 0 \; -R \end{bmatrix}^T \tag{3}$$

And the general formula for the velocity in the vertical frame is given by

$$V_V = \begin{bmatrix} v_\delta \; v_\tau \; -v_r \end{bmatrix}^T = \frac{\partial r_{cm}}{\partial t} + \omega_{V/R} \times r_{cm_V} \tag{4}$$

Thus, the kinematic equations can be obtained as

$$\dot{\delta} = \frac{v_\delta}{R}, \quad \dot{\tau} = \frac{v_\tau}{R\cos(\delta)}, \quad \dot{R} = v_r \tag{5}$$

As in the other cases, also Newton's Second Law will be used to set up the dynamic equations. And the total force applied to the vehicle should be first transferred into the vertical frame, as given by following expression:

$$\begin{aligned} F_V = \begin{bmatrix} F_{X_V} \\ F_{Y_V} \\ F_{Z_V} \end{bmatrix} &= F_{A_V} + F_{G_V} + F_{T_V} \\ &= T_{W\to V}(F_{A_W} + F_{T_W}) + F_{G_V} \end{aligned} \tag{6}$$

The dynamic equation can finally be obtained as the following form:

$$\begin{aligned} \dot{v}_\delta &= \frac{F_{X_V}}{m} - 2\Omega_E v_\tau \mathrm{s}(\delta) - \Omega_E^2 R\mathrm{s}(\delta)\mathrm{c}(\delta) - \frac{v_\tau^2 t(\delta) + v_\delta v_r}{R} \\ \dot{v}_\tau &= \frac{F_{Y_V}}{m} - 2\Omega_E(v_r \mathrm{c}(\delta) - v_\delta \mathrm{s}(\delta)) + \frac{v_\tau}{R}(v_\delta \mathrm{t}(\delta) - v_r) \\ \dot{v}_r &= -\frac{F_{Z_V}}{m} + 2\Omega_E v_\tau \mathrm{c}(\delta) + \Omega_E^2 R\mathrm{c}^2(\delta) + \frac{v_\tau^2 + v_\delta^2}{R} \end{aligned} \tag{7}$$

These final equations do not present any problem during simulation computations, except a singularity when the latitude is at ±π/2, which means the vehicle is over the north or south pole of the earth. The external forces applied to the vehicle including aerodynamic force $F_{A_W}$ (drag, lift and side force), gravitational force $F_{G_V}$ and propulsion force $F_{T_W}$ [1]. Due to the reason that the aiming of this article is on ground test of IMU/GNSS receiver, those external forces are actually not exerted on the moving platform.

## 2.2 Angular Motion

The dynamic equations for angular motion are obtained from the Euler's Law [7]:

$$\begin{bmatrix} \dot{p} \\ \dot{q} \\ \dot{r} \end{bmatrix} = I^{-1}\left(M_B - \begin{bmatrix} p \\ q \\ r \end{bmatrix} \times \left(I\begin{bmatrix} p \\ q \\ r \end{bmatrix}\right)\right) \tag{8}$$

where $M_B$ is the total external torque vector acting on the vehicle, expressed in the body reference frame. It consists of three parts, the aerodynamic torques due to control flaps, the aerodynamic force induced torques and the torques induced by thrust forces [8]:

$$\begin{aligned} M_B &= M_{A_B} + M_{F_B} + M_{T_B} \\ &= M_{A_B} + r_{ac} \times \left(T_{W\to B} F_{A_W}\right) + \sum_{i=1}^{n}\left(r_{T,i} \times F_{T,i_B}\right) \end{aligned} \tag{9}$$

The kinematic equation is dependent on the method of defining the rotation of the body. This can be performed using quaternions, Euler angles or aerodynamic angles. In this modeling, the quaternions are used, which are justified by the absence of singularities in the rotations [7], also for the convenient of mechanical layout formula for the inertial navigation system. The kinematic equation of angular moment is

$$\begin{bmatrix} p \\ q \\ r \end{bmatrix} = 2\begin{bmatrix} \dot{q}_1 q_4 - \dot{q}_4 q_1 + q_3\dot{q}_2 - q_2\dot{q}_3 \\ \dot{q}_2 q_4 - \dot{q}_4 q_2 - q_3\dot{q}_1 + q_1\dot{q}_3 \\ \dot{q}_3 q_4 - \dot{q}_4 q_3 + q_2\dot{q}_1 - q_1\dot{q}_2 \end{bmatrix} \tag{10}$$

Relations could be built up between the quaternions and the Euler angles or aerodynamic angles, based on the equivalent transformation matrices between different reference frames [7, 9]. Also, the aerodynamic torques are not introduced here that irrelevant to the receiver test. The interested readers could be referred to [1].

Equations (5), (7), (8) and (10) set up a complete six DOF dynamic model for the atmospheric re-entry vehicle flight, where 13 state variables are employed.

# 3 IMU/GNSS Integrated System Modeling

As shown in Fig. 1, the IMU/GNSS receiver is fixed in the moving vehicle. It is basically a strapdown inertial navigation system—SINS. By using the information from gyroscope output, a virtual mathematical platform is established. Then, the attitude transformation matrix is obtained, and finally, the measurements from accelerometer can be projected to navigation coordinate. The principle of SINS is given (Fig. 2).

## 3.1 Attitude Update Equation

Like angular motion equation in Sect. 2.2, there are some methods for SINS attitude update equation as quaternions, Euler angles or aerodynamic angles. Here, we choose quaternions due to the simplicity of computation. Considering the output of gyroscope is angle increment, a fixed time increment sampling algorithm is used for quaternion computation:

Suppose, we have a vector $\boldsymbol{\mu}$ and angle $\mu$, $\boldsymbol{\mu}$ is called equivalent rotation vector, if navigation frame $N$ can be coincidence with body frame $b$ when rotating of $\boldsymbol{\mu}$ with angle $\mu$.

Suppose the angle increment of three axes is $\Delta\theta_x$, $\Delta\theta_y$, $\Delta\theta_z$, define increment matrix $\Delta\Theta$ as

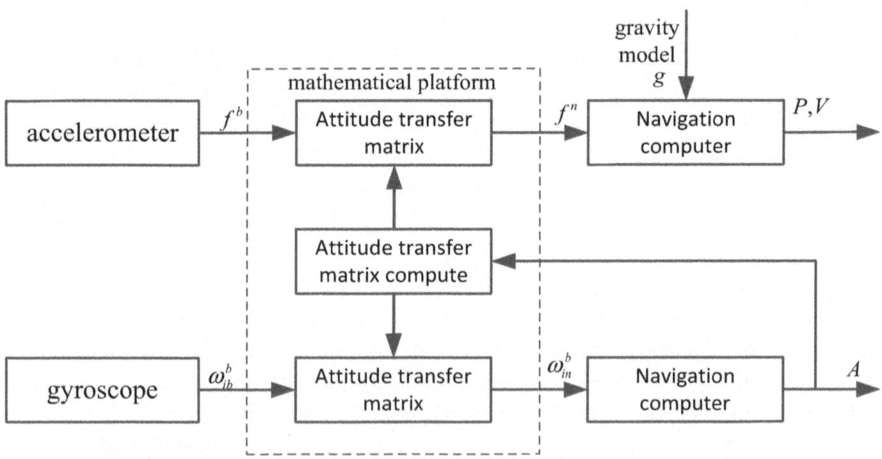


**Fig. 2** The principle of SINS

$$\Delta\Theta = \begin{bmatrix} 0 & -\Delta\theta_x & -\Delta\theta_y & -\Delta\theta_z \\ \Delta\theta_x & 0 & \Delta\theta_z & -\Delta\theta_y \\ \Delta\theta_y & -\Delta\theta_z & 0 & \Delta\theta_x \\ \Delta\theta_z & \Delta\theta_y & -\Delta\theta_x & 0 \end{bmatrix} \tag{11}$$

Discretization of quaternion update equation follows:

$$Q(t_{k+1}) = \left( I\cos\frac{\Delta\theta}{2} + \Delta\Theta\frac{\sin\frac{\Delta\theta}{2}}{\Delta\theta} \right) Q(t_k) \tag{12}$$

During real computation process, it can be handled by using expansion of Taylor series to required precision.

The first-order expansion:

$$Q(t_{k+1}) = \left( I + \frac{\Delta\Theta}{2} \right) Q(t_k) \tag{13}$$

The second-order expansion:

$$Q(t_{k+1}) = \left( I\left(1 - \frac{\Delta\theta^2}{8}\right) + \frac{\Delta\Theta}{2} \right) Q(t_k) \tag{14}$$

The third-order expansion:

$$Q(t_{k+1}) = \left( I\left(1 - \frac{\Delta\theta^2}{8}\right) + \left(\frac{1}{2} - \frac{\Delta\theta^2}{48}\right)\frac{\Delta\Theta}{2} \right) Q(t_k) \tag{15}$$

And the fourth order:

$$Q(t_{k+1}) = \left( I\left(1 - \frac{\Delta\theta^2}{8} + \frac{\Delta\theta^4}{384}\right) + \left(\frac{1}{2} - \frac{\Delta\theta^2}{48}\right)\frac{\Delta\Theta}{2} \right) Q(t_k) \tag{16}$$

## 3.2 *Velocity Update Equation*

Due to the reasons that the position and velocity input information of re-entry vehicle control system are spherical position variables $(R, \tau, \delta)$ and the Cartesian velocity variables $(v_\delta, v_\tau, v_r)$ (ground speed), here we choose geographical frame (see Appendix) as the navigation coordinate (subscript $n$) for SINS velocity update, which can be easily rotated from ground speed by $T_x(\pi)T_z(\pi/2)$. The principle of SINS mechanization is given in Fig. 3.

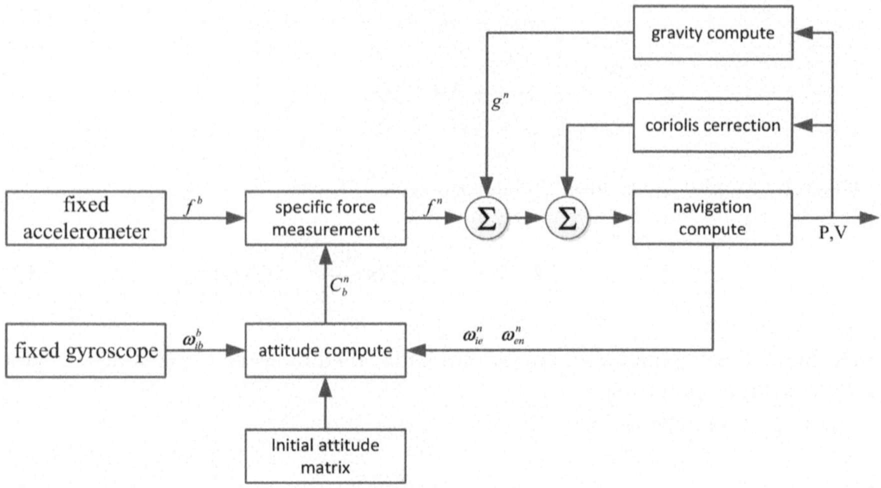


**Fig. 3** The principle of SINS mechanization

Transferring the ground speed from Eq. (4) as in vertical frame to geographical frame (navigation frame) as $V_e^n$, then we have

$$\dot{V}_e^n = C_b^n f^b - \left[2\omega_{ie}^n + \omega_{en}^n\right] \times V_e^n + g^n \tag{17}$$

where $f^b$ is specific force measurement vector from accelerometer, $C_b^n$ is attitude matrix, $\omega_{ie}^n$ is the earth rotation angular rate vector in n frame, $\omega_{en}^n$ is the relative to earth rotation angular rate vector in $n$ frame, $g^n$ is gravity vector in $n$ frame.

Finally, we have the velocity update equations in n frame as

$$\dot{\delta} = \frac{V_N}{R_e + h}, \quad \dot{\tau} = \frac{V_E}{(R_e + h)\cos(\delta)}, \quad \dot{h} = V_U \tag{18}$$

which is almost the same as Eq. (5). The $V_E$, $V_N$, $V_U$ are the ground speed of direction East, North, Up in n frame, $R_e$ is earth radius.

## 4 SINS Error Propagation Equations

The traditional gyroscope error models are [10]:

$$\begin{bmatrix} \varepsilon_x \\ \varepsilon_y \\ \varepsilon_z \end{bmatrix} = \varepsilon = \varepsilon_0 + \varepsilon_w + \delta \boldsymbol{K}_g \omega_{ib}^b + [\delta \boldsymbol{G} \times]\omega_{ib}^b$$

$$= \begin{bmatrix} \varepsilon_{x0} + \varepsilon_{xw} + \delta K_{gx}\omega_{ibx}^{b} - \delta G_{z}\omega_{iby}^{b} + \delta G_{y}\omega_{ibz}^{b} \\ \varepsilon_{y0} + \varepsilon_{yw} + \delta K_{gy}\omega_{iby}^{b} + \delta G_{z}\omega_{ibx}^{b} - \delta G_{x}\omega_{ibz}^{b} \\ \varepsilon_{z0} + \varepsilon_{zw} + \delta K_{gz}\omega_{ibz}^{b} - \delta G_{y}\omega_{ibx}^{b} + \delta G_{x}\omega_{iby}^{b} \end{bmatrix} \tag{19}$$

where $\boldsymbol{\varepsilon}_0$ denotes gyroscope constant drift, $\boldsymbol{\varepsilon}_w$ denotes gyroscope random drift, $\delta \boldsymbol{K}_g$ is scale factor error, $\delta \boldsymbol{G}$ is installation angle error.

Similarly, the accelerometer error model:

$$\begin{bmatrix} \nabla_x \\ \nabla_y \\ \nabla_z \end{bmatrix} = \nabla = \nabla_0 + \nabla_w + \delta \boldsymbol{K}_a \boldsymbol{f}^b + [\delta \boldsymbol{A} \times] \boldsymbol{f}^b$$

$$= \begin{bmatrix} \nabla_{x0} + \nabla_{xw} + \delta K_{ax} f_x^b - \delta A_z f_y^b + \delta A_y f_z^b \\ \nabla_{y0} + \nabla_{yw} + \delta K_{ay} f_y^b + \delta A_z f_x^b - \delta A_x f_z^b \\ \nabla_{z0} + \nabla_{zw} + \delta K_{az} f_z^b - \delta A_y f_x^b + \delta A_x f_y^b \end{bmatrix} \tag{20}$$

where $\nabla_0$ denotes accelerometer constant drift, $\nabla_w$ denotes random drift, $\delta \boldsymbol{K}_a$ is scale factor error, $\delta \boldsymbol{A}$ is installation angle error.

## 4.1 Attitude Error Equations

In inertial navigation systems, suppose the computed attitude matrix is $\tilde{\boldsymbol{C}}_b^n$, the real attitude matrix is $\boldsymbol{C}_b^n$, and the transformation between the two is

$$\tilde{\boldsymbol{C}}_b^n = \boldsymbol{B}\boldsymbol{C}_b^n \tag{21}$$

where $\boldsymbol{B}$ is transformation matrix from real $n$ frame to computed $n$ frame, i.e., the initial alignment attitude error of inertial systems. Usually, the misalignment angles are small that $\boldsymbol{B}$ can be represented as

$$\boldsymbol{B} = [I - \Psi] \tag{22}$$

where $\Psi$ is the skew symmetric matrix of three-axis misalignment angles $\delta\alpha$, $\delta\beta$, $\delta\gamma$:

$$\Psi = \begin{bmatrix} 0 & -\delta\gamma & \delta\beta \\ \delta\gamma & 0 & \delta\alpha \\ -\delta\beta & \delta\alpha & 0 \end{bmatrix} \tag{23}$$

Then, the attitude computation equation is

$$\Psi = I - \tilde{\boldsymbol{C}}_b^n \boldsymbol{C}_b^{nT} \tag{24}$$

The attitude error propagation equation can be obtained by differential as

$$\dot{\Psi} = -\dot{\tilde{C}}_b^n C_b^{nT} - \tilde{C}_b^n \dot{C}_b^{nT} \quad (25)$$

With some formula derivations, the error propagation equations in matrix form are

$$\begin{aligned}\dot{\Psi} &= -[I - \Psi]C_b^n[\widetilde{\Omega}_{ib}^b - \Omega_{ib}^b]C_b^{nT} \\ &+ \widetilde{\Omega}_{in}^n[I - \Psi]C_b^n C_b^{nT} - [I - \Psi]C_b^n C_b^{nT}\Omega_{in}^n\end{aligned} \quad (26)$$

where $\widetilde{\Omega}_{ib}^b$ denotes the measured angular rate matrix of vehicle, $\widetilde{\Omega}_{in}^n$ denotes the computed angular rate matrix in $n$ frame.

Let $\delta\Omega_{in} = \widetilde{\Omega}_{in} - \Omega_{in}$, $\delta\Omega_{ib} = \widetilde{\Omega}_{ib} - \Omega_{ib}$, omitting some small quantity, the matrix form of attitude error propagation equation is

$$\dot{\Psi} \approx \Psi\,\Omega_{in}^n - \Omega_{in}^n\Psi + \delta\,\Omega_{in}^i - \mathrm{C}_b^n\delta\,\Omega_{ib}^b\mathrm{C}_b^{nT} \quad (27)$$

With simplify, equation above can be rewritten as

$$\dot{\Psi} \approx -\boldsymbol{\omega}_{in}^n \times \Psi + \delta\,\boldsymbol{\omega}_{in}^n - C_b^n\Omega_{ib}^b \quad (28)$$

where $\Psi = \begin{bmatrix}\delta\alpha & \delta\beta & \delta\gamma\end{bmatrix}^T$, $\boldsymbol{\omega}_{in}^n \times = \Omega_{in}^n$, $\delta\boldsymbol{\omega}_{in}^n \times = \delta\Omega_{in}^n$, $\delta\boldsymbol{\omega}_{ib}^b \times = \delta\Omega_{ib}^b$.

## 4.2 Velocity Error Equations

The idea SINS velocity error equation is

$$\dot{V} = C_b^n f^b - \left[2\omega_{ie}^n + \omega_{en}^n\right] \times V + g_l \quad (29)$$

where $g_l$ is the local gravity vector, and

$$g_l = g - \omega_{ie} \times \left[\omega_{ie} \times r\right] \quad (30)$$

So, the computed velocity error is

$$\dot{\tilde{V}} = \tilde{C}_b^n \tilde{f}^b - \left[2\tilde{\omega}_{ie}^n + \tilde{\omega}_{en}^n\right] \times \tilde{V} + \tilde{g}_l \quad (31)$$

Differential of Eqs. (30) and (31), we get the propagation of velocity error equation:

$$\delta\dot{V} = -\Psi C_b^n f^b + C_b^n\delta f^b - \left[2\omega_{ie}^n + \omega_{en}^n\right] \times \delta V$$

$$- \left[2\delta\boldsymbol{\omega}_{ie}^{n} + \delta\boldsymbol{\omega}_{en}^{n}\right] \times \delta\boldsymbol{g} \tag{32}$$

Omitting the Coriolis error and gravity vector error, we get:

$$\delta\dot{\boldsymbol{V}} = \left[\boldsymbol{f}^{n}\times\right]\Psi + \boldsymbol{C}_{b}^{n}\delta\boldsymbol{f}^{b} \tag{33}$$

Conventional nonlinear extended Kalman filter is adopted for this IMU/GNSS Integrated System. The states variables include the inertial error of position, velocity, the bias of gyroscope and accelerometer. Vast literatures introduced the filter design that not provided here for concise [10].

## 5 On Ground Test of IMU/GNSS Receiver

The HiSGR receiver can process the received signal measurements to compute the navigation solutions. The core of navigation algorithm is achieved by real-time precise orbit determination software (RTPODs), which is from collaboration project with Shanghai Astronomical Observatory, Chinese Academy of Sciences [11–13]. RTPODs are a compact and portable software package optimized for real-time processing and are designed for use on embedded DSP systems. It makes use of a nonlinear filter as well as precise dynamic models for orbiting applications. The models used in RTPODs include a full EGM-96 70 by 70 gravity field, the DTM 94 atmospheric drag model [14], a solar radiation pressure model, earth orientation and polar motion models as well as a relativity model. In addition, RTPODs have the capability to utilize the reduced dynamic technique in which empirical accelerations are estimated in order to account for any dynamics left unmodeled [15]. Finally, the position, velocity and timing (PVT) point solution will be provided when four or more satellites are being tracked, via a single-shot nonlinear least-squares solver, and also provide onboard orbit determination capabilities under sparse observe conditions. When RTPODs operate in dual-frequency mode, more observables are used to compute the ionospheric delay and ionosphere-free pseudoranges for each satellite in view.

The RTPODs mainly operate in GNSS stand-alone navigation mode and can also be switched to integrated navigation mode that incorporates inertial measurements and on-line processed using ultra-tight coupled GPS/inertial Kalman filter for predicted orbit position, velocity and attitude, which is used to remove the Doppler effects to allow weak signal tracking in some scenarios.

The availability of the source code allows customization of RTPODs for space applications in difference altitude, and the whole software onboard can be totally reconfigured by using commanding and telemetry interface. So far, several modifications have been made for RTPODs, and Table 1 listed some verified software versions for future use.

Some of the RTPODs versions for orbiting applications, as low and high earth orbit, have been demonstrated on previous publications [4]. RTPODs-T is a special

**Table 1** Some RTPODs versions

| RTPODs versions | Applications |
|---|---|
| RTPODs-LEO | Low earth orbit |
| RTPODs-R | Reflected signal receiving |
| RTPODs-HEO | High eccentric orbit |
| RTPODs-GEO | Geostationary orbit |
| RTPODs-LUNAR* | Cislunar/orbiting moon |
| RTPODs-L2* | Helo Lagrange L2 |
| RTPODs-T | Test for ground vehicle |

*Under development

version that is used for on ground vehicle test of IMU/GNSS function, which is suitable for atmospheric re-entry navigation. Detail design of RTPODs-T (IMU/GNSS integrated SINS algorithm) has been introduced in above sections; here we provided some open field on ground test results.

The test is conducted in ASEIT working area from 14:15:00 to 15:15:00 Apr 8, 2019. Figure 1 demonstrated the assembling of HiSGR in platform with NovAtel's SPAN@ integrated navigation system, which is used for the comparison and evaluation of final solutions, and the definition of run car coordination and base station during running test are also provided.

Figure 4 demonstrated the reference trajectory from NovAtel's SPAN system, by using WAYPOINT software. The black lines and green dots showed the running route both in open ground and underground. Clearly, the SPAN system provided the smooth and precise positioning solution for commercial use. Figures 5 and 6 illustrated the post-processing trajectory from both NovAtel SPAN and HiSGR receiver. The bolded purple lines and green lines in Fig. 6 demonstrated the SPAN system output, both

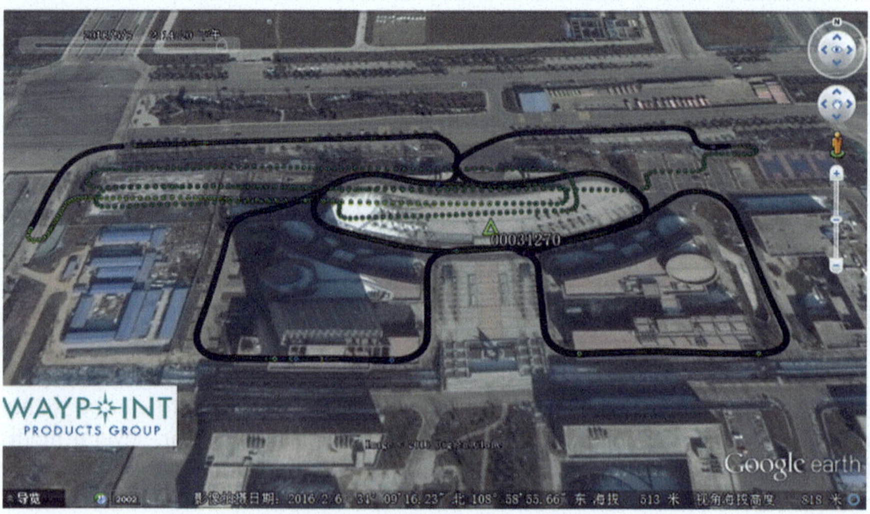


**Fig. 4** Moving trajectory from NovAtel SPAN system

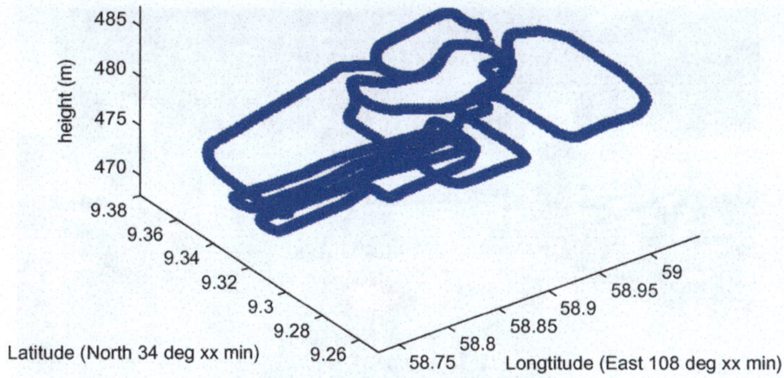


**Fig. 5** Processed trajectory from NovAtel SPAN system

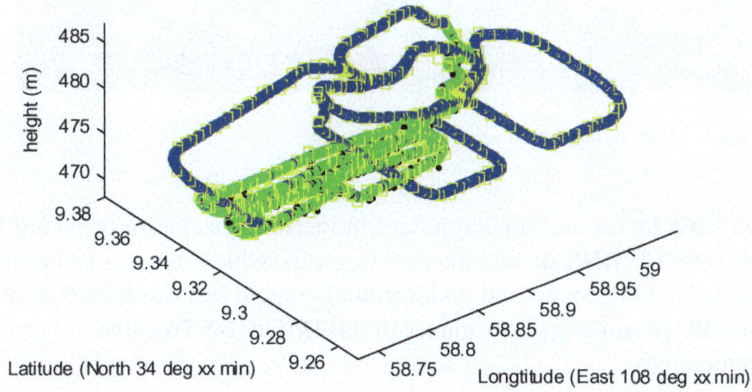


**Fig. 6** Trajectory comparison

on open field running and underground test, while the green squares and small black dots show the same output from HiSGR.

The running trajectory underground is intentional selected as the simulation of re-entry vehicle during blackout period, as in Fig. 7, surely, the receiver achieved quick signal reacquisition when out of underground garage, and HiSGR performs stable during the running test on open field. Position error is less than 10 meters during on ground test and achieved less than 20 meters during underground.

## 6 Conclusions

This article introduced the on ground test of a novel IMU/GNSS integrated navigation system for atmospheric re-entry applications. Dynamic and kinetic model of re-entry vehicle are introduced with proper coordinates, and dynamic and kinematic equations for angular motion are also provided by using Euler's Law and quaternions. The

**Fig. 7** Moving vehicle underground

sensor bias, scale factor and misalignment in inertial system are provided in detail, and the principle of SINS mechanization is given, which can easily be inserted to GNC closed loop. On ground and underground vehicle test results are provided that demonstrated the promising performance of this IMU/GNSS receiver for atmospheric re-entry applications.

**Acknowledgements** This article and the HiSGR project are supported by Nature Science Fund of Shanghai (No. 19ZR1426800), National Nature Science Fund of China (No. 91438107) and SJTU Global Strategic Partnership Fund (2019 SJTU-UoT No. WF610561702).

## Appendix

**R-frame**: Earth Centered Earth Fixed frame (index R), the center is in the center of the earth, $Z_R$ is pointing North, $X_R$ is pointing to the Greenwich meridian, and $Y_R$ completes the right-handed system.

**V-frame**: Vertical Reference frame (index $V$), the center is in the center of gravity of the body, $Z_V$ is pointing down collinear with the gravity vector, $X_V$ is pointing North, and $Y_V$ is pointing East.

**N-frame**: Geographical frame/Navigation frame, the center is in the center of mass of the vehicle, $Z_V$ is pointing up collinear with the negative gravity vector, $X_V$ is pointing East, and $Y_V$ is pointing North.

## References


1. Wu S-F, Costa RR, Chu Q-P, Mulder JA, Ortega G (2001) Nonlinear dynamic modeling and simulation of an atmospheric re-entry spacecraft. Aerosp Sci Technol 5(5):365–381
2. NASA Armstrong Fact Sheet: X-38 Prototype Crew Return Vehicle. https://www.nasa.gov/centers/armstrong/news/FactSheets/FS-038-DFRC.html
3. The First X-38 Technology Demonstrator (V-131) Shown with modifications to the rear to conform more to the shape of the Crew Return Vehicle (CRV). https://www.dfrc.nasa.gov/Gallery/Photo/X-38/HTML/EC99-45255-18.html
4. Wang X, Wang Y, Su Z, Meng Y, Gong D (2016) Design and test of HiSGR: a novel GNSS/INS ultra tight coupled receiver. J Commun Inf Netw 1(3):67–76
5. Mooij E (1997) The motion of a vehicle in a planetary atmosphere. Delft University Press
6. Laban M (1994) On-line aircraft aerodynamic model identification, Ph.D. Dissertation, TU Deflt, Delft University Press, The Netherlands. ISBN 90-6275-987-4
7. Regan FJ, Anandakrishnan SM (1993) Dynamics of atmospheric re-entry, AIAA education series, Washington DC, USA
8. Wu S-F, Chu Q-P (1999) The atmospheric re-entry spacecraft CRV/X-38 – aerodynamic modeling and analysis. Technical Report, Faculty of Aerospace Engineering, TU Delft, Delft, the Netherlands, December 1999
9. Mooij E (1994) The motion of a vehicle in a planetary atmosphere. Report LR-768, Faculty of Aerospace Engineering, Delft, The Netherlands
10. Fu M, Zheng X, Deng Z et al (2012) Transfer alignment theory and application. Science Press, Beijing, pp 23–34
11. Li PJ, Hu XG, Huang Y, Wang GL, Jiang DR (2012) Orbit determination for Chang'E-2 lunar probe and evaluation of lunar gravity models. Sci China Phys Mech Astron 55(3):514–522
12. Fan M, Hu XG, Dong G, Huang Y, Cao J (2015) Orbit improvement for Chang'E-5T lunar returning probe with GNSS technique. Adv Space Res 56(11):2473–2482
13. Huang Y, Fan M, Hu X, Li P (2015) The application of GNSS in the near-Earth navigation of China's lunar probe CE-5T1. Iau Gen Assembly 21
14. Berger C, Biancale R, Ill M, Barlier F (1998) Improvement of the empirical thermospheric model DTM: DTM94—a comparative review of various temporal variations and prospects in space geodesy applications. J Geodesy 72(3):161–178
15. Wu S-C, Yunck TP, Thornton CL (1991) Reduced-dynamic technique for precise orbit determination of low earth satellites. J Guid Control Dyn 14(1):24–30

**Xiao-Liang Wang** was born in 1981. He received B.S. and M.S. degrees in system control engineering from Northwestern Polytechnic University in 2003 and 2006, Xi'an, China, and received the Ph.D. degree major in navigation, guidance and control in School of Aeronautics and Astronautics from Shanghai Jiaotong University, Shanghai, China, in 2011. From 2011 to 2017, he worked as a Senior Engineer with Institute of Satellite Navigation and Intra-Satellite-Link Technology, Academy of Space Electronic Information Technology, Xi'an, China. He is devoting the GNSS applications in space and hosted as the principle leader of High Sensitive GNSS Receiver (HiSGR) project with many published articles during this period. He is currently with the School of Aeronautics and Astronautics in Shanghai Jiaotong University. His research interests include GNC technology of atmospheric re-entry vehicles. Design and evaluation of spaceborne GNSS receiver for multiple missions, weak GNSS signal acquisition and tracking, ultra-tight coupled GNSS/Inertial technology in space, autonomous real-time orbit determination onboard, nonlinear Kalman filter technology and nonlinear estimation and control of time-delayed systems.

**Shu-Fan Wu** was born in 1964. He received a Ph.D. degree major in Navigation, Guidance and Control in Nanjing University of Aeronautics & Astronautics (NUAA) in 1990. He worked as assistant/ associate professor in NUAA during 1990–1998, then as research fellow in faculty of aerospace engineering of TU Delft during 1998–1999 and as research fellow in Surrey space center of Surrey University of UK during 2000–2002, as senior/fellow spacecraft engineer in ESA/ESTEC in Netherlands during 2002–2013. He has been as CTO of Shanghai Engineering Center for Micro-satellites (SECM), Chinese Academy of Science during 2013–2017. Now, he is working as a chair professor at Shanghai Jiao Tong University. His research interests include navigation and control of spacecraft orbit and attitude, nonlinear dynamic modeling and robust control of atmospheric re-entry spacecraft.

**De-Ren Gong** received a Ph.D. degree major in Navigation, Guidance and Control in School of Aeronautics and Astronautics, Shanghai Jiao Tong University in 2012. Received M.S. and B.S. degree from Xi'an Jiao Tong University in 2006, 2003, Xi'an, China. He has been worked as a postdoctoral research fellow in Shanghai Jiao Tong University during 2012–2014 and currently works there as an associate research fellow. His research interests include guidance, navigation and control of spacecraft orbit and attitude, precise modeling of spacecraft dynamic and kinematics, nonlinear estimation and control of time-delayed systems and hardware-in-the-loop (HIL) simulation for distributed systems.

**Zhe Su** received a Ph.D. degree in circuits and system from institute of electronic engineer, Xidian University, Xi'an, China, in 2011. He is currently with Institute of Satellite Navigation and Intra-Satellite-Link Technology, Academy of Space Electronic Information Technology. His research interest includes autonomous navigation using X-ray pulse star technology, design and test of new generation GNSS signals and onboard payload.

# Stall Margin Enhancement of Aeroengine Compressor with a Novel Type of Alternately Swept Blades

**Chao Fang, Yizhi Zhang, Yidan Li and Xiaohua Liu**

**Abstract** Rotating stall, as a typical kind of flow instability in aero-compressor, could lead to disastrous consequences of aeroengine. Therefore, an effective method is perused to enhance the stall margin. Some of the previous researches focus on the holistically swept rotor. This paper concentrates on the impact of a novel type of axial swept blades on the aerodynamic behaviour of transonic axial-flow compressor rotors. A CFD package, which solves the Reynolds-averaged Navier–Stokes equations, is used to compute the complex flow field of the compressor. It is validated against the existing experimental data. Comparisons with experimental data indicate that the overall features of the rotor performance are calculated well by the numerical solution with acceptable accuracy. A number of new swept rotors were modelled based on the original blade, by axially moving the location of blade alternately. All the new rotors are simulated, and comparison of the results shows that the alternately swept rotor enhances the stall margin effectively. The stall margin of new rotors can reach up to 18.16%, while that of the original rotor is only 9.71%. More physical explanations on the stall margin improvement are given based on a detailed analysis of the flow field.

**Keywords** Aeroengine compressor · Stall margin · Alternate swept blade · Numerical simulation

Although this paper is partly inspired by the reference [15] and [18], the substance and the methods are different between this paper and these two references.

C. Fang · Y. Zhang · Y. Li · X. Liu (✉)
School of Aeronautics and Astronautics, Shanghai Jiao Tong University, Shanghai, China
e-mail: Xiaohua-Liu@sjtu.edu.cn

C. Fang
e-mail: fcchao0416@126.com

Y. Zhang
e-mail: Itszyz919@sina.com

Y. Li
e-mail: yidanli@sjtu.edu.cn



Z. Jing (ed.), *Proceedings of the International Conference on Aerospace System Science and Engineering 2019*, Lecture Notes in Electrical Engineering 622,
https://doi.org/10.1007/978-981-15-1773-0_4

## 1 Introduction

With the increase of compressor stage load, the secondary flow in the blade passages is more three-dimensional, the tip leakage is more serious, and the boundary layer and corner region are more easily separated [1]. Therefore, the requirement for stable operating margin of compressor is more stringent. There is a certain contradiction between pursuing higher stage pressure ratio and improving the stall margin of compressor. In order to enhance the stall margin of the compressor, the methods can be adopted by the compressor at present include variable guide vanes [2, 3], casing treatment [4], air bleeding [5], birotor, and trispool. Most of these methods have been applied in engineering and achieved good results.

Vo et al. [6] obtained the definition of the phenomena that lead to the onset of short length-scale (spike) rotating stall disturbances and got two conditions for the formation of spike disturbances [7, 8], both of which are relevant to the tip clearance flow. One is that the interface between the tip clearance and oncoming flows becomes parallel to the leading edge, which would lead to the spike. The second is the initiation of backflow, stemming from the fluid in adjacent passages, at the trailing-edge plane. The two criteria imply the method to delay the rotating stall.

The effects of swept rotor on the overall aerodynamic performance and compressor stability have been widely analysed in the published literature [9–11]. The result was investigated based on numerical and experimental analysis of an unswept rotor, a forward-swept rotor, and a backward-swept rotor by solving Reynolds-averaged Navier–Stokes equations. Hah et al. [12] displayed the phenomenon that the forward-swept rotor brings a larger stall margin than the unswept rotor and that the backward-swept rotor has the opposite effect on overall performance. Wadia et al. [13] obtained similar results and provided a physical description about reducing shock/boundary layer arising from a local low loading at the tip region of forward-swept rotor. Denton and Xu [14] observed the overall aerodynamic performance of swept rotor, and the results of backward-swept rotor and forward-swept rotor showed that the efficiency is not very remarkable. The common conclusion was obtained by He et al. [15] in four kinds of swept rotors simulation including forward axial swept, forward chordwise swept, backward axial swept, and backward chordwise swept which are modified at the position of tip blade section.

According to the research of Vo et al. [6], tip clearance flow will affect adjacent blade passages. The present paper concentrates on the impact of a novel type of alternately swept blades. We first introduce the design specification of test rotor, and the novel types were defined here. Secondly, the comparison of the numerical and experimental results of the original rotor would show the accuracy of the steady CFD model. After that, all the new swept rotors were simulated to investigate the effects of alternately swept on the overall performance and stall margin. Finally, the comparison of all simulation was presented, and the flow field would be analysed.

**Table 1** Design specification of rotor 37 [18]

| Parameter | Value |
|---|---|
| Blade number | 36 |
| Total pressure ratio | 2.106 |
| Hub-tip ratio | 0.7 |
| Aspect ratio | 1.19 |
| Tip solidity | 1.29 |
| Tip clearance | 0.0335 mm |
| Mass flow | 20.19 kg/s |
| Rotating speed | 17,188 rpm |
| Tip speed | 454 m/s |
| Adiabatic efficiency | 0.877 |

## 2 Rotor Geometries Definition

### *2.1 Original Rotor Type*

A transonic axial compressor rotor NASA rotor 37 was used as original rotor type in this paper. The rotor 37 was designed by NASA Lewis Research Centre. The design specification of this rotor is shown in Table 1. The rotor has 36 blades with the tip clearance of 0.335 mm. The operating rotating speed was 17,188 rpm, while pressure ratio and mass flow rate were 2.106 and 20.19 kg/s at operating point. The overall aerodynamic performance and detail flow field were provided by Moore and Reid [16]. Figure 1 Measurement stations show the measurement stations in this paper. Our flow results were undertaken for station 1 and 4 [17].

### *2.2 Swept Rotors*

We adopt rotor 37 as the original rotor and obtained some new rotors by axially moving the location of blade alternately. The definition of sweep is shown in Fig. 2 The definition of sweep. Three types of rotors were modelled based on rotor 37. Type 1 refers to the original rotor. Type 2 refers to the rotor forward axially moving the location of blades holistically for 0.001 m. Type 3 refers to the rotor which is obtained by forward axially moving the location of blades alternately for 0.001 m. Type 4 refers to the rotor which is obtained by forward axially moving the location of blades alternately for 0.002 m.

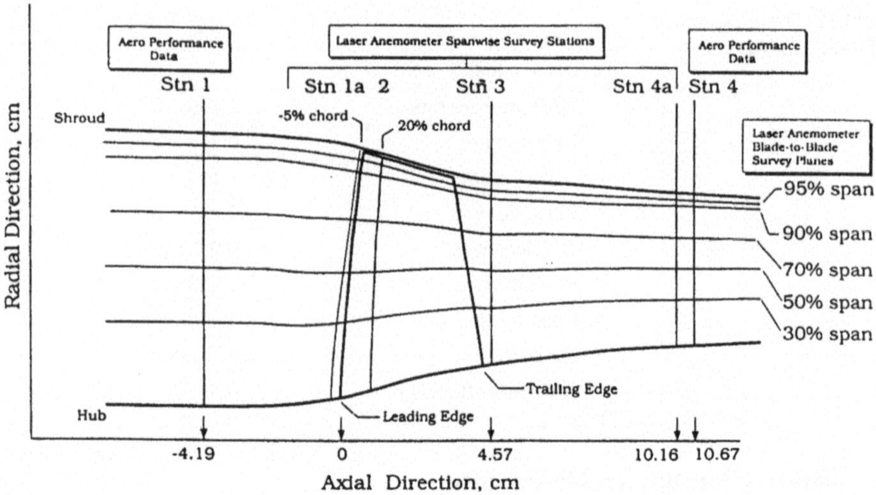


**Fig. 1** Measurement stations [19]

**Fig. 2** Definition of sweep

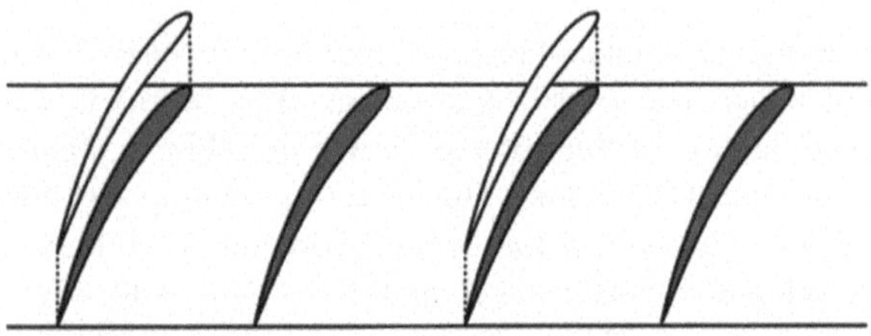

# 3 Model Validation

## *3.1 Computational Grid*

The multi-block structured grid, of about 1,300,000 cells, was shown in Fig. 3. Two passages were generated in one computational domain. An H-type grid was used for the main flow region, while a composite H/O-type grid was generated in tip clearance region. The whole grid consisted of 146 cells in azimuthal direction, 49 cells in spanwise direction, and 177 cells in streamwise direction. The cell width on solid walls was less than $1 \times 10^{-6}$.

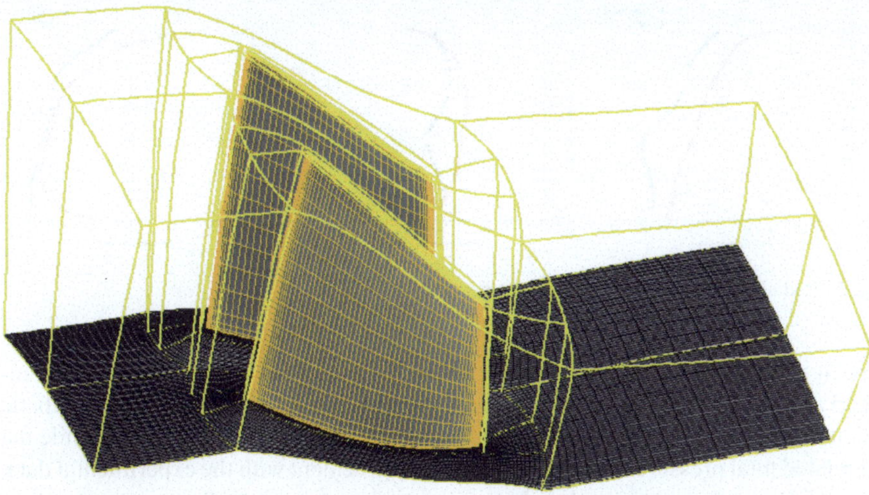

**Fig. 3** Computational grid for two passages

## 3.2 *Comparison Between the Experiment and Calculation*

For all the investigated rotor geometries, the flow field was computed in same conditions. The rotating speed was fixed to 17,188 rpm for all rotor types, and the inlet total pressure and total temperature were set to $p_{01} = 101325$ Pa and $T_{01} = 288.15$ K. The convergence criteria were fixed to $1 \times 10^{-6}$. Steady-state solutions were computed using S-A turbulence model [20]. The experimental data were investigated from AGARD test case report [19].

The experimental data and calculation results are presented in Fig. 4. The measured and calculated mass flow rate were normalized using the corresponding mass

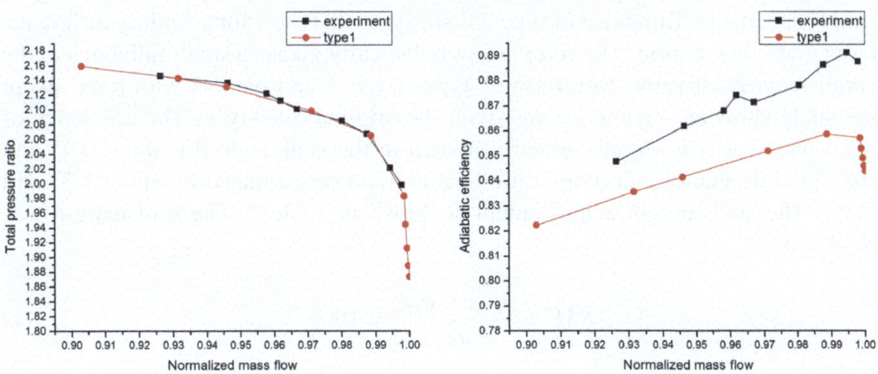


**Fig. 4** Performance maps for experiment and calculation of original type

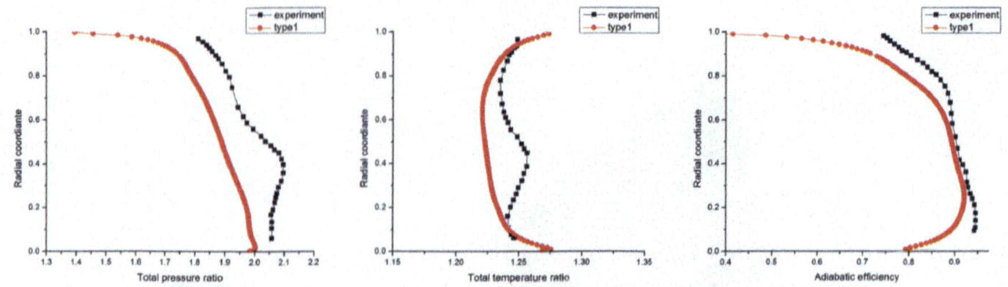


**Fig. 5** Radial results at station 4 for choke point

flow at choke point. Our model predicted well the choking point, which was computed a value of 20.915 kg/s against the experiment one of 20.93 kg/s. The adiabatic efficiency fell by slightly less (about 3.6%) than experiment in peak zone, while the calculated total pressure ratio achieved a great agreement with the experimental data. Figure 5 shows the spanwise distributions of pitch-averaged total pressure ratio, total temperature ratio, and adiabatic efficiency at station 4 for choke point. The calculated profiles get an acceptable result with the experimental data.

## 4 Simulation Results and Analysis

### *4.1 Overall Features*

All the novel types were simulated by solving the Reynolds-averaged Navier–Stokes equation with S-A turbulence model. For these alternate and unaltered rotors, we adopt the same computational setting including the boundary conditions and grid setting as we showed previously.

The performance comparisons with all novel rotor types are presented in Fig. 6. From Fig. 6, the performance of type 2 is similar with type 1 for a slightly difference in low mass flow region. The rotor swept holistically means a small influence in the overall features of rotor. But, the two types (type 3 and type 4) with rotor swept alternately show an obvious change with the original rotor type. The efficiency of type 3 and type 4 is slightly underestimated in the peak zone (by about 0.93 and 1.16%), while the total pressure ratio is slightly underestimated by about 1.53 and 2.82%. The stall margin enhancement is shown in Table 2. The stall margin was calculated by

$$SM' = \frac{q_{md} - q_{ms}}{q_{md}} \times 100\% \tag{1}$$

where $q_{md}$ is the mass flow rate of choke point and $q_{ms}$ is the mass flow rate of near-stall point.

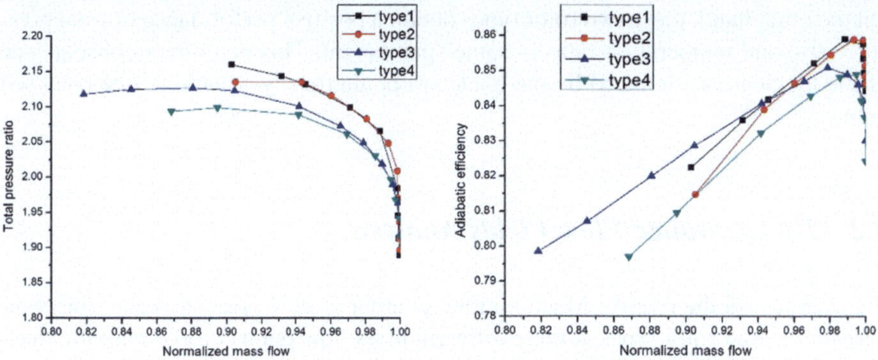


**Fig. 6** Performance maps of all rotor types

**Table 2** The stall margin for all types

| | $q_{md}$ (kg/s) | $q_{ms}$ (kg/s) | Stall margin (%) |
|---|---|---|---|
| Type 1 | 20.915 | 18.884 | 9.71 |
| Type 2 | 21.04 | 19.045 | 9.48 |
| Type 3 | 20.93 | 17.13 | 18.16 |
| Type 4 | 20.982 | 18.22 | 13.16 |

The stall margin of type 3 can reach up to 18.16%, while that of the original rotor type 1 is only 9.71%. Moreover, type 4 achieved a slightly higher stall margin by 13.16%. The novel types of rotor swept alternately obtained a great achievement in stall margin improvement.

Figure 7 compares the spanwise distribution of the total pressure ratio, total temperature ratio, and adiabatic efficiency of all rotor types at choking mass flow. Clearly, for all three performance quantities, type 1 is similar with these of type 2. The previous viewpoint about the influence of rotor swept holistically is approved by the spanwise distribution files. Between 30 and 70% span height, for both two alternate swept rotors, a slightly negative performance of adiabatic efficiency can be observed.

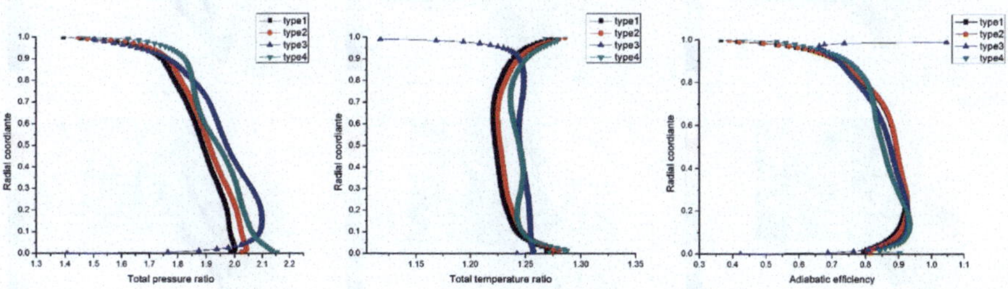


**Fig. 7** TRadial results of four types at choke point

On the other hand, these two rotor types obtain a positive performance of total pressure ratio and temperature ratio in same span height. This is a direct consequence of the tip clearance induced by the blade swept alternately, which will be analysed below.

## 4.2 *Tip Clearance Flow Fields Analysis*

Figure 8 shows the relative Mach number contour at 95% span. It is also the comparison for four rotor types at three different mass flow points (100% choking mass flow, 96% choking mass flow, near-stall point). For all rotor types, with the decrease of mass flow, the low Mach number area is expanded from the trailing edge to the blade suction surface. And this expanding process would be prevented by the shock wave before the leading edge. The phenomenon can be observed obviously in type

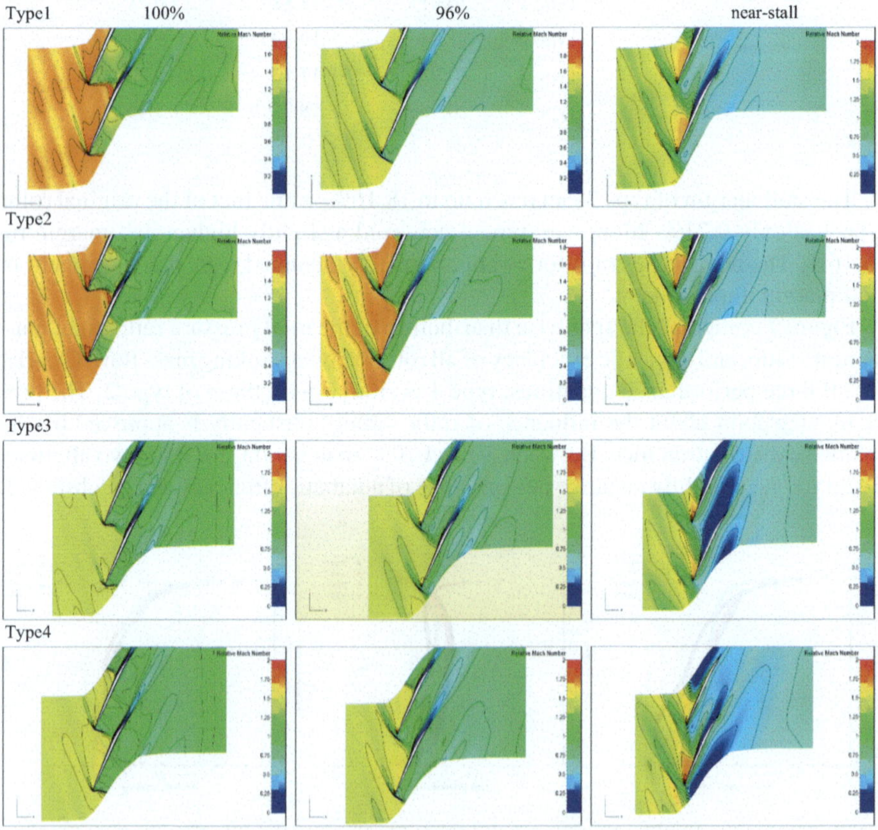


**Fig. 8** Comparison of relative Mach number at 95% span

3 and type 4 at near-stall point, which that the low Mach number areas are different in two passages. Meanwhile, these areas were prevented by the shock wave, too. The shock wave can reach a deeper position in passages for the alternate swept rotor types, which leads to that low Mach number area was kept in rearward position in one passage, while the low Mach number area deteriorated in adjacent passage. As we all know, the low Mach number areas can be considered as rotating stall areas. For the rotors swept holistically, the rotating stall would be expanded to whole flow field. Concurrently, the rotating stall cannot be expanded to whole field because the adjacent passage can stay in stable condition with the rotor swept alternately. It also leads to a decrease in adiabatic efficiency and total pressure ratio which can be observed in Fig. 6.

Figure 9 shows the Entropy at 95% span. It is also the comparison for four rotor types at three different points. According to the research of Vo et al. [6], the phenomena that the interface between the tip clearance and oncoming flows becomes

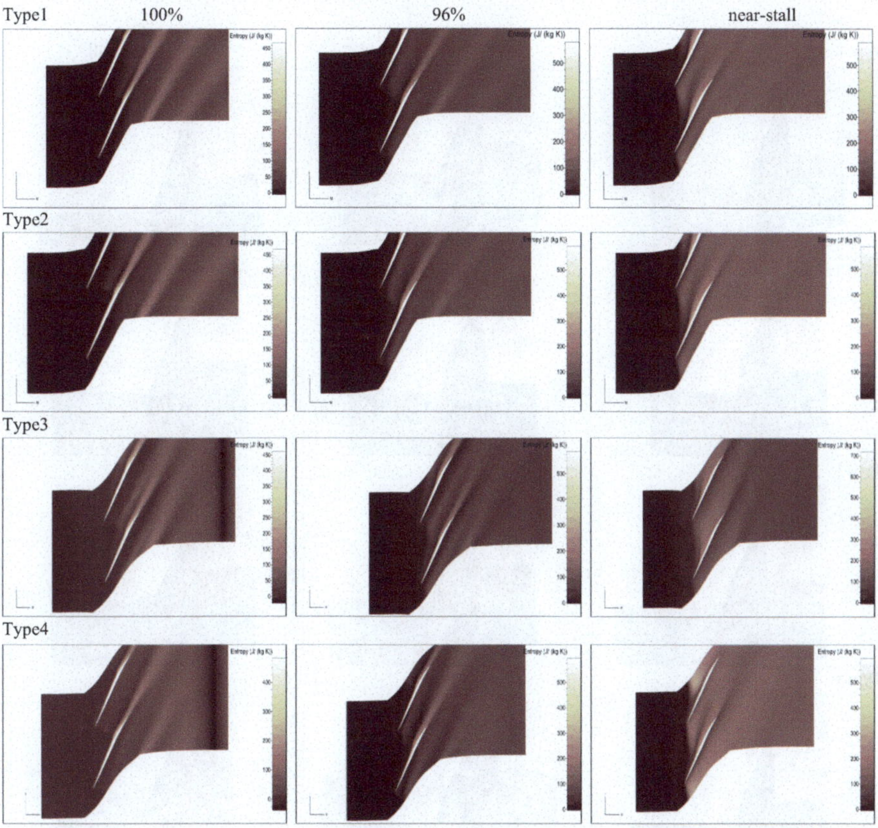


**Fig. 9** Comparison of Entropy at 95% span

parallel to the leading-edge plane lead to the onset of short length-scale (spike) rotating stall disturbances. This was demonstrated in Fig. 9. The processes of interface moving forward are different in adjacent passage for type 3 and type 4. It means that interface becomes parallel to the leading edge in one passage, while the interface in adjacent passage is still on the processing of moving forward. The interface moving was influenced by tip clearance which can be analysed in Fig. 10.

Figure 10 shows the tip clearance streamlines for these conditions. From the type 1 and type 2, the interface moving can be observed. The tip clearance was accumulated in the leading edge of blade. And the rearward position of the passage was shown as blank, which means that it is rotating stall, as shown in Figs. 8 and 9. For the types with rotor swept alternately, rotating stall was developed in a serious condition, while flow was still kept stable in suction surface of adjacent blade.

The distance between leading edge of blades was enhanced for the rotor swept alternately. Most of the tip leakage flows into the passages, which has a smaller effect on the tip clearance flow of adjacent blades. The tip clearance flow can interact on

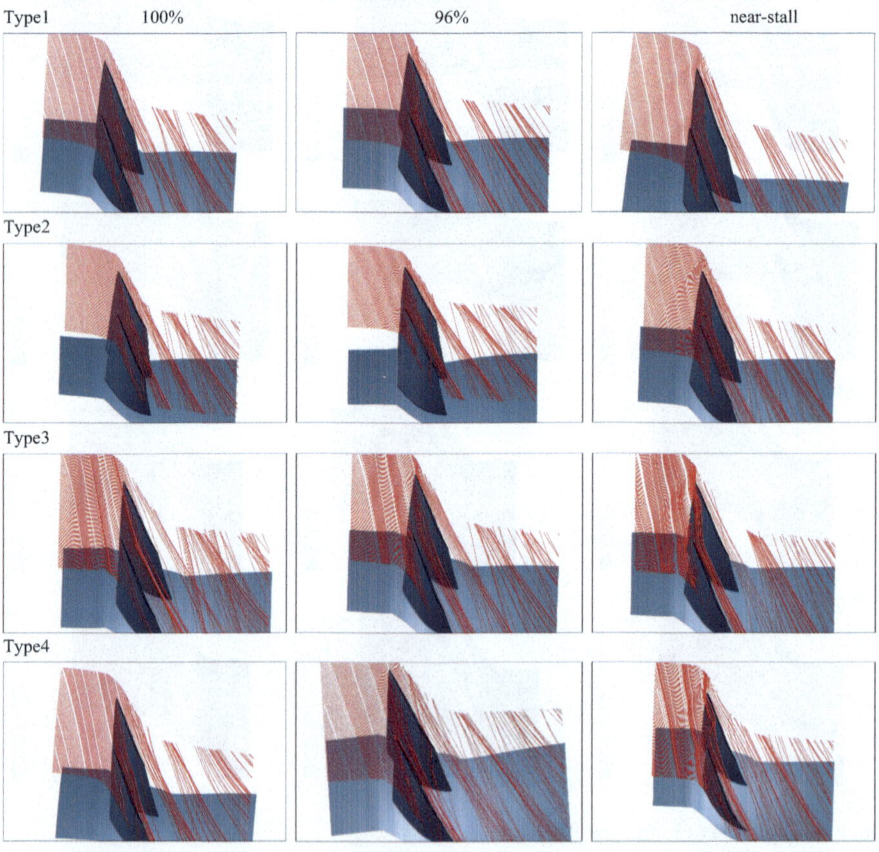


**Fig. 10** Comparison of streamlines in tip region

each other blades from Type 1 and Type 2. The phenomena lead to the stall margin improvement, which the flow accumulated in the leading edge had smaller influence on adjacent tip clearance. As shown in Fig. 8, the rotating stall cannot be expanded to whole flow field because the adjacent passage can stay in stable condition.

## 5 Conclusions

In this paper, the effects of rotor swept alternately or holistically on the stall margin are investigated. A novel rotor type is presented, which integrates an improved stall margin. Three new types are modelled based on the NASA rotor 37. The results are summarized as followed.

The rotors with swept alternately achieve a significant improvement in stall margin, while the rotor with swept holistically has no influence on overall performance and details of flow field. The stall margin of new rotor types can reach up to 18.16%, while that of the original rotor is only 9.71%. The efficiency of rotors with swept alternately is slightly underestimated in the peak zone (by about 0.93 and 1.16%), while the total pressure ratio is slightly underestimated by about 1.53% and 2.82%.

The flow fields are different which are shown between the rotors with swept alternately and other rotor types. For the types with rotor swept alternately, flow stability is shown different in different passage. Rotating stall was developed in a serious condition, while flow was still kept stable in suction surface of adjacent blade. It means the rotor can be operated at a lower mass flow rate. As expected, a decrease of adiabatic efficiency was obtained by these rotors swept.

The distance between leading edge of blades was enhanced for the rotor swept alternately. The rotor swept alternately reduces the interaction of tip clearance flow between adjacent blades.

**Acknowledgements** The first author greatly appreciates the support from China Scholarship Council. This work is also supported by Natural Science Foundation of China (No. 51576124, No. 51506126). The support from the United Innovation Centre (UIC) of Aerothermal Technologies for Turbomachinery is also acknowledged.

## References

1. Kai Z (2008) The investigation of the tip clearance flow of axial flow compressor with adjustable tip additional blades. Harbin Institute of Technology, Harbin
2. Paduano J, Epstein AH, Valavani L, Longley JP, Greitzer EM, Guenette GR (1991, June) Active control of rotating stall in a low speed axial compressor. In: ASME 1991 International Gas Turbine and Aeroengine Congress and Exposition. American Society of Mechanical Engineers, pp V001T01A036–V001T01A036

3. Day IJ (1991, June). Stall inception in axial flow compressors. In: ASME 1991 International Gas Turbine and Aeroengine Congress and Exposition. American Society of Mechanical Engineers, pp V001T01A034–V001T01A034
4. Houghton T, Day I (2011) Enhancing the stability of subsonic compressors using casing grooves. J Turbomach 133(2):021007
5. Rukavina J, Okiishi T, Wennerstrom A (1990) Stall margin improvement in axial-flow compressors by circumferential variation of stationary blade setting angles. In: 26th joint propulsion conference, p 1912
6. Vo HD, Tan CS, Greitzer EM (2008) Criteria for spike initiated rotating stall. J Turbomach 130(1):011023
7. Pullan G, Young AM, Day IJ, Greitzer EM, Spakovszky ZS (2015) Origins and structure of spike-type rotating stall. J Turbomach 137(5):051007
8. Tan CS, Day I, Morris S, Wadia A (2010) Spike-type compressor stall inception, detection, and control. Annu Rev Fluid Mech 42:275–300
9. Ahn CS, Kim KY (2002) Aerodynamic design optimization of an axial compressor rotor. ASME Paper GT-2002-30445
10. Denton JD, Xu L (1998) The exploitation of three-dimensional flow in turbomachinery design. Proc Inst Mech Eng Part C: J Mech Eng Sci 213(2):125–137
11. Yamaguchi N (1991) Secondary-loss reduction by forward-skewing of axial compressor rotor blading. 91-YOKOHAMA 8
12. Hah C, Puterbaugh SL, Wadia AR (1998, June) Control of shock structure and secondary flow field inside transonic compressor rotors through aerodynamic sweep. In: ASME 1998 International Gas Turbine and Aeroengine Congress and Exhibition. American Society of Mechanical Engineers, pp V001T01A132–V001T01A132
13. Wadia AR, Szucs PN, Crall DW (1997, June) Inner workings of aerodynamic sweep. In: ASME 1997 International Gas Turbine and Aeroengine Congress and Exhibition. American Society of Mechanical Engineers, pp V001T03A062–V001T03A062
14. Denton JD, Xu L (2002, January) The effects of lean and sweep on transonic fan performance. In: ASME Turbo Expo 2002: Power for Land, Sea, and Air. American Society of Mechanical Engineers, pp 23–32
15. He C, Ma Y, Liu X, Sun D, Sun X (2018) Aerodynamic instabilities of swept airfoil design in transonic axial-flow compressors. AIAA J 1878–1893
16. Moore RD, Reid L (1980) Performance of single-stage axial-flow transonic compressor with rotor and stator aspect ratios of 1.19 and 1.26 respectively, and with design pressure ratio of 2.05. NASA-TP-1659, E-138
17. Benini E, Biollo R (2006, January) On the aerodynamics of swept and leaned transonic compressor rotors. In: ASME Turbo Expo 2006: Power for Land, Sea, and Air. American Society of Mechanical Engineers, pp 283–291
18. He C, Sun D, Sun X (2018) Stall inception analysis of transonic compressors with chordwise and axial sweep. J Turbomach 140(4):041009
19. Dunham J (1998) CFD validation for propulsion system components (la validation CFD des organes des propulseurs) (No. AGARD-AR-355). Advisory Group for Aerospace Research and Development Neuilly-Sur-Seine (France)
20. Allmaras SR, Johnson FT (2012, July) Modifications and clarifications for the implementation of the Spalart-Allmaras turbulence model. In: Seventh International Conference on Computational Fluid Dynamics (ICCFD7), pp 1–11

# Improving the Cooling Air Supply System for the HPT Blades of High-Temperature GTE

**A. Minchenko, V. Nesterenko, I. Malinovsky and A. Revanth Reddy**

**Abstract** This paper describes the results of studies of the system for supplying cooling air to the HPT of high-temperature aviation bypass GTE. In the cooling cavity of the blade, a dividing partition is installed, which allows cold air to be supplied to the front cooling cavity of the blade, taken out of the high-pressure compressor, and to the rear cavity—air with lower pressure and temperature, taken from the intermediate stage of the compressor. Air cooled by the working blades of a GTE is fed into the tubes of a U-shaped air-to-air heat exchanger blown with air from the outer contour of this GTE. The results of the studies showed that the temperature of the air taken from the compressor in the AtA HE can be reduced by 110°–240°, depending on the geometric dimensions of the tubes and the configuration of the AtA HE. Problems to be solved: minimization of pressure losses in the external circuit of a gas turbine engine, development of methods for constructively increasing the intensity of air temperature reduction in tubular AtA HE and schemes for the optimal supply of this air to the inlet of cooled propeller blades. A tubular row-type AtA HE was designed, with micro-heat transfer intensifiers installed on the inner surface of small-sized thin-walled tubes, cylindrical or oval, into which cooled air drawn after the compressor or another, colder, but with lower pressure from its intermediate stage, is supplied. The system of cooling air cutoff, in the channels for supplying the rear cavity of the working blade of the turboprop engine on the cruising mode of GTE operation, implemented in the blades of the turbine rotor with a vortex matrix, is considered. In conclusion, the work presents recommendations on the design methodology of these units in modern and future aviation gas turbine engines.

**Keywords** Gas turbine engine · High-pressure turbine · Heat exchanger

---

A. Minchenko (✉) · V. Nesterenko · I. Malinovsky · A. Revanth Reddy
School of Propulsion Engineering, Moscow Aviation Institute (National Research University), Moscow, Russia
e-mail: acd-motor@mail.ru



Z. Jing (ed.), *Proceedings of the International Conference on Aerospace System Science and Engineering 2019*, Lecture Notes in Electrical Engineering 622,
https://doi.org/10.1007/978-981-15-1773-0_5

## Abbreviations

| | |
|---|---|
| GTE | Gas turbine engine |
| LPT, HPT | High-/low-pressure turbine |
| LPC, HPC | High-/low-pressure compressor |
| ND | Nozzle diaphragm |
| RB | Rotor blade |
| AtA HE | Air-to-air heat exchanger |
| ALV | Auto-lock valve of cooling air |
| BP/BPR | Bypass/bypass ratio |
| A/B | Afterburner |

## 1 Introduction

The development trends of modern civil and military aircraft engines are directed along the path of forcing the main operating parameters of the thermodynamic cycle: the degree of increase in air pressure in the compressor $\pi_c^* \pi^*{}_c$, air temperature behind the compressor $T_c^*$ and the temperature of the gas in front of the turbine $T_{gas}^*$. When creating gas turbines for promising gas turbine engines, one of the main problems is to ensure high efficiency and operational reliability of their thermally loaded parts and assemblies with increased parameters of the operating cycle. Existing practice has shown that the growth rate of the strength characteristics of materials used to manufacture parts of modern turbines does not keep pace with the rate of forcing the parameters of the thermodynamic cycle of a gas turbine engine. Table 1 shows the main parameters of current and future jet engines. From the data of Table 1, it can be seen that the increase in $T_{gas}^*$ in front of the turbine is accompanied by the simultaneous requirement to increase its resource, which is proof of the task relevance of improving its cooling system.

**Table 1** Key parameters of modern and perspective jet engines

| Generation | III (1965–1975 year) | IV (1985–1995 year) | V (2000–2015 year) | VI (2025–2030 year) |
|---|---|---|---|---|
| Construction type | Turbojet engine (A/B) | BP turbofan engine, Turbojet engine A/B | BP turbofan engine, Turbojet engine A/B | BP turbofan engine (A/B) |
| $T_{gas}^*$, K | 1450 | 1650 | 1850–1950 | 2100–2300 |
| $m/\pi_{\kappa\sum}$ | 0–2/15–20 | 2–6/25–35 | 8–12/35–45 | ≥12/50–60 |
| $C_R$, (kg/kg h) | 0.70–0.80 | 0.63–0.65 | 0.53–0.54 | 0.44–0.47 |
| Resource (cold/hot part; thousand cycles) | 5/– | 25/17 | 40/20 | 50/25 |

The main share—up to 70% of the cost of air for cooling the turbine is associated with satisfaction the requirements for reliability and ensuring the required service life of the blades and HPT disk, as the most loaded parts for which the reduction of long-term strength is crucial. Numerical studies of the thermodynamic model of a cooled gas turbine and experimental data from real aviation GTE have shown that at present, every 100 K increases in the value of $T^*_{gas}$ force an additional 2.0–3.3% of the air to be drawn from the compressor for cooling the turbine, which leads to reduced engine performance in general. Consequently, an extremely important and urgent task is to reduce the airflow for cooling the turbine, without which a further increase in the cycle parameters does not significantly improve the efficiency and thrust of the GTE. The creation of efficient cooling and thermal protection systems for components and parts of modern, high-temperature gas turbines is relevant and is one of the most important tasks, which ensures further improvement of modern and future aircraft engines.

The solution to this complex task can be achieved in the following ways:

- lowering the temperature of the cooling air taken from the compressor;
- the use of low-pressure air with less energy for its compression, which contributes to reducing the cost of cooling;
- effective distribution of cooling air through the cooling channels from its place of extraction to the entrance into the cavity of the turbine rotor blade;
- intensification of heat transfer in the cooling system channels;
- reduction of air leakage from the cooling system;
- regulation of airflow for cooling elements of a gas turbine;

## 2 HPT and Its Cooling System

The modern airborne system of the aviation jet engines solves a large number of very different tasks. In addition to ensuring the required level of turbine components cooling, it is necessary to ensure the optimum value of the axial force acting on the thrust bearing of the rotor, cooling the main rotor bearings, supercharging the rotor support labyrinth system, in order to prevent oil from being released into the flow part of the jet engine, etc. Therefore, the design of the air system of an aircraft engine should be carried out taking into account the whole complex of factors affecting the efficiency of the engine as a whole [1].

Figure 1 shows a structural diagram of a modern turbine, in the bypass of which AtA HE is installed with a section of three tubes, in cross section having an inline configuration [2].

Figure 2a shows the complex dependence of the change [3] of the cooling intensity of the RB of the HPT on the magnitude of the gas temperature in front of the turbine $T^*_{gas}$ and the magnitude of its cooling intensity:

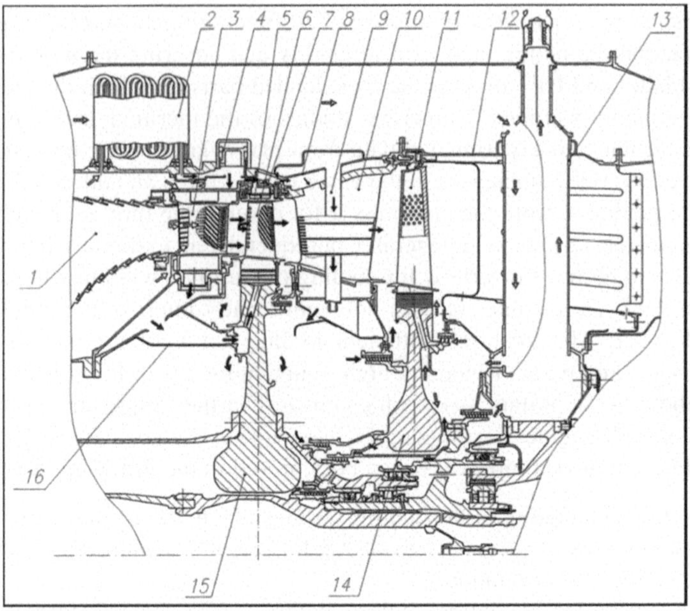


**Fig. 1** The composition of the nodes and the scheme of the turbine cooling system of a two-shaft turbofan engine with a small degree of BPR: 1—a combustion chamber; 2—cross-current AtA HE with *U*-shaped tubes; 3—cooled ND of the HPT; 4—ALV; 5—RB of HPT; 6—outer ring; 7—cellular inserts; 8—rim of the ND; 9—bypass cooling air into the wheel space cavity; 10—ND blade of the LPT; 11—RB of LPT; 12—the case of a back support of the turbine; 13—LPT power rack; 14—LPT disk; 15—HPT disk; 16—swirling device

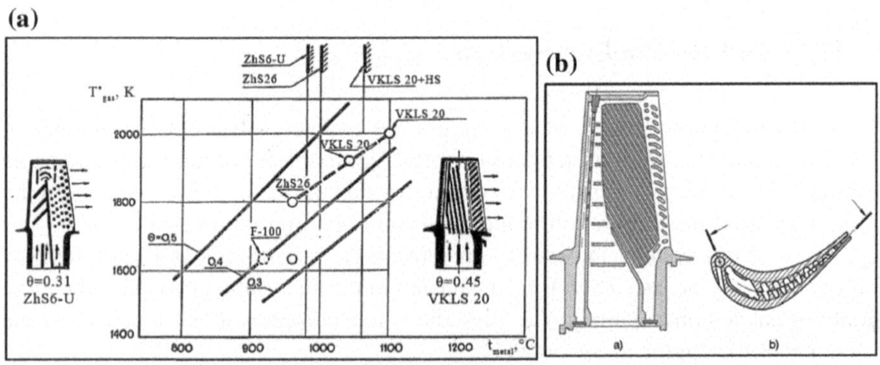


**Fig. 2** **a** Effect of $T^*_{gas}$ before the HPT, the magnitude of the intensity of cooling and the characteristics of the metal on the temperature of the HPT blade. **b** Show a screw-channel RB of a HPT, with cooling intensity is $\theta = 0.54$

$$\theta = \frac{T_{gas} - T_{blade}}{T_{gas} - T_{cooling\ air}} \quad (1)$$

$$\theta = \frac{1700 - 1165}{1700 - 710} = 0.54 \quad (2)$$

$$\theta = \frac{1700 - 1120}{1700 - 626} = 0.54 \quad (3)$$

Below, on dependences (2) and (3), for example, the results of two calculations are shown, showing the magnitude of the decrease in temperature of the turbine blade depending on the change in temperature of the cooling air. When the intensity of its cooling is $\theta = 0.54$, a decrease in the cooling air temperature by 84° leads to a decrease in the temperature of the blade wall by 45°, their ratio to:

Unlike Fig. 1, Fig. 3 shows the cooling system of the HPT with AtA HE. In HE1, the cooling air is supplied due to the HPC, then this air $G_6$ enters through the twisting grate under the covering disk, cools its upper part and, with a temperature increased by 60°, is directed into the internal cavity of the RB.

Air $G_{17}$ from the unloading cavity for the HPC is fed to the entrance to the HE2 and then sent to the wheel space cavity.

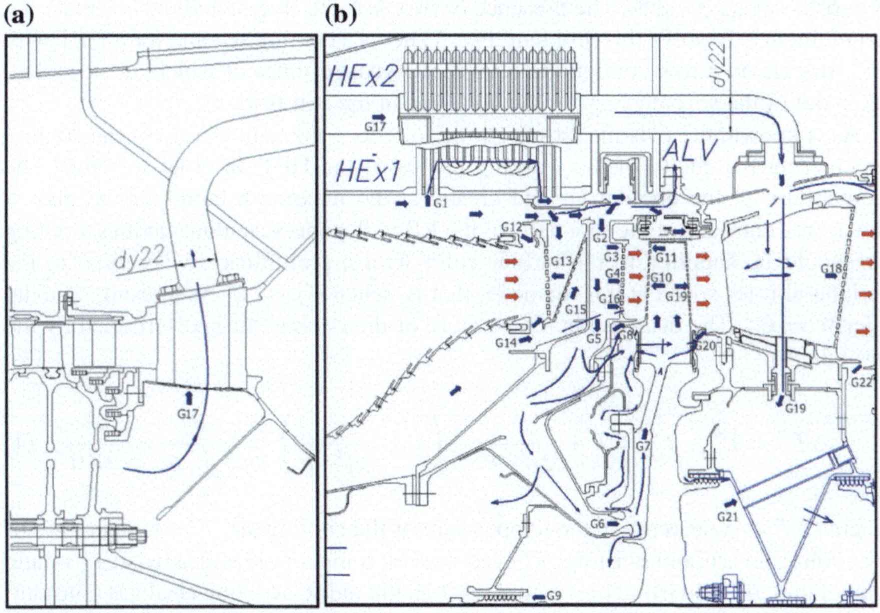


**Fig. 3** **a** Scheme of selection of cooling air supplied to the disk wheel space of HPT; **b** scheme of the cooling system of the turbofan engine with a RB having a one-sided supply of cooling air. In the blade, under the cover plate, cooling air is supplied due to the pressure build-up. This system includes two AtA HE, where the cooled air taken out from the HPT is supplied to HE1, and the air from the compression cavity is recycled to HE2

Table 2 shows the results of calculating the values of the flow rates of cooling air, its pressures and temperatures, which characterize the features and efficiency of the consider cooling system. The disadvantage of the studied cooling system is an increase in the temperature of the cooling air due to its low peripheral speed, compared with the diametric size corresponding to the base of the turbine blade, as well as due to its heating from the disk. The level of this increase in temperature corresponds to about half the magnitude of the decrease in its temperature in AtA HE.

Figure 4 presents a new cooling scheme for a modernized HPT, also with AtA HE and ALV, which can be used in the designs shown in Figs. 1 and 3.

As can be seen from Fig. 4, in this HPT there are two supplies of cooling air entering the internal cavity of the working blade: air $G_6$ is fed into the front cavity 4 of the RB HPT, which enters through the twisting lattice 2, which is located as high as possible; $G_{18}$ air enters the rear cavity 6 through another twisting lattice 7 and this air is cut off at cruising mode of engine operation by means of a ALV installed in the manifold under HE2, in order to save fuel consumption due to a decrease in air supply to cool the turbofan.

For the jet engines with a low BPR, the level of the outer contour of the outer loop with the AtA HE units is of great importance. One of the requirements for AtA HE of promising GTE is the minimum hydraulic resistance $\Delta\sigma_{\text{AtA}} = \text{k} \cdot \sigma_{\text{II}}$, where: $\sigma_{\text{II}} = 4\text{–}6\%$; $\Delta\sigma_{\text{AtA}} \leq 2.5\%$. The presence of two AtA HE does not allow to satisfy the requirements. To solve this problem, Fig. 5 shows a cooling scheme for a HPT with one AtA HE in a row configuration, in the U-shaped tubes of which air is cooled, taken out of the seventh stage of the HPC and at the exit from it.

As is known, the twisting lattice allows to reduce the temperature of the cooling air entering the cavity of the working blades of the HPT. In order to reduce the temperature of the cooling air and create, at the maximum radius of the disk, a shock-free entrance of the root parts of the RB to the interscapulum cavities, cooling air should be supplied to the turbine rotor with a preliminary twist equal to the peripheral rotor speed at this diameter, that is, when $C_{1u} = U$. As a result, an axial inlet $W = C_a$. The decrease in temperature of the cooling air is determined by the formula (4):

$$\Delta T^* = T_c^* - T_{c_a}^* = T + \frac{c^2}{2\frac{k}{k-1}R} - \left(T + \frac{c_a^2}{2\frac{k}{k-1}R}\right) = \frac{u^2}{2\frac{k}{k-1}R} = \frac{u^2}{2010} \quad (4)$$

where: $\Delta T^*$—a decrease in the temperature of the cooling air; $T_c^*$—temperature of the cooling air without twisting; $T_{c_a}^*$—cooling air temperature with a twist; $T$—static temperature; $C$—swirling flow rate; $k$—adiabatic index; $R$—universal gas constant; $C_a$—axial speed of the cooling air; $u$—circumferential speed of the rotor in the place of installation of the twisting lattice.

**Table 2** Results of calculating the values of the flow rates of cooling air

| Local cooling airflow rates (Fig. 4) | | $G_1$ | $G_2$ | $G_3$ | $G_4$ | $G_5$ | $G_6$ | $G_7$ | $G_8$ | $G_9$ | $G_{10}$ | $G_{11}$ | $G_{12}$ |
|---|---|---|---|---|---|---|---|---|---|---|---|---|---|
| Parameters | $G$ (%) | 5.982 | 5.982 | 0.67 | 1.882 | 2.882 | 1.976 | 0.726 | 1.321 | 0.132 | 1.367 | 1.982 | 0.11 |
| | $P$ (Pa) | 39.32 | 32.05 | 29.78 | 30.07 | 30.95 | 30.95 | 29.58 | 17.68 | 17.58 | 21.21 | 39.28 | 21.21 |
| | $T$ (K) | 610 | 505 | 505 | 505 | 515 | 515 | 515 | 540 | 455 | 635 | 610 | 659 |
| Local cooling airflow rates (Fig. 4) | | $G_{13}$ | $G_{14}$ | $G_{15}$ | $G_{16}$ | $G_{17}$ | $G_{18}$ | $G_{19}$ | $G_{20}$ | $G_{21}$ | $G_{22}$ | $G_{23}$ | $G_{24}$ |
| Parameters | $G$ (%) | 1.53 | 3.616 | 2.592 | 0.652 | 5.141 | 3.241 | 2.59 | 0.778 | 1.009 | 0.353 | 0.415 | 0.565 |
| | $P$ (Pa) | 39.12 | 39.75 | 34.73 | 37.93 | 26.53 | 20.71 | 19.28 | 10.67 | 9.054 | 11.36 | 6.2 | 5.994 |
| | $T$ (K) | 615 | 610 | 631 | 617 | 574 | 496 | 568 | 586 | 543 | 511 | 518 | 552 |

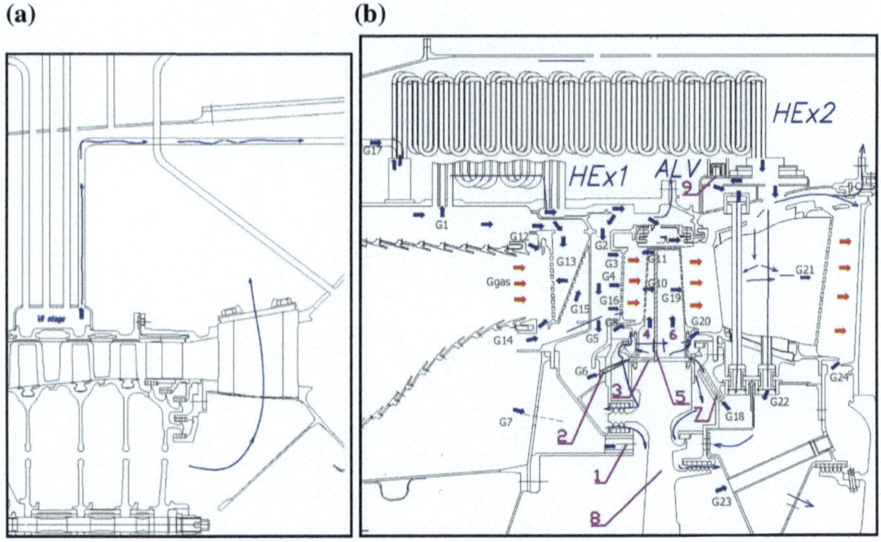


**Fig. 4** **a** Scheme of selection of the cooling air supplied to the rear cavity of the RB of the HPT; **b** scheme of the cooling system of a HPT with a RB having a radial partition 5, ALV and two AtA HE, where HE1 supplies air drawn from the ALV, and HE2 supplies air with a lower temperature taking of the seventh step of the HPC

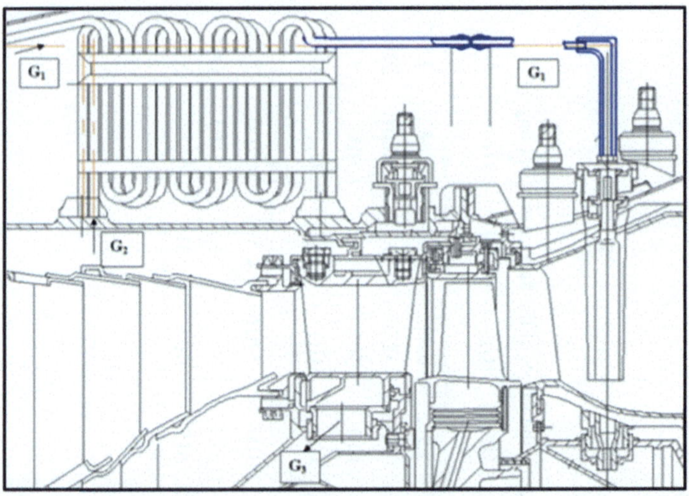


**Fig. 5** Constructive scheme with one AtA HE, in which the air is cooled $G_1$ taken from the intermediate stage and $G_2$, taken behind the last stage of the HPC

## 3 Constructive Ways to Improve the Efficiency of HPT Tubular AtA HE

Consider the features of the choice of design and parameters of AtA HE HPT with different diameter of thin-walled tubes. A known structural scheme and parameters of a tubular AtA HE cross-current installed on the body of the combustion chamber in the internal circuit of the jet engines (Fig. 1). In these schemes, there is a cooling air cutoff system on the cruise mode of the engine in order to increase its efficiency. The initial outer diameter of the tubes of the studied AtA HE is 5.0 mm, the wall thickness is 0.3 mm, the number of tubular modules is $N_m = 64$. The number of tubes in one module is six. The total number of tubes is 384, the length of one tube in the sweep is $L = 679$ mm. All tubular modules are evenly spaced around the circumference on the annular body of the jet engine bypass. Let us present the operational parameters of the AtA HE: for cooled air, inside the tubes—flow rate $G_1 = 6.45$ kg/s, gas inlet pressure $P_{11} = 23 \times 10^5$ Pa, gas inlet temperature $T_{11} = 774$ K; in the heated air of the external circuit, the flow rate is $G_2 = 40$ kg/s, the gas pressure at the inlet is $P_{21} = 3.66 \times 10^5$ Pa, the gas temperature at the inlet is $T_{21} = 438$ K. In an experimental study of AtA HE in the engine, it was found that the loss of the total pressure of the cooled air, in general, amounted to 8.5%. In the bypass—the total pressure loss from the AtA HE installation was 1.6%. The change in the efficiency of heat exchange of this AtA HE, estimated by the parameter ε, is determined by the initial value of the cooled air temperature decrease $\Delta T_{\text{AtA}} = 120°$ and its estimated maximum value, $\Delta T_{\text{maxAtA}} = 180°$. Then we get the value of ε equal to:

$$\varepsilon = \frac{\Delta T_{\text{AtA}}}{\Delta T_{\text{max AtA}}} = 0.67 \quad (5)$$

To achieve such heat transfer efficiency, we consider the effect of changing two design parameters of this AtA HE, diametric size of the tubes 4.0–6.0 mm and their length—from 0.1 to 1.2 m, on the reduction level $\Delta T_{\text{AtA}}$ [4]. Figure 6 shows the graphs of the dependence $\Delta T = f\ (L;\ d)$ for cross-flow AtA HE tubes obtained by calculation in the ANSYS CFX system.

In Fig. 6, curves 1, 2, and 3 are smooth straight tubes, distinguished by their outer diameter—4.0 mm, 5.0 mm, and 6.0 mm, respectively; curve 4 is made with six bends of the tube, a U-shaped tube with a diameter of 6.0 mm, with heat transfer intensifiers installed on the inner smooth surface of these tubes, with projections 0.3 mm high, arranged with a step of 5.0 mm; curves 5 and 6 are *U*-shaped tubes with six turns, diameter 5.0 mm, smooth and with similar intensifiers of heat exchange on their inner surface. Analysis of the dependences shows that tubes of small diameter heat up faster. However, to reduce the temperature of the same amount of cooled air, the number of tubes of small diameter should be increased, which affects the increase in their mass, slightly increasing, by about 10%.

Figure 7a, b show the streamlines and the structure of the flow of cooled air in different parts of the *U*-shaped tubes, when the flow is turned and in straight sections.

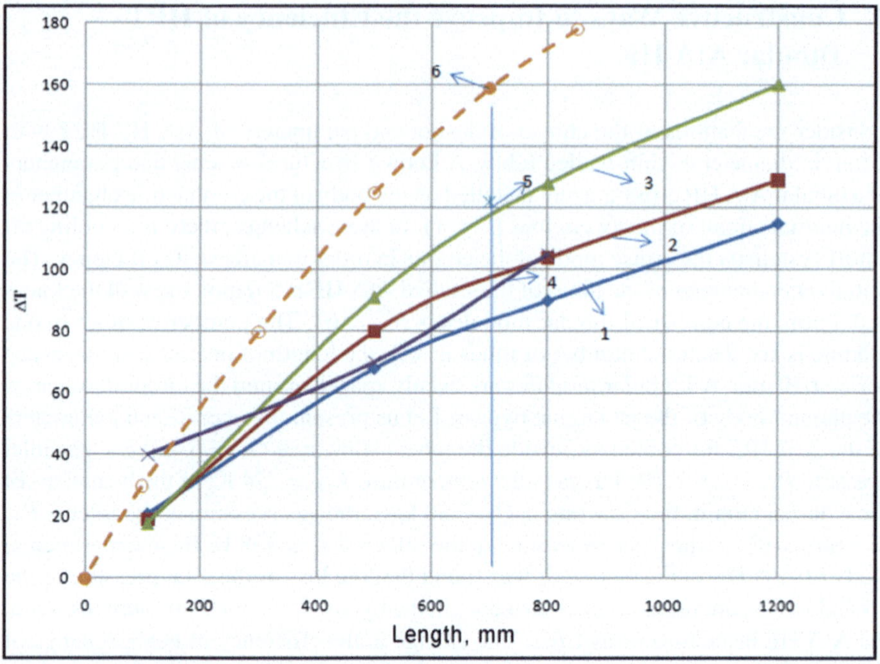


**Fig. 6** Calculated graphs of the dependence $\Delta T = f\ (L;\ d)$ for cross-flow inertial current tubes, calculated in the ANSYS CFX

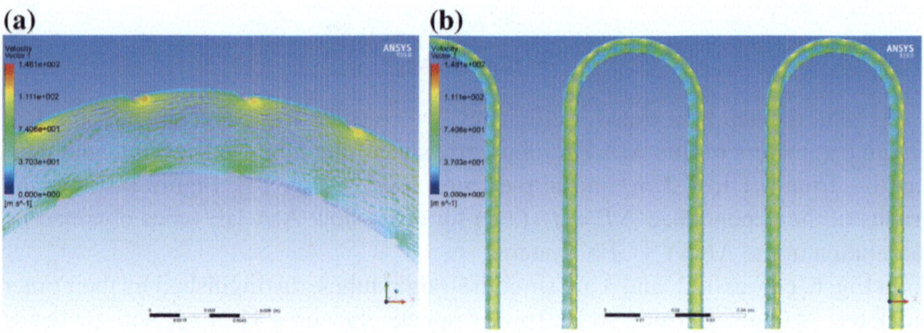


**Fig. 7** The structure of the flow of cooled air: **a** in the radius of the channel of AtA HE; **b** in its two *U*—shaped cylindrical channels

The protrusions localize the spread of the thickened boundary layer; therefore, the flow velocity adjacent to the cooled tube walls increases, which proportionally affects the increase in the heat exchange rate of this AtA HE.

## 4 Conclusion

1. In a bypass jet engine, a tubular inline AtA HE installed in its outer contour must have rows of tubes oriented along the axis of the engine. These tubes can be supplied with cooled air taken from different stages of high-pressure boiler with different temperatures, as well as cooling other elements of the construction of the HPT.
2. The use of heat exchange intensifiers in the form of annular or screw protrusions with a height of not more than 0.3 mm, installed in straight or U-shaped cylindrical tubes with diametric dimensions of 4.0–6.0 mm, allows increasing the cooling rate of hot air flowing through these tubes, about 50–60%, and for the investigated lengths of tubes increases $\Delta T_{\text{cooling}}$ up to values of 160–180°.
3. The loss of pressure of the cooled air in the AtA HE tubes, when using annular heat transfer intensifiers with a height of not more than 0.3 mm, should be no more than 15–20%.

## References


1. Nesterenko VG, Nesterenko VV, Asadollahi Gokhiekh A et al (2014) Research and analysis of the efficiency of air cooling systems for blades of high-pressure turbines of gas turbine engines. Aerosp Eng Technol 7:83–93
2. Hronin DV (1989) Design and design of aircraft gas turbine engines: textbook. Mashinostroenie, 368 p
3. Revant Reddy A, Nesterenko VG (2018) Constructive methods for improving the critical nodes of the cooling system of modern high-temperature high-pressure turbine of aviation gas turbine engines. Sci Tech Bull Volga Reg 5:73–77
4. Kalinin EK (1998) Effective heat exchange surfaces. Energoatomizdat, 408 C

# Investigation into the Wear Process of Laser Cladding from TiC Multilayer Coating for GTE Shrouded Blade Platforms

**P. Kleimenov, L. Lesnevskiy, M. Lyakhovetskiy, A. Ionov and E. Marchukov**

**Abstract** The object of the study was to create a protective coating Ti-6Al-4V/TiC for applying the possibility technology by the laser metal deposition (LMD) method. One of the most suitable areas for LMD is to repair worn and/or damaged parts. It is well known that during GTE operation the fan blades are subjected to periodically varying forces causing vibrational bending and torsional stresses. One of the methods currently used to reduce the magnitude of the vibration stress is to use a shroud blade platform that significantly changes the vibration characteristics of the individual blades and the overall blade ring. The purpose of this study was to obtain information on the resistance to micro-motion of titanium carbide coatings obtained by LMD and also to consider the possibility of applying this technique to modernize and repair coatings for GTE shrouded blade platforms. For the study of fretting wear (dry alternating friction), a special friction machine «3308» was used. During the test, the run-in process of the sample and the functional dependence of the type were examined in detail. The wear and friction time of the optical viewing system and the measurement of the friction zone without removing the sample. The evaluation of the properties of the coating material was carried out at various stages of running-in wear. The experimental device for fretting wear and fretting corrosion significantly improved the possibility and contact pressures up to 1.0 GPa (with the influence condition of the desired corrosive environment that is easy to implement). As a result of research, new information was obtained on the wear rates of coatings under fretting wear conditions fans shrouded blade platforms characteristic and low-pressure compressors of gas turbine engines. The results of the study can be used to increase the design and technical development of the GTE platform anti-friction problem and to improve the application of hardened coatings on GTE components.

---

P. Kleimenov (✉) · L. Lesnevskiy · M. Lyakhovetskiy · A. Ionov · E. Marchukov
School of Propulsion Engineering, Moscow Aviation Institute (National Research University), Moscow, Russia
e-mail: nerofm@yandex.ru

 
Z. Jing (ed.), *Proceedings of the International Conference on Aerospace System Science and Engineering 2019*, Lecture Notes in Electrical Engineering 622,
https://doi.org/10.1007/978-981-15-1773-0_6

## Abbreviations and Acronyms

| | |
|---|---|
| AM | Additive manufacturing |
| LMD | Laser metal deposition |
| GTE | Gas turbine engine |
| UMC | Unmelted carbide |

# 1 Introduction

It is well known that during GTE operation the fan blades are subjected to periodically varying forces causing vibrational bending and torsional stresses. One of the methods currently used to reduce the magnitude of the vibration stress is to use a platform that significantly changes the vibration characteristics of the individual blades and the overall crown of the blade. Providing zero clearance between the contact surfaces of the shrouded blade platforms in working condition eliminates the most dangerous first form of flexural vibrations of the blades. Therefore, the timely preservation of the geometry of the platform is an important condition for fan reliability and resources [1].

During the operation of the engine in the stationary mode, the oscillating motion of the blades and the blade ring causes a slight movement of the contact surface of the platform. In the transient mode of operation, the contact surfaces also move relative to each other but under variable contact pressure conditions experimental studies of the electrical activity of the contact surfaces have important practical implications. Since the titanium alloy used to manufacture the fan blades has extremely low resistance under micro-motion conditions, it is practically impossible to operate the blade without the reinforcing coating on the contact surface of the shroud blade platform [2].

The development of modern technology is becoming more and more strict on the requirements of materials and the improvement of wear resistance parts are the top priority of many industries. A promising solution to this problem is to apply a composite coating material on components that are subject to intense wear. The creation and implementation of innovative structural materials with high physical and mechanical properties are currently relevant. Among structural materials, carbide titanium is widely used as a composite coating material composed of alloy titanium and carbide with a mass fraction of 20–70%. According to their characteristics, they occupy an intermediate position between steel and carbide.

In the Russian practice for the hardening of the contact surfaces, the soldering grains technique of hard alloy and detonation spraying of hard alloys based on tungsten carbide were previously used. Foreign manufacturers of GTE for the shroud blade platforms protection (compressors, turbines) use mainly plasma spraying technology using high-power plasma torches (Gator-Gard®, Teleflex Inc.; Plasmadyne SG-100/80 from Tafa Inc.). Tungsten carbide powder containing 17% cobalt (Metco

73, AMDRY 9830, from Sulzer Metco AG) is used as a material for the coating. This paper investigates the possibility of applying a promising technology to form a protective coating by LMD method. The application of this technology may in the long-term allow forming a shroud blade platform on a blade body during manufacture and recovery after operation in one technological process with the application of a wear-resistant coating. As a wear-resistant material, we consider a composite mixture of titanium with titanium carbide in different ratios [3].

Considering LMD technology and taking into account that some parts are too small, repairing such part using the conventional process could result in damage of the parts especially because of the large heat-affected zone. Laser metal deposition process produces very low heat-affected zone because of the laser that is used as the energy source and the properties of laser that made it to be applied only to the needed area as a result of its high directionality and coherency. The success of using LMD for effective repair can also be attributed to the rapid cooling of the process that prevents the melt pool from staying too long which is responsible for the low heat-affected zone. The high resolution that is also achievable with the laser beam makes it possible to create miniaturize parts and repair of such parts with high precision using the laser metal deposition process. Another unique property of LMD process is its capability to produce a new part on an existing part with high metallurgical integrity. These important characteristics of the LMD process have helped to position the technology in product remanufacturing [4–6]. Design modification is also made possible with this technology without having to start from scratch and without having to turn an existing machine or equipment into scrap. This additive manufacturing technology gives designers the flexibility to modify an existing design with ease.

The influence of laser power on microstructure, microhardness and wear resistance properties of laser metal deposited Ti6Al4V/TiC composite was studied by Mahamood et al. [7]. Ti6Al4V/TiC composites were deposited on a Ti6A4V substrate in order to increase the wear resistance of the base material. The composition ratio of 50 W% Ti64 and 50 W% TiC was deposited at various laser powers from 0.4 to 3.2 kW, at the same time all other process parameters remained constant. Each of the powders was placed in different hoppers directly from the powder feeder, and two powders were simultaneously applied. It was found that the microstructures in the field zone consist of unmelted carbide (UMC), resolidified carbide and dendritic TiC. It was revealed that the microhardness and wear resistance properties change with a change in the laser power. As a result, the study showed that the optimal laser power should be 2 kW for the set of processing parameters considered in the study. Dongdong et al. [8] investigated the effect of the addition of TiC particles on the resulting laser spraying properties of pure Inconel 625 alloy and TiC/Inconel 625 composites. The results of the study showed that the inclusion of TiC particles significantly changed the microstructure of the Ni–Cr matrix, and it should be noted that the central zone of the deposited areas dominated by columnar dendrites and cellular dendrites. Thus, with the addition of nano-TiC particles, more and more columnar dendrites were visible in the microstructure. The addition of nano-TiC particles led to the formation of thinner columnar dendrites with well-developed secondary dendritic branches. The increase in the number of particles of micro-TiC leads to the fact that columnar dendrites become coarser and degenerated, and the growth of secondary dendrite decreases. As a result, it was found that cellular dendrites are purified from TiC particles. The addition of TiC nanoparticles also led to a significant improvement in microhardness, tensile properties and wear resistance properties.

## 2 Experimental Procedures

*LMD machine*

The set of installations consists of (see Fig. 1)

- 3000 W ytterbium fiber laser in the optical range of λ ~ 1070 nm;
- complex control cabinets;
- control panel;
- KUKA robot;
- powder feeder;
- a fridge;
- optical components with powder nozzles.

The laser is designed as a separate stand. To transmit the output radiation, the fiber optic cable exits the rack and terminates at the optical connector at the entrance of the optics. The laser contains software and hardware tools that can be used to create laser machining programs. In order to achieve coordinated control of the components of the robot system of the composite (feeder, laser source, robotic manipulator) as well as input and monitoring of process parameters, the control

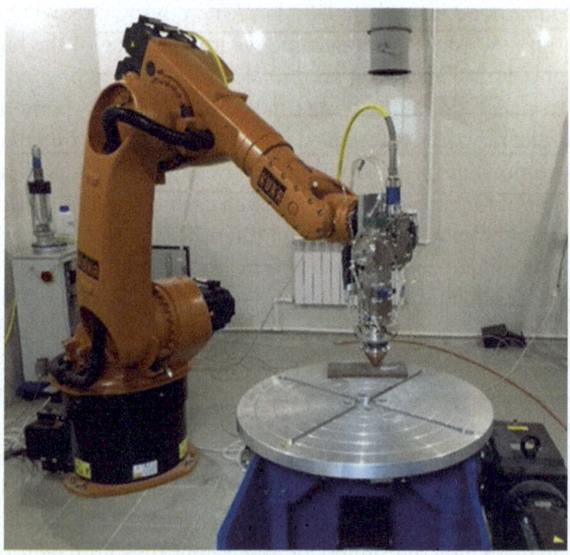

**Fig. 1** LMD set of installations

panel of the control cabinet is used. The controller controls the operation of the robot system according to a previously developed program.

During the operation of the complex system, the control module solves the following tasks:

- regulates the flow of transport and shielding gas;
- supplying a carrier gas to the powder feeder;
- providing shielding gas for the optics;
- the operation mode remote control of the powder feeder (setting the metering disk, the speed of the agitator);
- coordinates the matching of various components of the installation;
- when transferring control to the robot, the module transmits information from the robot to the laser source;
- emergency stop installation of components and release information about the failure in an emergency.

The KUKA robot is designed to provide a predetermined motion path of the optical focusing system relative to the substrate surface and a layer-by-layer configuration of the components on the 3D model. A powder feeder is used to deliver the powder material to an optical system with a powder nozzle and to ensure accurate dosing according to specified parameters. The refrigerator is designed for continuous water cooling of the laser system and optical system during LMD process. The optics provides collimation and focusing of the laser radiation to melt the powder material supplied through the powder nozzle to the focus of the laser radiation on the surface of the component structure.

*LMD Process*

Laser deposition was performed on a robotic laser composite using a fiber laser (KUKA robot + rotator MTS-250; fiber laser LS-4K). A laser nozzle with a multi-jet powder feed (YC-50 head for surface treatment) ensures that a roll of about 4 mm width and 1–1.2 mm height is applied in one pass. The overlapping layer thickness is 50% and for experienced formulations is 2–2.5 mm. Thus, the desired layer thickness is achieved in a small number of thermal cycles that heat the surrounding metal to a high temperature above 800 °C. The supply of the powder material was carried out using a dispenser having two flasks.

As the filler, a metal powder obtained by a method of atomizing a molten gas is used. All powders used have a fractional size of 40–150 microns. The powder mixture of Ti-6Al-4V/TiC in a different proportion by volume is welded onto the sample site in the working size at the mode (see Table 1). Substrate is the titanium alloy Ti-6Al-4V. The titanium-based powder is a self-fluxing alloy having high corrosion resistance and wear resistance.

Two types of titanium carbide are used as the hardened phase: cut and agglomerated carbides. The agglomerated titanium carbide in the cobalt ribbon is a finely divided carbide which is combined into a spherical particle using a cobalt base. This carbide fuses into the matrix and forms a continuous uniform transition without lattice distortion and microdefect formation.

**Table 1** Ratio of powder mixture and application features

| № | Ti-6Al-4V | WC-8Co | % | Σ | Number of layers | H |
|---|---|---|---|---|---|---|
| 1 | 0.6 | 1.4 | 30/70 | 2 | 2 | 0.6 |
| 2 | 0.5 | 1.5 | 25/75 | 2 | 2 | 0.5–0.6 |
| 3 | 0.4 | 1.6 | 20/80 | 2 | 2 | 0.5 |
| 4 | 0.3 | 1.7 | 15/85 | 2 | 2 | 0.4–0.5 |
| 5 | 0.2 | 1.8 | 10/90 | 2 | 2 | 0.3–0.4 |

A powder mixture Ti-6Al-4V/TiC and chipped titanium carbides are deposited at the workplace indicated. The characteristics work of laser consists of the main parameters of coating operating mode (see Table 2).

*The Samples Coating Process*

1. The surfacing was carried out on experimental cylindrical samples of size 10 × 6.8 mm. The modes and the percentage of titanium carbides were varied. To eliminate cracks, preheating of the blanks to a temperature of 450 °C was used.

   In the process of applying coatings on experimental cylindrical samples of 10 × 6.8 mm in size, no cracks or pores were found, and the carbides remained undissolved. Figures 2 and 3 show the macrostructure of the fusion zone with the substrate for samples with a ratio of 30/70 and 10/90. The fusion is even;

**Table 2** Coating operating mode

| Parameter | Value |
|---|---|
| Power (%) | 36–38 |
| Moving rate (mm/s) | 5 |
| Step (mm) | 2.5 |
| Substrate | Ti-6Al-4V |

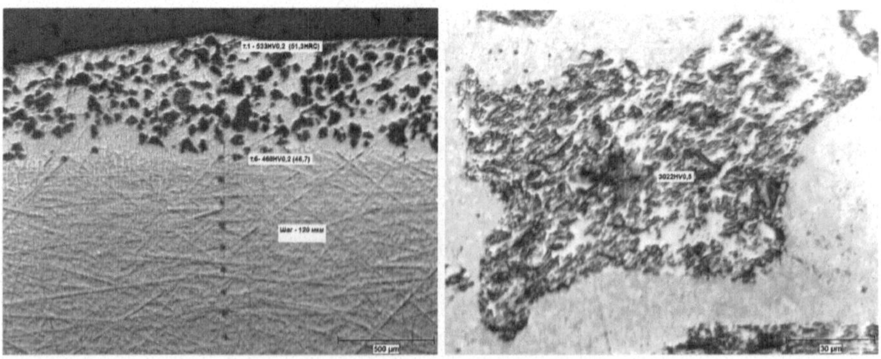


**Fig. 2** Fusion zone macrostructure with the samples substrate in 30/70 ratio

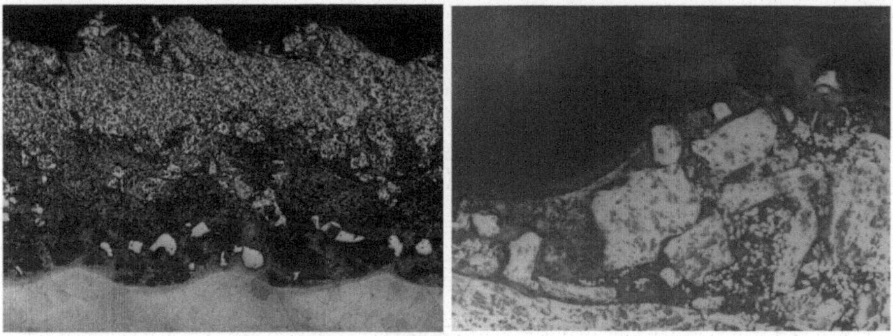

**Fig. 3** Fusion zone macrostructure with the samples substrate in 10/90 ratio

the carbides are evenly distributed throughout the thickness and length of the coating. The surfacing was carried out in two passes; the coating thickness was 1.2 mm. The hardness of the coating was at different values of the titanium carbide content from 30 to 40 HRC. With an increase in the carbide phase content at the deposited layer, an increase in hardness and brittleness occurs; therefore, the optimal content of Ti-6Al-4V/TiC (10/90 by volume) is found in sample No. 5, since this sample has the most uniform distribution of carbides in thickness and length of the coating.

2. As well a sample was made from alloy Ti-6Al-4V, that the working surface was strengthened by induction soldering with powder solder based on titanium (Vpr-16) of WC-8Co hard alloy grains with granulation 180–250 μm.

*Friction and Wear Test*

In order to study fretting wear (dry alternating friction), a specialized friction machine «3308» was used. The friction machine scheme is shown in Fig. 4. On the shaft—1, fixed between the two racks—2 the body, there are two levers—3, the specimens at the ends of which are mounted in a tapered sleeve (see Fig. 5). Under the action of the leaf spring—5, a frictional contact with a force "P" is formed by means of a loading unit between the samples—4. It can be smoothly adjusted by means of a spring displacement unit—6, by changing springs of different thicknesses load range. The reciprocation of the samples—4 relative to each other is ensured by the rotation of the shaft—7 with an eccentricity "e" fixed, while the outer bearing race—8 is in constant contact with the lower arm.

At the first stage of the study, preliminary tests were conducted to select the sample shape and test procedures. For the test, a contact plane scheme was chosen, implemented on a cylindrical sample. After application, the laser-coated samples were tested without further processing. At the same time, while soldering the hard alloy WC-8Co, the end working surface of the samples was processed by grinding to form a flat surface. To fix the cylindrical specimen on the friction machine, a high-precision clamping cartridge was developed and manufactured.

Before conducting validation tests for testing the test procedure, selecting test intervals and obtaining information about the behavior of the tested coatings under the

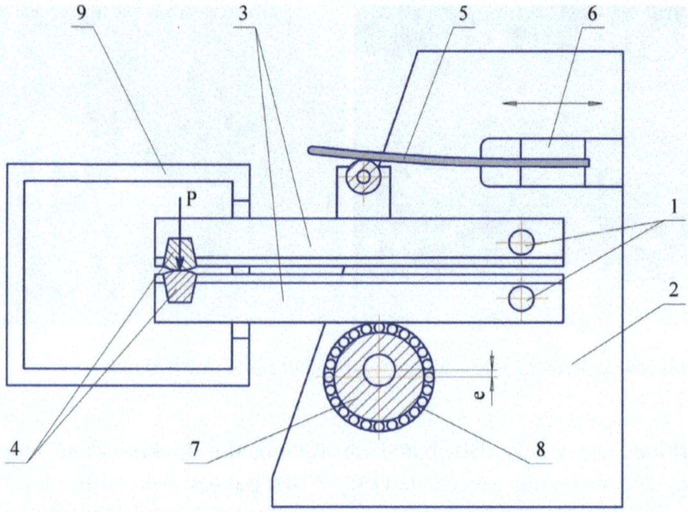


**Fig. 4** Scheme specialized friction machine «3308», 1—axis lever; 2—friction machine shell plates; 3—levers; 4—specimens; 5—flat springs; 6—spring displacement units; 7—eccentric pusher drive shafts; 8—outer pusher bearing sleeves; 9—volume modeling friction environment

**Fig. 5** Standard test coating sample

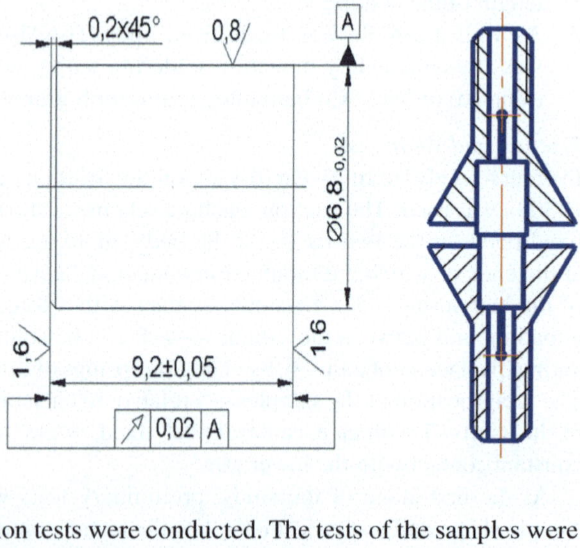


given test conditions, evaluation tests were conducted. The tests of the samples were carried out under the following conditions: external environment—air; sample temperature is room temperature; contact pressure in a friction pair—1.3 kgf/mm$^2$; displacement in a friction pair—0.7 mm; contact form in the friction zone—degenerate cylinders; frequency—10 Hz; test time—360 min.

The wear of the samples was evaluated in two ways: During the experiment, the relative convergence of the samples was measured using a Keyence LC2450 laser sensor with the LC2400 controller, and before and after the tests, the samples were weighed on a ViBRA H220CE analytical balance.

## 3 Results and Discussion

The results of evaluation of wear on the approximation of samples in the friction process are presented in Fig. 6 of which clearly shows that the smallest wear resistance (closest approach) in the process of fretting wear showed a coated sample with a component ratio of 30/70 (Ti-6Al-4V/TiC) and the best three samples with a ratio of components 10/90, 15/85 and 25/75. In this case, the reference point was a sample with a coating obtained by soldering tungsten carbide grains and its wear turned out to be higher compared to the coatings obtained by welding. The results of the study showed that the inclusion of TiC particles significantly changed the microstructure of the Ni–Cr matrix in the central zone of the deposited areas dominated by columnar dendrites and cellular dendrites. A similar process was also observed with a 20/80 coverage of 270 min. For steel specimens, the wear rate throughout the entire process after the burn-in period was almost the same.

Figure 7 shows the results of weight wear of the coatings. The best indicators are observed for coatings with ratios of 15/85, 20/80 and 25/75, a slightly worse value of mass carryover for a coating of 10/90, while the worst indicators are for a coating of 30/70 and WC-Co. In these results, it is worth making some corrections for different density of coatings due to different ratios and densities of the applied components of the composite coating.

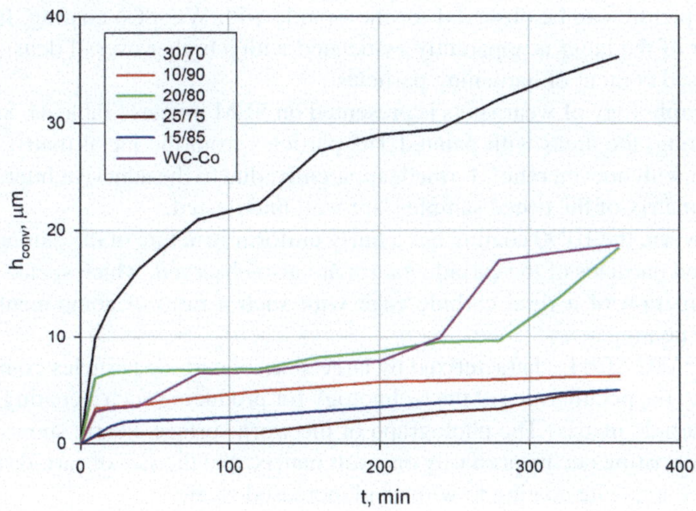


**Fig. 6** Convergence wear graph for tested samples

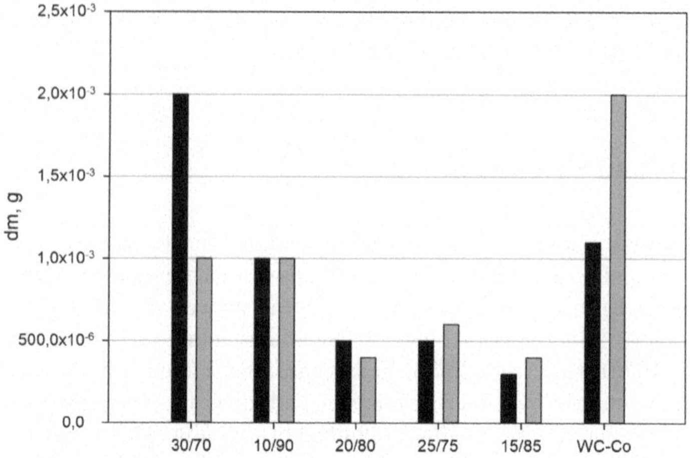


**Fig. 7** Weight wear diagram for tested samples

Table 3 shows photographs of wear spots obtained on samples with coatings of 30/70, 10/90 and WC-8Co. As can be seen, the process of wear on the 30/70 coating covered the entire area of the samples; you can also see zones of adhesive wear. Such a process may be due to the fact that the hardening particles, TiC, are in the coating in the unbound state and are located in the metal matrix Ti, as can be seen in Fig. 2, which leads to low wear resistance. The sample of the coating obtained with a component ratio of 10/90 has a different structure (Fig. 3): Two interpenetrating matrices of metallic titanium and carbide particles were formed. This led to a lower rate of wear of the latter, as evidenced by the smaller area of the wear spot (Table 3). A similar picture can be observed for the sample with WC-8Co coating. Increased mass wear of the latter is apparently associated with a higher overall density due to the increased content of hardening particles.

The morphology of wear spots is presented on SEM images (Table 4). So for the 30/70 coating, the areas with painted TiC particles from the metal matrix, as well as the area with uneven relief, formed, apparently, due to the adhesive interaction of the metal matrix of the tested samples, are well manifested.

When worn, the 10/90 coating has a fairly uniform structure of the damaged area, and no dyed particles of the carbide matrix are also observed, which speaks in favor of the formation of a rigid carbide cage with such a ratio of components of the composite coating.

Coating WC-8Co is characterized by large sizes of carbide particles compared to TiC. Due to the peculiarities of the technology for producing such a coating, there is no rigid carbide matrix. The photograph of the worn surface clearly shows that the wear of the coating occurs primarily on a soft matrix, and the size of carbide particles allows the composite coating to withstand increased wear.

**Table 3** Images of worn samples by special camera

| Applied material and application method | Image of sample number 1 paired | Image of sample number 2 paired |
|---|---|---|
| Ti-6Al-4V/*TiC* (30%/70%)—*LMD method* | 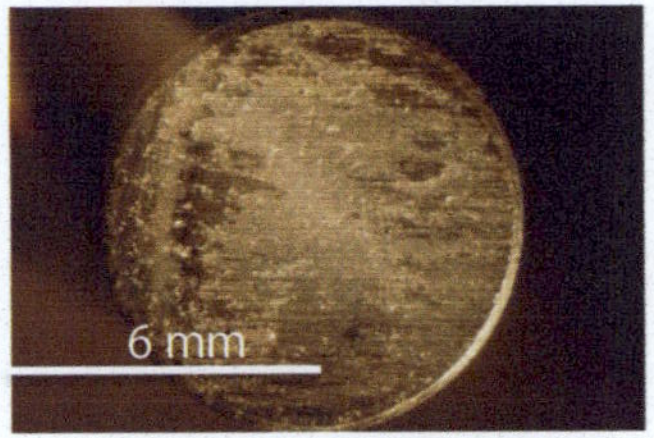  |   |
| Ti-6Al-4V/*TiC* (10%/90%)—*LMD method* | 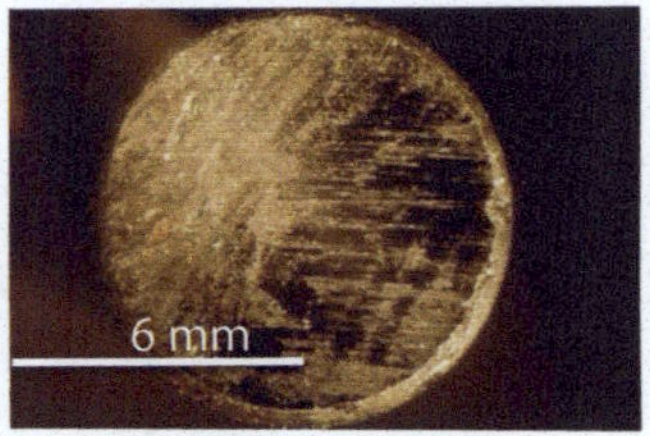  | 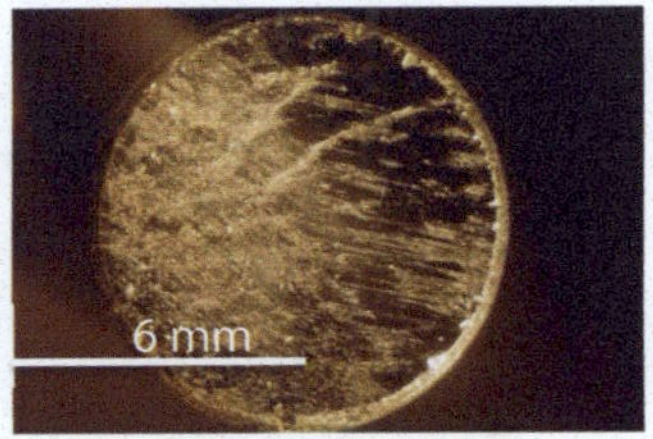  |
| WC-8Co—*technology soldering grains of hard alloy* |   |   |

**Table 4** Morphology and phase composition of the unworn and worn area by SEM

| Applied material and application method | Morphology and phase composition of the *Unworn* area | Morphology and phase composition of the *Worn* area |
|---|---|---|
| Ti-6Al-4V/*TiC* (30%/70%)—*LMD method* | 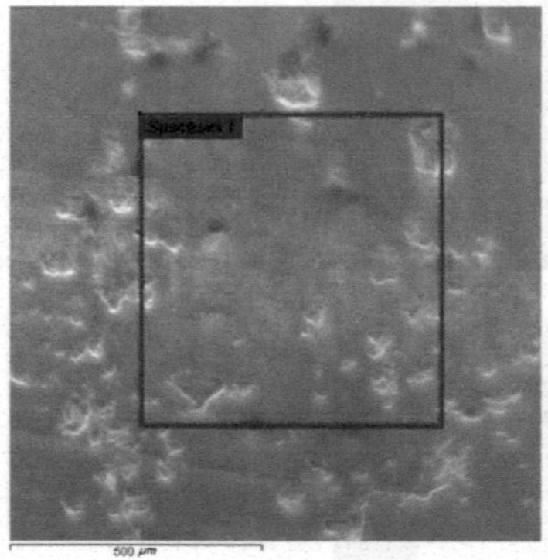 | 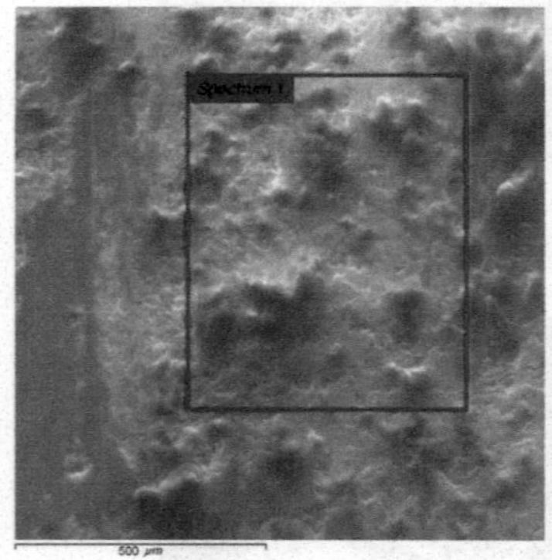 |

(continued)

**Table 4** (continued)

| Applied material and application method | Morphology and phase composition of the *Unworn* area | Morphology and phase composition of the *Worn* area |
|---|---|---|
| Ti-6Al-4V/*TiC* (10%/90%)—*LMD method* | 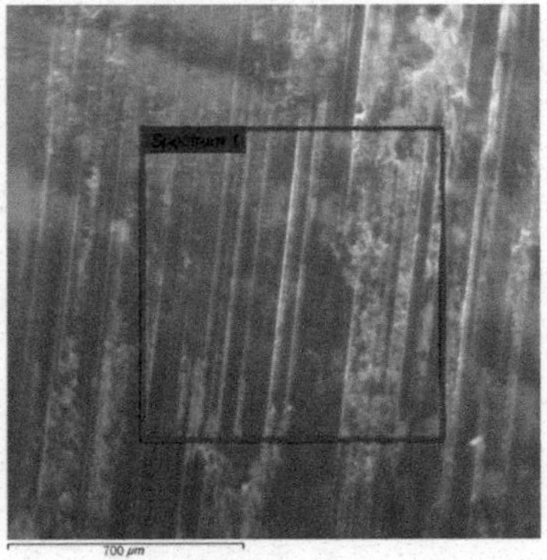  | 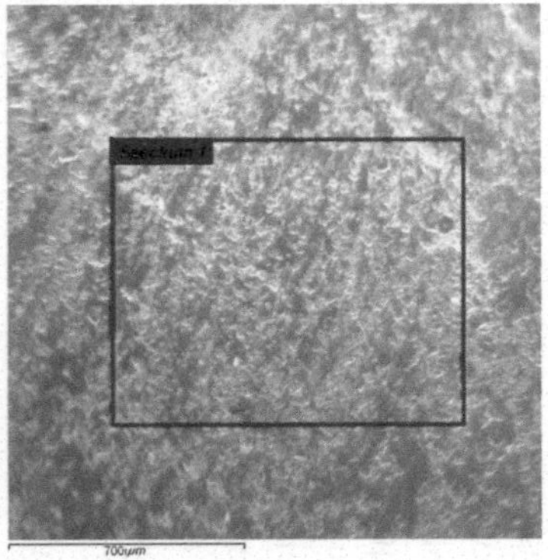  |
| WC-8Co—*technology soldering grains of hard alloy* | 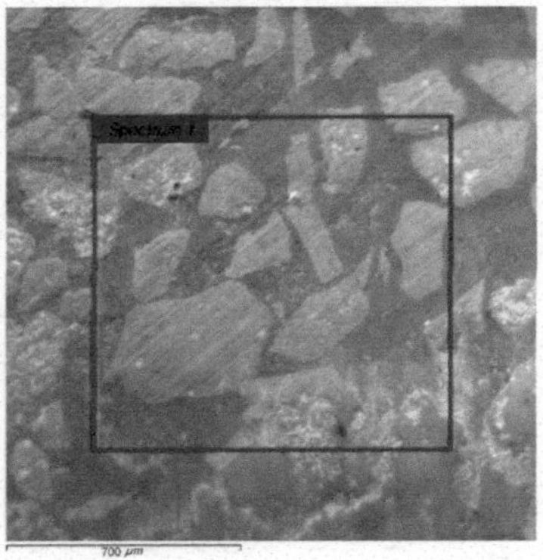  | 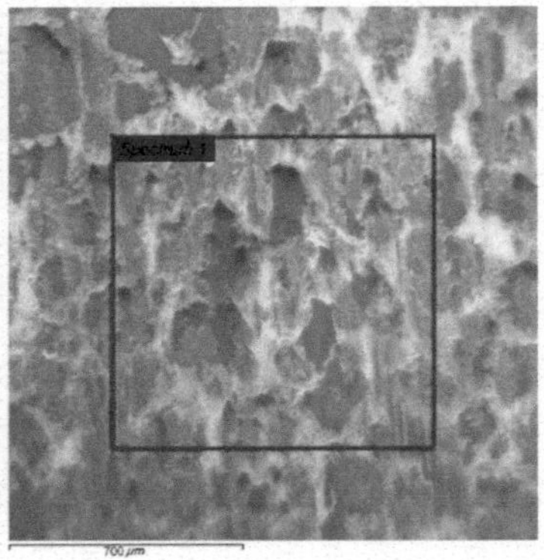  |

## 4 Conclusions and Further Work

Studies have shown that the use of direct laser synthesis technology allows, when properly selected components of the composite coating to obtain interpenetrating carbide and metal matrices, the wear resistance whether is higher compared to the coating obtained by the method of induction brazing of tungsten carbide grains.

To solve the problem of the applicability of a TiC coating, in our opinion, additional studies are needed to answer the question about the behavior of the developed compositions of composite coatings during long-term tests, as well as what is the critical contact pressure, including under thermal load.

As a result of the research, new information was obtained on the wear rates of coatings of two different types under the conditions of fretting wear typical for the fans shroud blade platforms and low-pressure compressors of gas turbine engines. The differences in the mechanisms of wear of ceramic coatings based on LMD technology for a powder mixture of Ti-6Al-4V/TiC and induction soldering of hard-grained WC-8Co grains are revealed.

The research results can be used in the design and technological development of the issue of increasing the fretting resistance of GTE retaining shelves and improving the technology of applying hardening coatings to GTE components.

## References


1. Holmberg K, Matthews A (2009) Coatings tribology: properties, mechanisms, techniques and application in surface engineering. Elsevier, 560c
2. Boitsov BV, Lesnevskiy LN, Lyakhovetskiy MA, Petuhov YuV, Pruss EM, Troshin AE, Ushakov AM (2013) Improving the safety of Aluminum press molds by protecting them from wear and corrosion by the microarc oxidation method. J Mach Manuf Reliab 42(6):479–485
3. Simchi A, Petzoldt F, Pohl H (2001) Direct metal laser sintering: material considerations and mechanisms of particle bonding. Int J Powder Metall 37(2):49–61
4. Kamrani AK (2014) Direct laser deposition for re-manufacturing of components. In: 2014 World Automation Congress (WAC), Waikoloa, HI, pp 1–5. https://doi.org/10.1109/wac.2014.6935640
5. Gao J, Chen X, Zheng D (2010) Remanufacturing oriented adaptive repair system for worn components. In: Proceedings of responsive manufacturing e Green Manufacturing ICRM, 5th International Conference, Ningbo, China, p 13e18. https://doi.org/10.1049/cp.2010.0406
6. Bremer C (2000) Adaptive strategies for manufacturing and repair of blades and blisks. In: Proceedings of ASME Turbo Expo 2000: 45th ASME international gas turbine & aeroengine technical congress, Munich, 8–11 May 2000
7. Mahamood RM, Akinlabi ET, Shukla M, Pityana S (2014) Characterization of laser deposited Ti6A4V/TiC composite. Lasers Eng 29(3–4):197–213
8. Dongdong G, Cao S, Lin K (2016) Laser metal deposition additive manufacturing of TiC reinforced Inconel 625 composites: influence of the additive TiC particle and its starting size. J Manuf Sci Eng 139(4):041014. https://doi.org/10.1115/1.4034934

# Comparison of Deconvolution Algorithms of Phased Microphone Array for Sound Source Localization in an Airframe Noise Test

**Jiayu Wang and Wei Ma**

**Abstract** Nowadays, Phased microphone arrays have a powerful capability for acoustic source localization. The conventional beamforming constructs a dirty map of source distributions from array microphone pressure signals. Compared with conventional beamforming, deconvolution algorithms, such as DAMAS, CLEAN-SC, NNLS, FISTA and SpaRSA, can significantly improve the spatial resolution but require high computational effort. The performances of these deconvolution algorithms have been compared using simulated applications and experimental applications with simple sound source distributions. However, these comparisons are not carried out in experimental applications with complex sound source distributions. In this paper, the performances of five deconvolution algorithms (DAMAS, CLEAN-SC, NNLS, FISTA and SpaRSA) are compared in an airframe noise test, which contains very complex sound source distributions. DAMAS and CLEAN-SC achieve better spatial resolution than NNLS, FISTA and SpaRSA. DAMAS need more computational effort compared with CLEAN-SC. In addition, DAMAS can significantly reduce computational run time using compression computational grid. DAMAS with compression computational grid and CLEAN-SC are thus recommended for source localizations in experimental applications with complex sound distributions.

**Keywords** Microphone array · Beamforming · Deconvolution algorithms · Airframe noise

## 1 Introduction

Nowadays, with the improvement of people's living standards, the demand for quiet and a comfortable environment is getting stronger and stronger. At the same time, the interest for acoustic source localization has been increasingly growing. Beamforming

---

J. Wang · W. Ma (✉)
School of Aeronautics and Astronautics, Shanghai Jiao Tong University,
Shanghai 200240, People's Republic of China
e-mail: mawei@sjtu.edu.cn



Z. Jing (ed.), *Proceedings of the International Conference on Aerospace System Science and Engineering 2019*, Lecture Notes in Electrical Engineering 622,
https://doi.org/10.1007/978-981-15-1773-0_7

with arrays of microphones is indispensable for the localization of sound sources on moving objects, on flying aircraft, on high-speed trains, on motor cars in motion, on open rotors like helicopter and wind turbine rotors [1].

In order to better represent the distributed noise (such as the aerodynamic noise from the airframe) in microphone array measurements, the deconvolution algorithms are required. Deconvolution algorithms reconstruct a clean map of source distributions from a dirty map by iteratively deconvolution and thus can significantly improve the spatial resolution [2–5].

For that purpose, a variety of deconvolution algorithms have been developed in the last several decades. In 1974, the non-negative least-squares (NNLS) was introduced by Lawson and Hanson [6], where all calculations are accelerated by spectral procedures. In 1998, Dougherty and Stoker [7] first applied CLEAN algorithm in sound source localization. Sijtsma [8] extended CLEAN to CLEAN-SC, which based on spatial source coherence. Unlike other deconvolution algorithms, it does not use the point spread function. It works well in combination with cross spectra matrix (CSM) diagonal removal [9]. At the 2004 AIAA/CEAS Aeroacoustic Conference, a breakthrough of deconvolution algorithms in acoustic microphone array technology was reported by Brooks and Humphreys, and they applied DAMAS algorithm in acoustic-array measurements [10, 11]. Between 2005 and 2006, Brooks and Humphreys extended it to three-dimensional acoustic image [12] and for coherent acoustic sources [13]. Unfortunately, DAMAS usually requires high computational effort in most situations. In 2017, Ma and Liu [14–16] published their work on compression computational grid, where they can successfully improve the efficiency of DAMAS via compression computational grid that only contains the significant grid points and does not contain the redundant grid points.

On many occasions, it has turned out that it is necessary to have a better assessment of these deconvolution algorithms using a set of common data sets. Many researchers have drawn the attention of the comparison of these deconvolution algorithms using simulated applications and experimental applications with simple sound source distributions. Ehrenfried et al. [17] have applied the three deconvolution algorithms DAMAS2 [18], the Fourier-based NNLS and Richardson-Lucy (RL) [19, 20] to reconstruct the source distributions from the dirty map with a line array and a one-dimensional region of interest. Herold et al. [21] compared several deconvolution algorithms using data from an aeroacoustic measurement of NACA 0012 airfoil positioned in an open jet. Recently, Bahr et al. [22] compared several common microphone phased array processing techniques applied to two open datasets. However, the comparison of DAMAS, CLEAN-SC, NNLS, FISTA and SpaRSA haven't been carried out in experimental applications with complex sound source distributions.

The main purpose of this paper is to determine which deconvolution algorithms (DAMAS, CLEAN-SC, NNLS, FISTA and SpaRSA) can play a better role for source localization in experimental applications with complex sound distributions. The benchmark test DLR1 is used to assess these deconvolution algorithms. The DLR1 benchmark test consists of a test configuration with a Dornier-728 semispan (or half) model in the high-lift configuration in the cryogenic wind tunnel at the

DLR Cologne site (Kryo-Kanal Koeln, DNW-KKK) [23]. One goal is to carry out the application of several deconvolution algorithms in aeroacoustic measurements. Another objective is to carry out the comparison of these deconvolution algorithms.

The rest of this paper is organized as follows. Conventional beamforming and deconvolution algorithms are illustrated in Sect. 2. Experimental applications with complex source distributions are carried out in Sect. 3. A discussion is given in Sect. 4. Finally, conclusions are presented in Sect. 5.

## 2 Conventional Beamforming and Deconvolution Algorithms

Conventional beamforming is simple and robust. Conventional beamforming can construct a dirty map of source distributions from array microphone pressure signals. However, its main disadvantages include poor spatial resolution particularly at low frequencies and poor dynamic range due to side-lobe effects [16].

Brooks and Humphrey [10, 11] proposed DAMAS algorithm, which is an iterative algebraic deconvolution algorithm. DAMAS aims at solving the convolution equation by a Gauss–Seidel procedure, replacing nonphysical negative solutions by zero and achieving more accurately quantify position and strength of acoustic sources.

In order to remove side lobes actually measured beam patterns. Sijtsma [8] launched CLEAN-SC. CLEAN-SC can successfully extract absolute sound power levels from the source plots [8]. CLEAN-SC basically performs a decomposition of the CSM into coherent components. Unlike other deconvolution algorithms, it does not use the point spread function. It works well in combination with CSM diagonal removal.

NNLS [6] algorithm aims directly at the minimization of the square sum of the residuals. A gradient-type procedure is used to solve the NNLS problem.

FISTA [24] deconvolution algorithm is ideally suitable for image reconstruction from indirect and possibly under-sampled data and can achieve high computational simplicity.

Sparse reconstruction by separable approximation (SpaRSA) is an iterative deconvolution algorithm to a minimum of the objective function and suitable for solving large-scale optimization problems involving the sum of a smooth error term and a possibly non-smooth regularizer [25]. SpaRSA has far-reaching effects in the field of imaging. More details for SpaRSA can be found in [25].

## 3 Experimental Applications

In this section, five different deconvolution algorithms (DAMAS, CLEAN-SC, NNLS, FISTA and SpaRSA) are applied to benchmark test DLR1. This acoustic-array measurement [23] performed in a cryogenic wind tunnel at the DLR Cologne

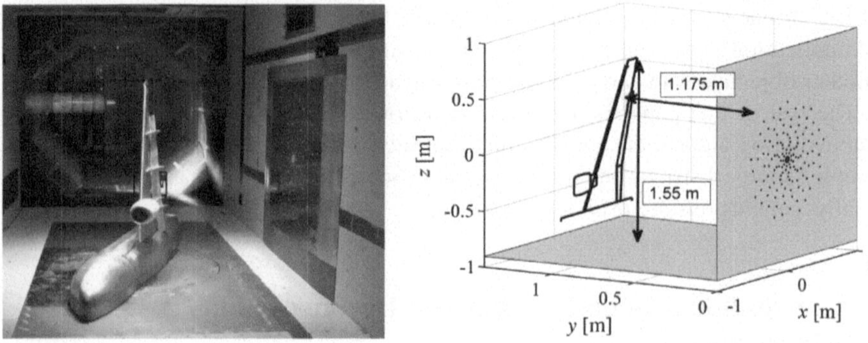


**Fig. 1** Photo of the test section with the array mounted on the sidewall (left) and the Dornier 728 half-model in the center, view in flow direction. Drawing of the measurement setup (right) [23]

**Table 1** Information of data processing

| | |
|---|---|
| Angle of attack | 3° |
| Mach number | 0.25 |
| Temperature | 290 K |
| Grid resolution | 20 mm |
| The computational grid | $53 \times 73$ with 3869 points in $x$-$y$ plane |
| Plotting scale | 20 dB |
| Number of iterations | 2000 |

site, Kryo-Kanal Koeln (DNW-KKK) using a 9.24% Dornier 728 half-model. This cryogenic wind tunnel is a continuous-flow low-speed wind tunnel (DNW-KKK) with a 2.4×2.4 m closed-wall test section. By injection of liquid nitrogen, this wind tunnel can be operated in the range of $100 < T < 300$K at Mach number up to 0.38. Figure 1 indicates the test setup. A spiral microphone array consisting of 144 microphones is designed and constructed for this measurement. The array is mounted by the compressed laminated wood onto the sidewall, and the Dornier 728 half-model is located in the center of the test section. The model of scale 1:9.24 is configured in a landing configuration and has a mean aerodynamic chord length of 0.353 m and a half-span width of 1.44m. More details for this measurement can be found in [23]. Recently, a comparison of microphone phased array methods applied to the benchmark test DLR1 was carried out by Bahr et al. [22].

The data processing is carried out on an Intel Core i5-4210H 2.90GHz processor with MATLAB. The information of this data processing is listed in Table 1. The choice of angle of attack, Mach number, temperature and plotting scale is for the sake of comparison with the deconvolution result of Bahr et al. [22]. The computational grid used in this paper is the same as that in the benchmark. The rotation of the computational grid is the same as that introduced in Bahr et al. [22]. In the calculation of the dirty map of conventional beamforming, CSM is directly from the benchmark test data. In the process of calculating the steering vector, the influence of Mach number is taken into account. There are two different experimental applications, in which frequencies are $f = 8496$Hz and $f = 6300$Hz, respectively, plotted in Figs. 2 and 3. The computational run time of these experimental applications is listed in Table 2.

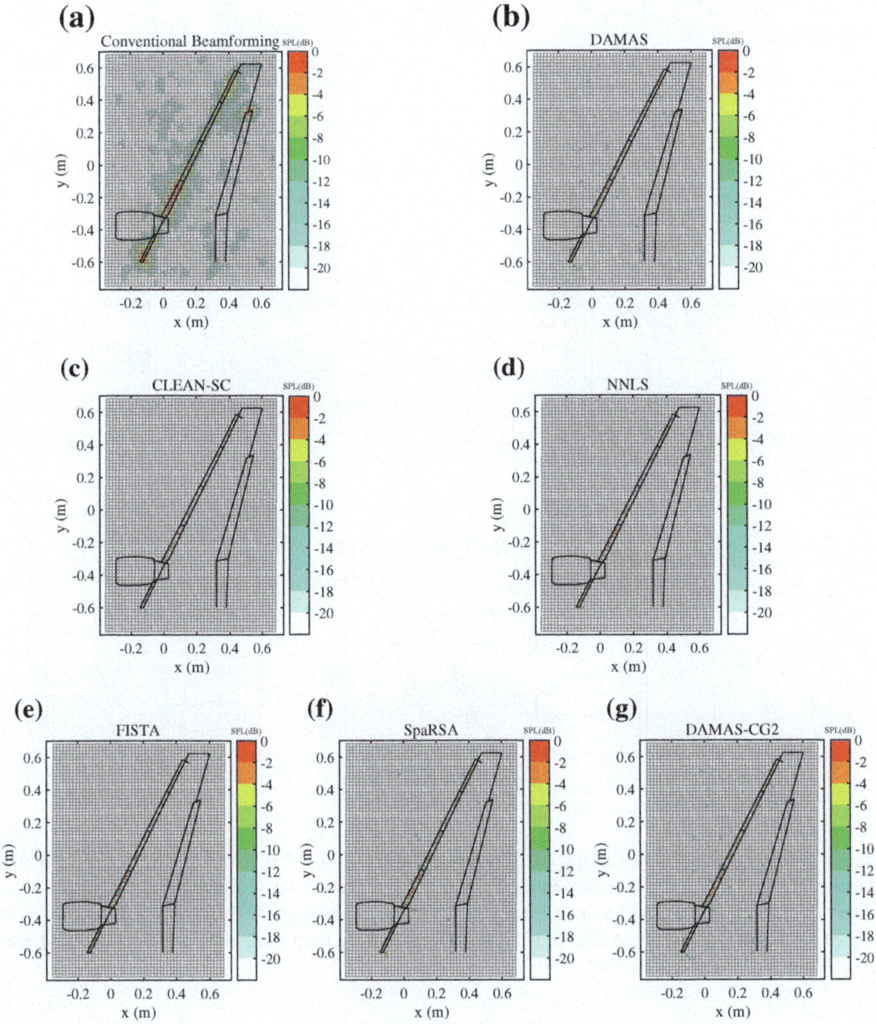


**Fig. 2** Sound pressure level maps of Benchmark test DLR1. $f = 8496$ Hz. The location of the Dornier 728 half-model is sketched in the background. **a** Dirty map of conventional beamforming. **b** Deconvolved map of DAMAS on original grid. **c** Deconvolved map of CLEAN-SC. **d** Deconvolved map of NNLS. **e** Deconvolved map of FISTA. **f** Deconvolved map of SpaRSA. **g** Deconvolved map of DAMAS on compression grid

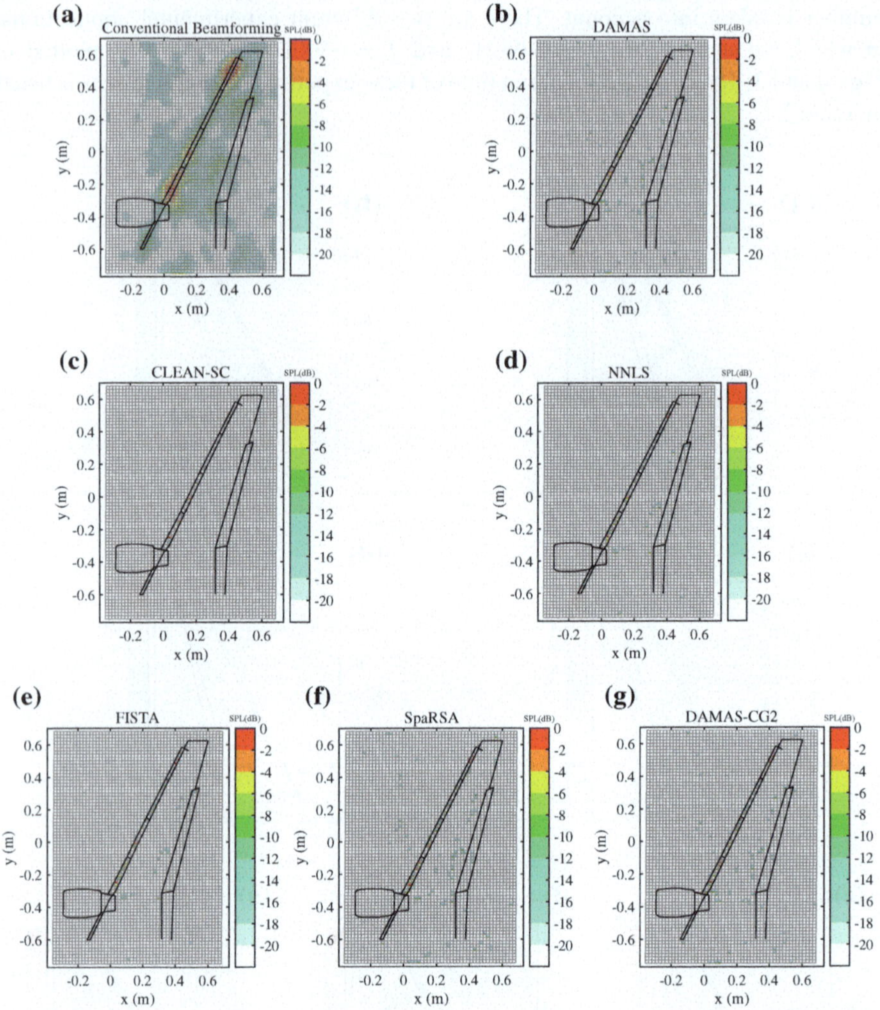


**Fig. 3** Sound pressure level maps of Benchmark test DLR1. $f = 6300$Hz The location of the Dornier 728 half-model is sketched in the background. **a** Dirty map of conventional beamforming. **b** Deconvolved map of DAMAS on original grid. **c** Deconvolved map of CLEAN-SC. **d** Deconvolved map of NNLS. **e** Deconvolved map of FISTA. **f** Deconvolved map of SpaRSA. **g** Deconvolved map of DAMAS on compression grid

**Table 2** Computational run time of different deconvolution algorithms

| Deconvolution algorithms | Computational run time | |
|---|---|---|
| | Case 1 (s) | Case 2 (s) |
| DAMAS | 1416 | 1623 |
| CLEAN-SC | 51 | 33 |
| NNLS | 1255 | 1456 |
| FISTA | 1264 | 1470 |
| SpaRSA | 1370 | 1790 |
| DAMAS-CG2 | 237 | 501 |

## *3.1 Spatial Resolution*

### 3.1.1 Case 1, $f = 8496$Hz

The deconvolution results of $f = 8496$Hz case is very similar to the recent deconvolution results of Bahr et al. [16, 22].

The dirty map of conventional beamforming shown by Fig. 2a is quite similar to the source maps of conventional beamforming contributed by DLR has shown in Fig. 4a in [22]. There are two main noise sources when $f = 8496$Hz. One is the slat noise near the leading edge. The other one is the flap side edge noise [16].

Figure 2b indicates the clean map of DAMAS on original grid which is very similar to the DAMAS result contributed by DLR in Fig. 5h in [16, 22]. The slat noise and the flap noise are well distributed in this clean map, due to significantly improved spatial resolution.

The clean map of CLEAN-SC shown in Fig. 2c is also quite similar to the CLEAN-SC result contributed by DLR in Fig. 4c in [22]. This clean map can achieve high spatial resolution containing exactly one source on every slat and one source on the flap side edge.

In Fig. 2d–f, NNLS, FISTA and SpaRSA are applied to the benchmark DLR1 that shows only slightly fewer sources. However, the peak level is nearly as same as the result of DAMAS and CLEAN-SC.

Compared with Fig. 2a, the clean map of DAMAS with compression computational grid based on conventional beamforming (denoted by DAMAS-CG2 [14]) shown in Fig. 2g can retain the spatial resolution of DAMAS on original grid.

### 3.1.2 Case 2, $f = 6300$Hz

In order to carry out this comparison in different frequencies, in the second application, frequency is set as 6300Hz. In this case, there is only one high region dominated the dirty map of conventional beamforming. That is the slat noise near the leading edge. As shown in Fig. 3b, c and g, DAMAS, CLEAN-SC and DAMAS-CG2 can identify separately most slat noise sources. Figure 3d–f shows the deconvoluted results of NNLS, FISTA and SpaRSA. The spatial resolutions are a little slighter than that of DAMAS, CLEAN-SC and DAMAS-CG2.

### *3.2 Computational Run Time*

The computational run time of different deconvolution algorithms in Case 1 and Case 2 is shown in Table 2. The computational efficiency of CLEAN-SC is obvious in both two cases.

In Case 1, the computational run time of DAMAS, NNLS, FISTA and SpaRSA are 1416 s, 1255 s, 1264 s and 1370 s, respectively, while that of CLEAN-SC is only 51 s.

In Case 2, the computational run time of DAMAS, NNLS, FISTA and SpaRSA are 1623 s, 1456 s, 1470 s and 1790 s, respectively, while that of CLEAN-SC is only 33 s.

DAMAS-CG2 can successfully reduce computational run time. The computer run time of DAMAS-CG2 in both two cases is only 237 s and 501 s, respectively.

## 4 Discussion

In this paper, in order to compare the performance of five different deconvolution algorithms (DAMAS, CLEAN-SC, NNLS, FISTA and SpaRSA) in experimental applications with complex sound distributions, we apply these deconvolution algorithms to the benchmark DLR1.

The following statements are only valid for the angle of attack $\alpha = 3°$, $Ma = 0.25$, $T = 290\,\text{K}$, $f = 8496\,\text{Hz}$ & $f = 6300\,\text{Hz}$, the grid resolution $d_{xy} = 20\,\text{mm}$ and plotting scale 20 dB.

When applied to aeroacoustic measurements with complex sound source distributions, these deconvolution algorithms are observed to behave as follows:

- **Conventional Beamforming**
  The result of conventional beamforming is a dirty map, and all sound sources are visible. However, it has poor spatial resolution and side-lobe effects.
- **DAMAS and CLEAN-SC**
  DAMAS and CLEAN-SC can significantly improve spatial resolution. Noise sources can be well distributed.
- **NNLS, FISTA and SpaRSA**
  NNLS, FISTA and SpaRSA can only distinguish slightly fewer source.
- **Computational Efficiency**
  CLEAN-SC is much faster than DAMAS, NNLS, FISTA and SpaRSA. However, DAMAS-CG2 can achieve similar spatial resolution with DAMAS. The compu-

tational run time of DAMAS-CG2 is longer than CLEAN-SC. While compared with other three deconvolution algorithms, DAMAS-CG2 can greatly reduce the computational run time.

## 5 Conclusion

Five different deconvolution algorithms (DAMAS, CLEAN-SC, NNLS, FISTA and SpaRSA) are compared through applications to the benchmark DLR1. The spatial resolution and computational run time are selected as the main criteria in this comparison.

In terms of spatial resolution, every deconvolution algorithms described above can distribute the dominant sound source. DAMSA and CLEAN-SC can successfully improve the spatial resolution and achieve higher spatial resolution than NNLS, FISTA and SpaRSA.

In terms of computational run time, CLEAN-SC is much faster than the other four deconvolution algorithms described above. DAMAS with compression computational grid can significantly reduce computational run time.

In order to obtain higher spatial resolution and greatly improve the computational efficiency, DAMAS with compression computational grid and CLEAN-SC are thus recommended for source localization in experimental applications with complex sound distributions. For the future investigations, it will be of interest to improve the computational efficiency of NNLS, FISTA and SpaRSA.

**Acknowledgements** The authors would like to thank Dr. Thomas Geyer of BTU Cottbus-Senftenberg, Germany for providing the login information of the DLR1 benchmark test. This work was supported by China Scholarship Council and the Natural Science Foundation of China (Grant NO. 51506121).

## References

1. Michel U (2006) History of acoustic beamforming. In: Proceedings of 1st Berlin beamforming conference2006
2. Frieden BR (1972) Restoring with maximum likelihood and maximum entropy. J Opt Soc Am 62(4):511–518
3. Banham MR, Katsaggelos AK (1977) Digital image restoration. IEEE Signal Process Mag 14(2):24–41
4. Gull SF, Daniell GJ (1978) Image reconstruction from incomplete and noisy data. Nature 272(5655):686–690
5. Narayan R, Nityananda R (1986) Maximum entropy image restoration in astronomy. Ann Rev Astron Astrophys 24(1):127–170
6. Lawson CL, Hanson RJ (1995) Solving least squares problems. Math Comput 30(135):665
7. Dougherty RP, Stoker RW (1998) Sidelobe suppression for phased array aeroacoustic measurements. In: 4th AIAA/CEAS aeroacoustics conference
8. Sijtsma P (2009) CLEAN based on spatial source coherence. Int J Aeroacoustics 6(4):357–374

9. Sarradj E, Herold G, Sijtsma P, Merino Martinez R, Geyer TF, Bahr CJ, Porteous R, Moreau D, Doolan CJ (2017) A microphone array method benchmarking exercise using synthesized input data. In: 23rd AIAA/CEAS aeroacoustics conference
10. Brooks TF, Humphreys WM (2004) A deconvolution approach for the mapping of acoustic sources (DAMAS) determined from phased microphone arrays. In: 10th AIAA/CEAS aeroacoustics conference
11. Brooks TF, Humphreys WM (2006) A deconvolution approach for the mapping of acoustic sources ( DAMAS) determined from phased microphone arrays. J Sound Vib 294(4):856–879
12. Brooks TF, Humphreys WM (2005) Three-dimensional applications of DAMAS methodology for aeroacoustic noise source definition. In: 11th AIAA/CEAS aeroacoustics conference
13. Brooks TF, Humphreys WM (2006) Extension of DAMAS phased array processing for spatial coherence determination (DAMAS-C). In: 12th AIAA/CEAS aeroacoustics conferences
14. Ma W, Liu X (2017) DAMAS with compression computational grid for acoustic source mapping. J Sound Vib 410:473–484
15. Ma W, Liu X (2017) Improving the efficiency of DAMAS for sound source localization via wavelet compression computational grid. J Sound Vib 395:341–353
16. Ma W, Liu X (2018) Compression computational grid based on functional beamforming for acoustic source localization. Appl Acoust 134:75–87
17. Ehrenfried K, Koop L (2007) Comparison of iterative deconvolution algorithms for the mapping of acoustic sources. AIAA J 45(7):1–19
18. Dougherty RP (2013) Extensions of DAMAS and benefits and limitations of deconvolution in beamforming. In: 11th AIAA/CEAS aeroacoustics conference
19. Lucy LB (1974) An iterative technique for the rectification of observed distributions. Astron J 79(6):745–754
20. Richardson WH (1972) Bayesian-based iterative method of image restoration. J Opt Soc Am 62(1):55–59
21. Herold G, Geyer TF, Sarradj E (2017) Comparison of inverse deconvolution algorithms for high-resolution aeroacoustic source characterization. In: 23rd AIAA/CEAS aeroacoustics conference
22. Bahr CJ, Humphreys WM, Ernst D, Ahlefeldt T, Spehr C, Pereira A, Leclre Q, Picard C, Porteous R, Moreau D, Fischer JR, Doolan CJ (2017) A comparison of microphone phased array methods applied to the study of airframe noise in wind tunnel testing. In: 23rd AIAA/CEAS aeroacoustics conference
23. Ahlefeldt T (2013) Aeroacoustic measurements of a scaled half-model at high reynolds numbers. AIAA J 51(12):2783–2791
24. Beck A, Teboulle M (2009) A fast iterative shrinkage-thresholding algorithm for linear inverse problems. SIAM J Imaging Sci 2(1):183–202
25. Wright SJ, Nowak RD, Figueiredo MAT (2009) Sparse reconstruction by separable approximation. IEEE Trans Signal Process 57(7):2479–2493

# Analytical Solutions to a Rijke Tube System with Periodic Excitations Through a Semi-analytical Approach

**Jianzhe Huang**

**Abstract** Thermoacoustic instability problems are widely existed in many real-world applications such as gas turbines and rocket motors. A Rijke tube is a typical thermoacoustic system, and it is difficult to analyze such a system due to the nonlinearity and time delay. In this paper, a set of nonlinear ordinary differential equations with time delay which represent a Rijke tube system will be studied. The state space of such a tube system is consisted of velocity and pressure, and the periodic motion can be discretized based on an implicit midpoint scheme. Through Newton-Raphson method, the node points on the periodic motion will be solved, and the analytical solution of such a periodic motion for Rijke tube system can be recovered using a set of Fourier representations. According to the theory of discrete maps, the stability of the periodic motion will be obtained. Finally, specific system parameters will be adopted in order to carry out numerical studies to show different periodic motions for such a tube system. The analytical bifurcations which show how period-1 motion involves to period-m motion and then becomes chaos will be demonstrated. With such a technique, some interesting nonlinear phenomena will be explained analytically, which will be of great help to understand and control such a Rijke tube system.

**Keywords** Thermoacoustic · Periodic motions · Chaos · Semi-analytic · Bifurcation

## 1 Introduction

In the 1800s, Rijke [1] found that an open-ended vertical tube could produce loud sound by placing hot metal gauze in the lower half, and such a tube is now named as Rijke tube. But he did not explain why hot gauze in the upper half of the tube did not produce sound. Rayleigh [2] explained the phenomena observed by Rijke, and a criterion was proposed for the development of heat-driven oscillations. For vertical Rijke tube, it is difficult to use for quantitative testing due to the acoustic coupling. In

J. Huang (✉)
Shanghai Jiao Tong University, Shanghai 200240, China
e-mail: huangdeng9@sjtu.edu.cn

 
Z. Jing (ed.), *Proceedings of the International Conference on Aerospace System Science and Engineering 2019*, Lecture Notes in Electrical Engineering 622,
https://doi.org/10.1007/978-981-15-1773-0_8

1964, Friedlander et al. [3] studied the horizontal Rijke tube, and the sound pressure level varying with relative heater position for different tube length was discussed experimentally. Bisio and Rubatto [4] designed a feedback control system to achieve the active control of noise for a horizontal Rijke tube. In 2008, Balasubramanian and Rujith [5] modeled the heat release rate of the heating element for Rijke tube using a modified form of King's law, and algebraic growth of oscillations was induced by the non-normality of the thermoacoustic system. For horizontal Rijke tube, the governing equations for the fluid flow are too stiff to be solved through computational fluid dynamics due to the small Mach number of the steady flow and the small thickness of the heat source. The numerical model for horizontal Rijke tube was proposed in [6]. Then Mariappan and Sujith [7] used Galerkin method to simulate the acoustic zone and CFD technique to model the hydrodynamic zone. It showed that the bifurcation results were different with and without the global-acceleration term. Juniper et al. [8] studied a horizontal Rijke tube model with radius of the upstream and downstream ducts varied, the perturbation method was adopted to obtain the solutions of such a system and weakly nonlinear analysis has been carried out. For such a Rijke tube system with strong nonlinearity, it is difficult to calculate the solution analytically since traditional analytic methods for nonlinear system are only valid for weak nonlinear problems. Generalized harmonic balance method [9] is an ideal tool for getting the approximate solutions which correlate with numerical simulation very well for strong nonlinear systems, and it can also deal with nonlinear dynamical system with time-delayed term [10]. However, it will take a great effort to derive the equations for such a Rijke tube model, since the degrees of freedom could be high when more mode shapes have been included [11]. The discrete implicit maps method [12] is a newly developed technology for solving various dynamical systems with strange nonlinear terms. With such a method, Wang and Huang [13] gave an analytical solution for periodic motions for a damped mobile piston system in a high-pressure gas cylinder with P control, and there was no computational error to calculate the bifurcation points since the dimension of the Jacobian matrix did not increase with the truncated order of harmonics for periodic solution increases. Guo and Luo [14] adopted the discrete implicit maps method to investigate the parametrically driven pendulum for which the nonlinear term is a triangular function, and the complex bifurcation diagrams of analytic solutions for periodic motions were presented. For time-delayed systems, Luo and Xing [15] studied the periodically forced hardening Duffing oscillator with time delay through the discrete implicit method, and the time-delay effects on period-1 to chaos for such a dynamical system have been discussed.

In this paper, a horizontal Rijke tube with periodical acoustic source will be studied. The acoustic momentum and energy equations for such a Rijke tube will be converted into a set of ordinary differential equations through Galerkin transformation. The time-delayed effect will be introduced due to the heat release rate fluctuations of the heat source for such a horizontal Rijke tube. The periodic motion will be discretized into a finite number of node points with constant time interval, and one node point maps to another one based on implicit midpoint scheme. The time-delayed term will be represented by the normal states due to the periodicity, and the node points can be solved through Newton-Raphson method. From the node points, the analytic solution of periodic motion will be recovered using a set of finite Fourier series. Then the harmonic amplitudes varying amplitude of periodic acoustic excitation will be illustrated. Some numerical simulations will be given to discuss the dynamic behaviors of such a horizontal Rijke tube with periodical acoustic source.

## 2 Model and Semi-analytic Method

For a horizontal Rjike tube, the governing equations can be given in Eqs. (1) and (2) by assuming that the system is symmetric, Mach number of the mean flow is small, and a perfect open-open acoustic boundary condition at both ends of the tube.

$$\gamma M\frac{\partial u'}{\partial t}+\frac{\partial p'}{\partial x}=0 \tag{1}$$

$$\frac{\partial p'}{\partial t}+\gamma M\frac{\partial u'}{\partial x}+\zeta p'=\frac{\gamma MK}{2}\left(\sqrt{\left|\frac{1}{3}+u'_f(t-\tau)\right|}-\frac{1}{3}\right)\delta(x-x_f) \tag{2}$$

where $u'$ and $p'$ are the non-dimensional acoustic velocity and pressure, respectively; $x$ is the non-dimensional axial distance measured from the inlet; $t$ is the non-dimensional time; $M$ is the Mach number of the steady-state flow; $\gamma$ is the adiabatic index; $\zeta$ is the damping coefficient; $K$ is the non-dimensional heater power. The subscript $f$ means the position of heater inside the Rijke tube. Therefore, the Dirac delta function at the right-hand side of Eq. (2) indicates that the heat effect only exists at the location of the heater.

Represent velocity and pressure by Fourier series, and it gives

$$u'=\sum_{i=1}^{N}\cos(i\pi x)U_i(t) \tag{3}$$

$$p'=\gamma M\sum_{i=1}^{N}\sin(i\pi x)P_i(t) \tag{4}$$

Then substitute Eqs. (3) and (4) into Eqs. (1) and (2), and perform the Galerkin transform. Then the aforementioned partial differential equations become

$$\dot{U}_i+i\pi P_i=0 \tag{5}$$

$$\dot{P}_i-i\pi U_i+\zeta P_i=K\left(\sqrt{\left|\frac{1}{3}+u'_f(t-\tau)\right|}-\frac{1}{3}\right)\sin(i\pi x_f) \tag{6}$$

In Ref. [16], the damping coefficient $\zeta$ for -$i$th mode is expressed as

$$\zeta = \zeta_i = c_1 i^2 + c_2\sqrt{i} \tag{7}$$

Then Eq. (6) can be rewritten as

$$\dot{P}_i - i\pi U_i + \zeta_i P_i = K\left(\sqrt{\left|\frac{1}{3} + u'_f(t-\tau)\right|} - \frac{1}{3}\right)\sin(i\pi x_f) \tag{8}$$

Introduce a periodic excitation into the system, Eq. (8) then becomes

$$\dot{P}_i - i\pi U_i + \zeta_i P_i = K\left(\sqrt{\left|\frac{1}{3} + u'_f(t-\tau)\right|} - \frac{1}{3}\right)\sin(i\pi x_f) + Q_i\sin(\Omega t) \tag{9}$$

Define $\mathbf{x} \equiv \{U_1, U_2, \ldots, U_N\}^{\mathrm{T}}$ and $\mathbf{y} \equiv \{P_1, P_2, \ldots, P_N\}^{\mathrm{T}}$, then Eqs. (5) and (9) can be written in the matrix form as

$$\dot{\mathbf{x}} = \begin{bmatrix} \dot{U}_1 \\ \dot{U}_2 \\ \vdots \\ \dot{U}_N \end{bmatrix} = \begin{bmatrix} -\pi P_1 \\ -2\pi P_1 \\ \vdots \\ -N\pi P_N \end{bmatrix} = -\mathbf{k}\mathbf{y} \tag{10}$$

$$\dot{\mathbf{y}} = \begin{bmatrix} \dot{P}_1 \\ \dot{P}_2 \\ \vdots \\ \dot{P}_N \end{bmatrix} = \begin{bmatrix} -\zeta_1 P_1 + \pi U_1 \\ -\zeta_2 P_2 + 2\pi U_2 \\ \vdots \\ -\zeta_N P_1 + N\pi U_1 \end{bmatrix} + K\left(\sqrt{\left|\frac{1}{3} + u'_f(t-\tau)\right|} - \sqrt{\frac{1}{3}}\right)\begin{bmatrix} \sin\pi x_f \\ \sin 2\pi x_f \\ \vdots \\ \sin N\pi x_f \end{bmatrix}$$

$$= -\,\mathbf{c}\mathbf{y} + \mathbf{k}\mathbf{x} + K\left(\sqrt{\left|\frac{1}{3} + u'_f(t-\tau)\right|} - \sqrt{\frac{1}{3}}\right)\mathbf{f}_1 \tag{11}$$

where

$$\mathbf{k} = \begin{bmatrix} \pi & 0 & 0 & \cdots & 0 \\ 0 & 2\pi & 0 & \cdots & \vdots \\ 0 & 0 & \ddots & \vdots & 0 \\ \vdots & \cdots & 0 & (N-1)\pi & 0 \\ 0 & \cdots & 0 & 0 & N\pi \end{bmatrix}$$

$$\mathbf{c} = \begin{bmatrix} \zeta_1 & 0 & 0 & \cdots & 0 \\ 0 & \zeta_2 & 0 & \cdots & \vdots \\ 0 & 0 & \ddots & \vdots & 0 \\ \vdots & \cdots & 0 & \zeta_{N-1} & 0 \\ 0 & \cdots & 0 & 0 & \zeta_N \end{bmatrix}$$

$$\mathbf{f}_1 = \left[ \sin \pi x_f \; \sin 2\pi x_f \; \cdots \; \sin N\pi x_f \right]^T$$

For a period-$m$ motion of such a horizontal Rijke tube, the trajectory is discretized into $mL$ partitions. The mapping $P_k$ ($k = 0, 1, \ldots, mL - 1$), which maps from one node point with state vector $(\mathbf{x}_k, \mathbf{y}_k, \mathbf{x}_{k-\mathrm{int}(\tau/h)-1}, \mathbf{x}_{k-\mathrm{int}(\tau/h)}, \mathbf{x}_{k-\mathrm{int}(\tau/h)+1})$ to another node point with state vector $(\mathbf{x}_{k+1}, \mathbf{y}_{k+1}, \mathbf{x}_{k-\mathrm{int}(\tau/h)}, \mathbf{x}_{k-\mathrm{int}(\tau/h)+1}, \mathbf{x}_{k-\mathrm{int}(\tau/h)+2})$, is expressed by

$$\mathbf{x}_{k+1} = \mathbf{x}_k - \frac{1}{2} h\mathbf{k}(\mathbf{y}_k + \mathbf{y}_{k+1}) \tag{12}$$

$$\mathbf{y}_{k+1} = \mathbf{y}_k + h\left[ -\mathbf{c}\frac{\mathbf{y}_k + \mathbf{y}_{k+1}}{2} + \mathbf{k}\frac{\mathbf{x}_k + \mathbf{x}_{k+1}}{2} + K\left( \sqrt{\left|\frac{1}{3} + V\right|} - \sqrt{\frac{1}{3}} \right)\mathbf{f}_1 \right. \\ \left. + \mathbf{Q} \sin \Omega\left( t_k + \frac{h}{2} \right) \right] \tag{13}$$

where

$$V = \frac{1}{2}\left[ u'_f(t_k - \tau) + u'_f(t_{k+1} - \tau) \right] \tag{14}$$

$$u'_f(t_k - \tau) = \left\{ \mathbf{x}^T_{k-\mathrm{int}(\tau/h)-1} + \left[ 1 - \frac{\tau}{h} + \mathrm{int}\left(\frac{\tau}{h}\right) \right]\left( \mathbf{x}^T_{k-\mathrm{int}(\tau/h)} - \mathbf{x}^T_{k-\mathrm{int}(\tau/h)-1} \right) \right\}\mathbf{f}_2 \tag{15}$$

$$\mathbf{Q} = [Q_1, Q_2, \ldots, Q_N]^{\mathrm{T}}$$

and where

$$\mathbf{f}_2 = \left[ \cos \pi x_f \; \cos 2\pi x_f \; \cdots \; \cos N\pi x_f \right]^T$$

$$\mathbf{x}_k \equiv \mathbf{x}(t = t_k), \mathbf{y}_k \equiv \mathbf{y}(t = t_k) \quad \text{and} \quad h = 2\pi/(L\Omega).$$

Due to the periodicity, it has $\mathbf{x}_0 = \mathbf{x}_{mL}$ and $\mathbf{y}_0 = \mathbf{y}_{mL}$. Categorize the node points into a single set, one have

$$\mathbf{z} \equiv \{\mathbf{x}_0, \mathbf{y}_0, \mathbf{x}_1, \mathbf{y}_1, \cdots, \mathbf{x}_{mL-1}, \mathbf{y}_{mL-1}\}^T \tag{16}$$

To solve those node points $\mathbf{z}$, one gives the initial approximations $\mathbf{z}^*$. Rearrange Eqs. (12) and (13), and it gives

$$\mathbf{g}_{1k} = -\mathbf{x}_{k+1} + \mathbf{x}_k - \frac{1}{2}h\mathbf{k}(\mathbf{y}_{k+1} + \mathbf{y}_k) \tag{17}$$

$$\begin{aligned}\mathbf{g}_{2k} = -\mathbf{y}_{k+1} + \mathbf{y}_k + h\Bigg[-\mathbf{c}\frac{\mathbf{y}_k + \mathbf{y}_{k+1}}{2} + \mathbf{k}\frac{\mathbf{x}_k + \mathbf{x}_{k+1}}{2} \\ +K\left(\sqrt{\left|\frac{1}{3} + V\right|} - \sqrt{\frac{1}{3}}\right)\mathbf{f}_1 + \mathbf{Q}\sin\,\Omega\left(t_k + \frac{h}{2}\right)\Bigg]\end{aligned} \tag{18}$$

Define $\mathbf{g} \equiv \{\mathbf{g}_{10}, \mathbf{g}_{20}, \mathbf{g}_{11}, \mathbf{g}_{21}, \ldots, \mathbf{g}_{1L}, \mathbf{g}_{2L}\}^{\mathrm{T}}$, and put such initial approximates into Eqs. (17) and (18) to calculate $\mathbf{g}^*$ for the first iteration. According to Newton-Raphson algorithm, the increment of the approximations for the next iteration can be obtained as

$$\left[\frac{\partial \mathbf{g}}{\partial \mathbf{z}}\right]\Bigg|_{\mathbf{z}=\mathbf{z}^*} \Delta\mathbf{z} = \mathbf{g}^* \tag{19}$$

Then the approximates for the next iteration can be obtained as $\mathbf{z}^{**} = \mathbf{z}^* + \Delta\mathbf{z}$, and the iteration terminates until $\|\Delta\mathbf{z}\| < \varepsilon$ where $\varepsilon$ is the preset convergence criterion.

For the $k$-th node point, the velocity $u'_f(t_k)$ and pressure $u'_f(t_k)$ at the position of heater can be computed with Eqs. (3) and (4). The analytic solution of velocity and pressure at the position of heater for period-m motion for such a horizontal Rijke tube can be represented by

$$u'_f = a_{01}^{(m)} + \sum_{l=1}^{\infty} b_{1l/m}\cos(l\Omega t/m) + c_{1l/m}\sin(l\Omega t/m) \tag{20}$$

$$p'_f = a_{02}^{(m)} + \sum_{l=1}^{\infty} b_{2l/m}\cos(l\Omega t/m) + c_{2l/m}\sin(l\Omega t/m) \tag{21}$$

where

$$\begin{aligned} a_{01}^{(m)} &= \frac{1}{L}\sum_{s=0}^{mL-1} u'_f(t_s) \\ b_{1l/m} &= \frac{2}{L}\sum_{s=0}^{mL-1} u'_f(t_s)\cos(l\Omega t_s/m) \\ c_{1l/m} &= \frac{2}{L}\sum_{s=0}^{mL-1} u'_f(t_s)\sin(l\Omega t_s/m) \end{aligned}$$

$$\begin{aligned}
a_{02}^{(m)} &= \frac{1}{L}\sum_{s=0}^{mL-1} p'_f(t_s)\\
b_{2l/m} &= \frac{2}{L}\sum_{s=0}^{mL-1} p'_f(t_s)\cos(l\Omega t_s/m)\\
c_{2l/m} &= \frac{2}{L}\sum_{s=0}^{mL-1} p'_f(t_s)\sin(l\Omega t_s/m)
\end{aligned} \tag{22}$$

The amplitude for the $l$-th order of harmonic for the velocity and pressure at the position of heater then can be calculated as

$$A_{l/m}^{(1)} = \sqrt{b_{1l/m}^2 + c_{1l/m}^2} \text{ and } A_{l/m}^{(2)} = \sqrt{b_{2l/m}^2 + c_{2l/m}^2} \tag{23}$$

## 3 Simulation and Discussions

In this section, the specific system parameters are chosen, which are tabulated in Table 1. In the simulation, only the first mode shape for the velocity and pressure is included, and the frequency of the excitation is assumed as $\Omega = 1.2$. Figure 1 gives the bifurcation diagram of velocity and pressure at the heater for such as horizontal Rijke tube by varying the amplitude of excitation from 0 to 4. The steady-state response is periodic and the period is one period of excitation when the amplitude of excitation is greater than 1.736. The steady-state response suddenly becomes chaotic as the amplitude of excitation continues to decrease. In order to understand how chaotic motion for such a Rijke tube model forms as the amplitude of excitation decreases, the zoomed plots for velocity and pressure are illustrated in Fig. 1c, d. It can be found that the period-1 motion jumps to another period-17 motion at $\Omega = 1.736$, and then it becomes chaotic at $\Omega = 1.72$.

In Fig. 2, the plots of node points of velocity and pressure at the heater varying excitation amplitude at the phase $t = 0, 2\pi/\Omega, 4\pi/\Omega$ … which are obtained by

**Table 1** System parameters for horizontal Rijke tube

| Parameter symbol | Value |
|---|---|
| $c_1$ | 0.1 |
| $c_2$ | 0.06 |
| $x_f$ | 0.3 |
| $\tau$ | 0.2 |
| $K$ | 0.6 |
| $\gamma$ | 1.4 |
| $M$ | 0.6 |

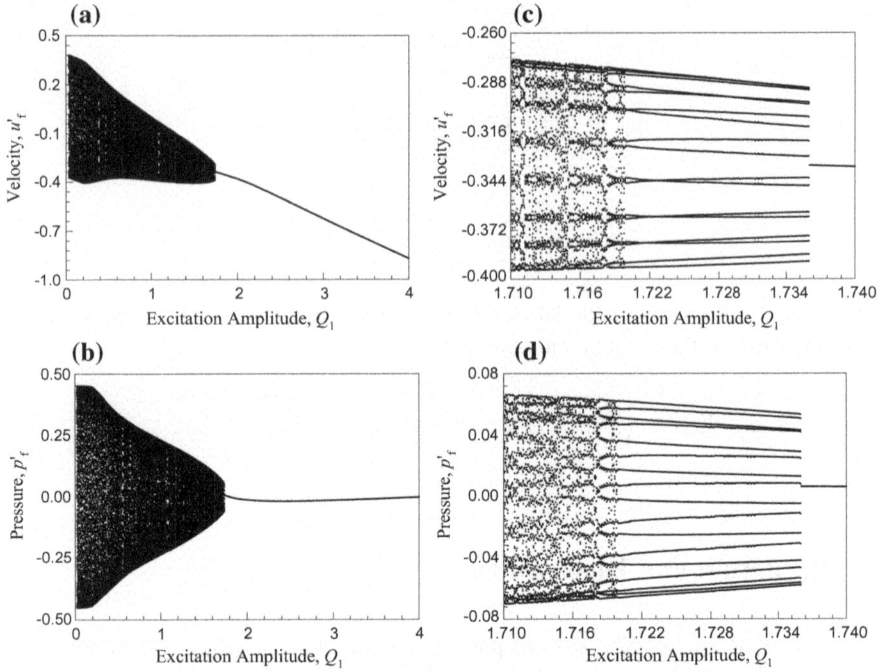


**Fig. 1** Bifurcation for Rijke tube by varying excitation amplitude $Q_1$: **a** velocity at the heater, **b** pressure at the heater, **c** zoomed plot for velocity and **d** zoomed plot for pressure

discrete implicit maps method are shown. The analytic solution of period-1 motion exists in the range of $Q_1 = [0, 4]$. The harmonic amplitudes of velocity and pressure for period-1 motions varying excitation amplitude are demonstrated in Figs. 3 and 4, respectively. For harmonic amplitude of velocity which is shown in Fig. 3, the constant term is non-zero which indicates that average value of velocity oscillation is not zero for $Q_1 = [0, 4]$. It increases from zero at first, and then it drops around $Q_1 = 1.8$. The first order of harmonic increases almost linearly as the amplitude of excitation increases. For the higher harmonics, the quantity levels of $A_2^{(1)}$ and $A_3^{(1)}$ are both $10^{-2}$, and the amplitude is almost zero when the excitation amplitude is small. When the order continues to increase, the range for which the harmonic amplitude is close to zero increases. For harmonic order increases to 10, the quantity level of harmonic amplitude decreases to $10^{-4}$. The quantity level of harmonic amplitude of pressure also drops asymptotically with the order of harmonic increases which is illustrated in Fig. 4. But for constant term of pressure at the heater, $a_{02}$ is always zero.

For the analytical solution **z** which is obtained from discrete implicit maps method, the numerical simulation can be carried out with initial conditions from the analytical solution **z** of period-1 motion for Rijke tube. For the numerical simulation with $Q_1 = 1.73$ which is shown in Fig. 5, the initial velocity and pressure at the heater are

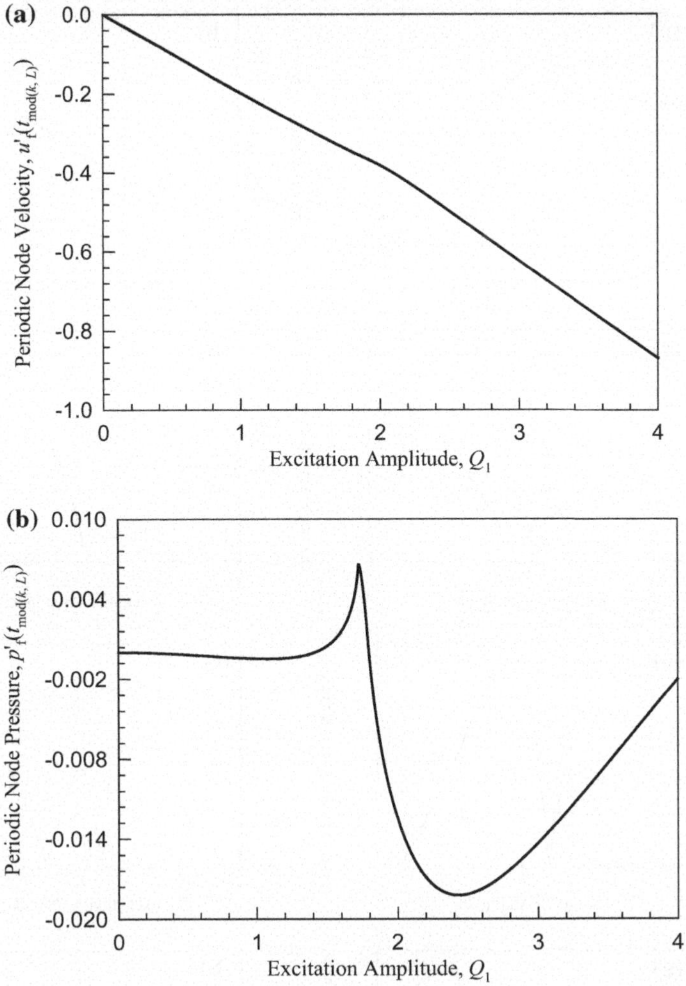


**Fig. 2** Node points of period-1 motion at phase $t = t_{\mathrm{mod}(k,\,L)}$ $(k = 0, 1, 2 \ldots)$ varying excitation amplitude $Q_1$: **a** velocity at the heater, **b** pressure at the heater

chosen as $u'_f = -0.334259$, $p'_f = 0.00653288$, respectively. In the phase plane, the solid curve is the numerical simulation. The green filled circle is the initial position, and the blue hollow circles are the time-delayed information before $t = 0$. The red fork symbols are the analytical solution. For the first 20 periods, the trajectory from numerical simulation sticks to analytical solution of period-1 motion which is plotted in Fig. 5a–c. From Fig. 1, the steady-state response for $Q_1 = 1.73$ is a period-17 motion. In order to investigate how motion involves from period-1 motion to period-17 motion, the Poincare maps of numerical simulation for $Q_1 = 1.73$ from 0 to 100 periods, from 100 to 200 periods, and from 300 to 400 periods are drawn in

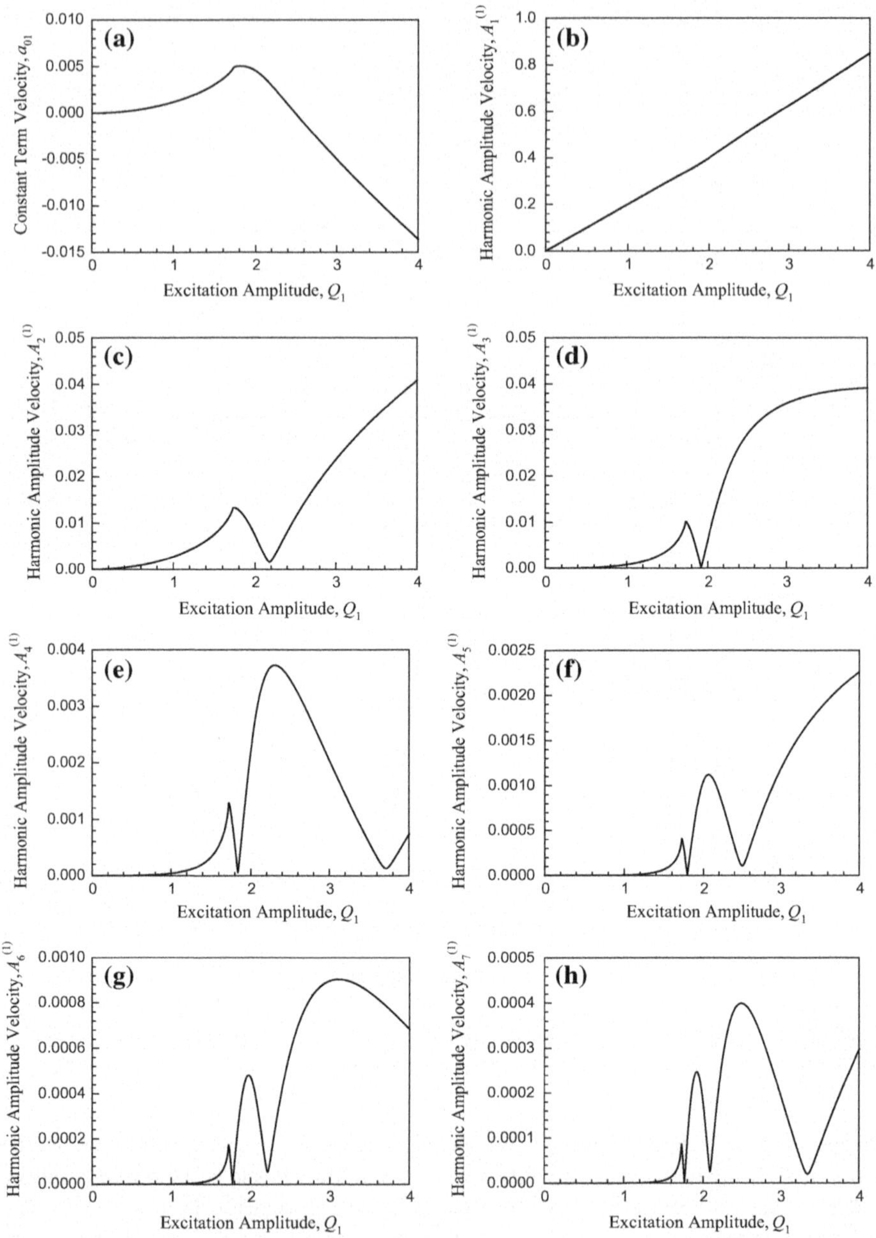


**Fig. 3** Harmonic amplitude varying excitation amplitude $Q_1$ for velocity at heater: **a** constant term $a_{01}$, **b–l** $A_k^{(1)}$ (k = 1, 2,…, 10)

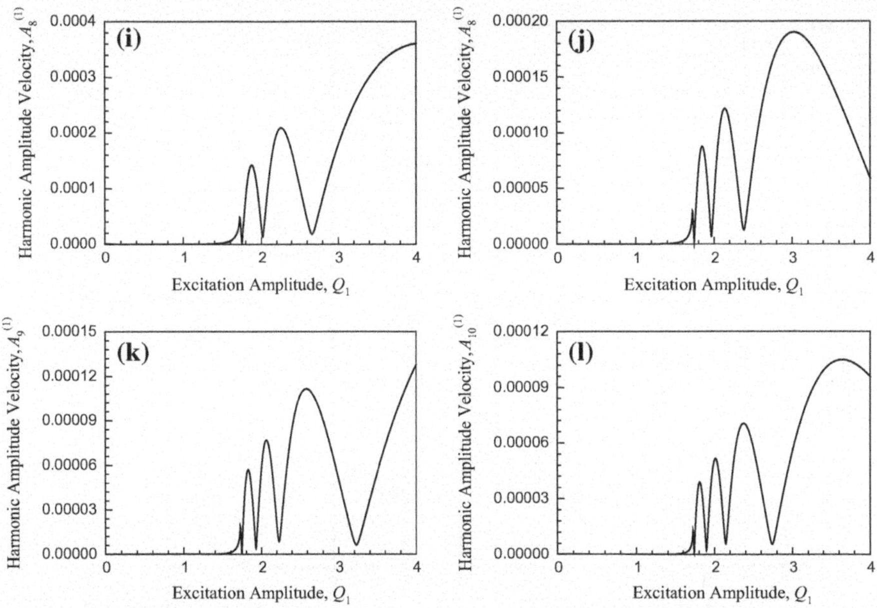


**Fig. 3** (continued)

Figs. 5d–g, respectively. It can be seen that the motion sticks on the orbit of period-1 solution for the first 100 periods, and then it gradually leaves that orbit. After hundreds of periods of transient process, it runs to the stable period-17 orbit after 300 periods of simulation time.

For $Q_1 = 2.00$, the stable period-1 motion is simulated with initial velocity and pressure at the heater $u'_f = -0.380952$, $p'_f = -0.0124891$, which is demonstrated in Fig. 6. The numerical simulation correlates with the analytical solution very well, and it never leaves such an orbit. In Figs. 6d, e, the harmonic amplitudes of velocity and pressure are shown. The quantity level of harmonic amplitude of velocity drops to $10^{-6}$ when the order increases to 20. But the harmonic amplitude of pressure drops slower as the order of harmonic increases, and quantity level still stays around $10^{-6}$ as $k \to 50$.

## 4 Conclusions and Future Work

In this paper, the analytic solution of periodic motion for a horizontal Rijke tube model with periodic excitation has been obtained through discrete implicit approach. The procedures to discretizing the periodic motions and solving for the analytic solutions have been described in details. The analytic solutions of period-1 motion varying excitation amplitude have been presented. It shows that the solution of period-1

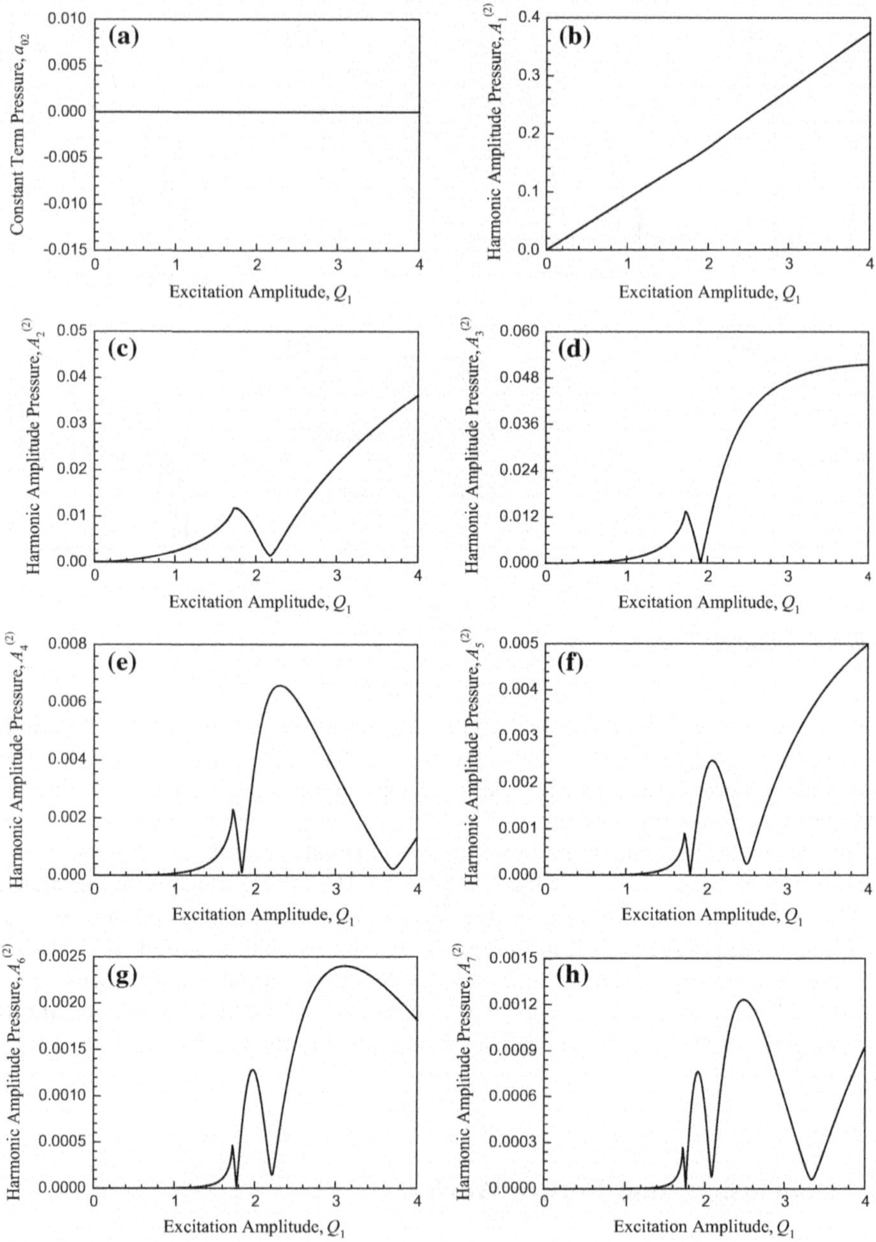


**Fig. 4** Harmonic amplitude varying excitation amplitude $Q_1$ for pressure at heater: **a** constant term $a_{02}$, **b–l** $A_k^{(2)}$ (k = 1, 2,…, 10)

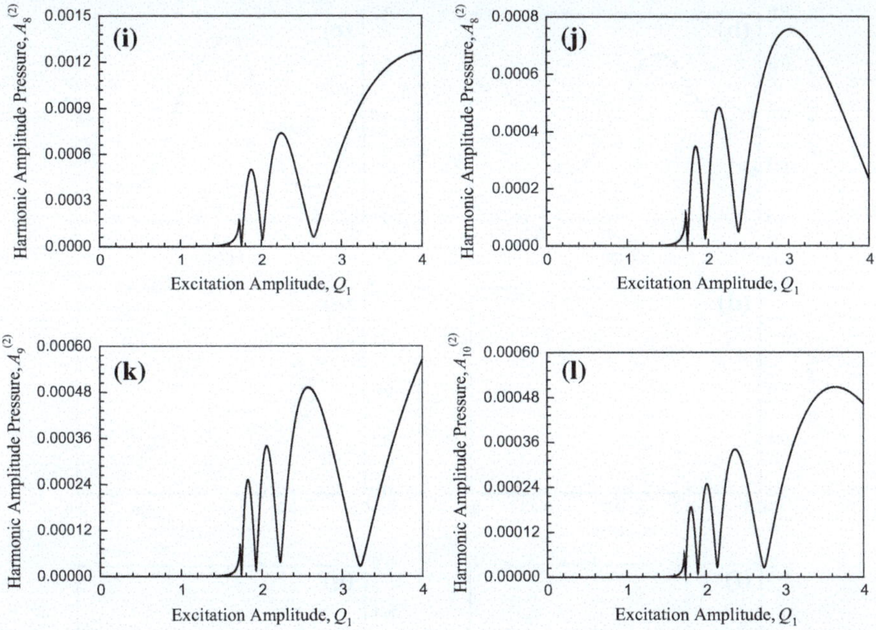


**Fig. 4** (continued)

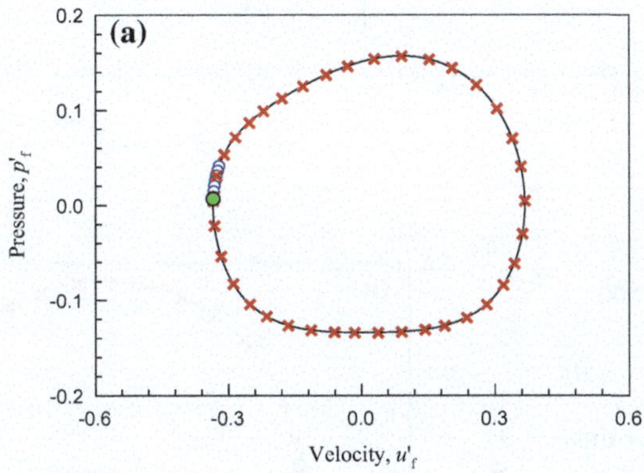


**Fig. 5** Numerical simulation for $Q_1 = 1.73$ ($u'_f = -0.334259$, $p'_f = 0.00653288$ at $t = 0$): **a** phase plane, **b** time trajectory of velocity, **c** time trajectory of pressure for the first 20 periods; **d** Poincare map for 0–100 periods, **e** Poincare map for 100–200 periods, **f** Poincare map for 200–300 periods, **g** Poincare map for 300–400 periods

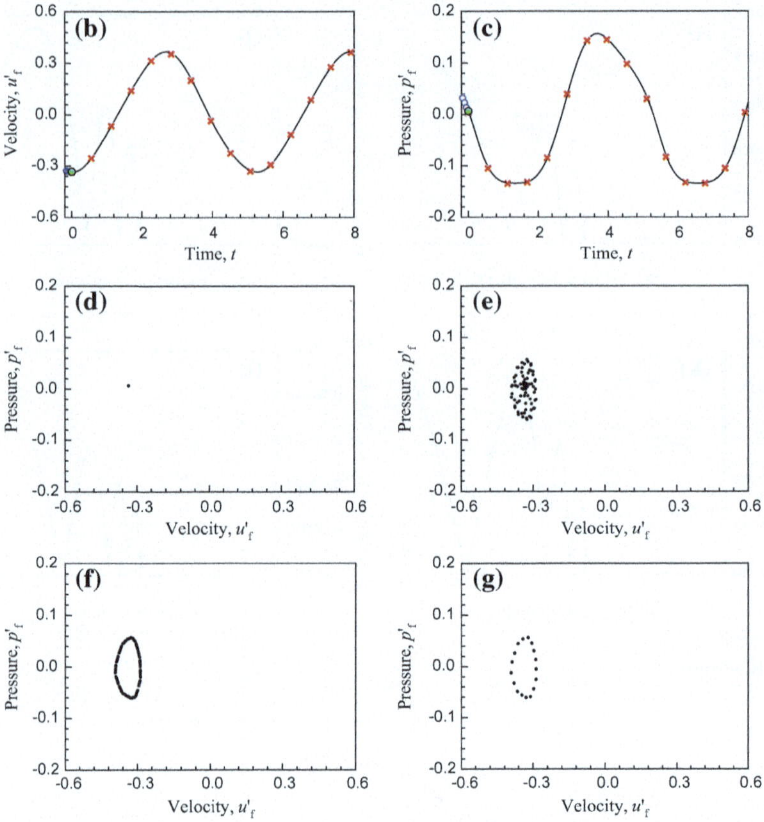


**Fig. 5** (continued)

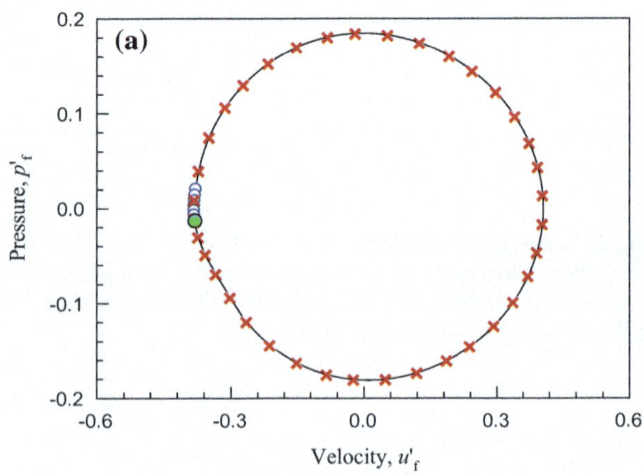


**Fig. 6** Numerical simulation for $Q_1 = 2.00$ ($u'_f = -0.380952$, $p'_f = -0.0124891$ at $t = 0$): **a** phase plane, **b** time trajectory of velocity, **c** time trajectory of pressure, **d** harmonic amplitudes of velocity, **e** harmonic amplitudes of pressure

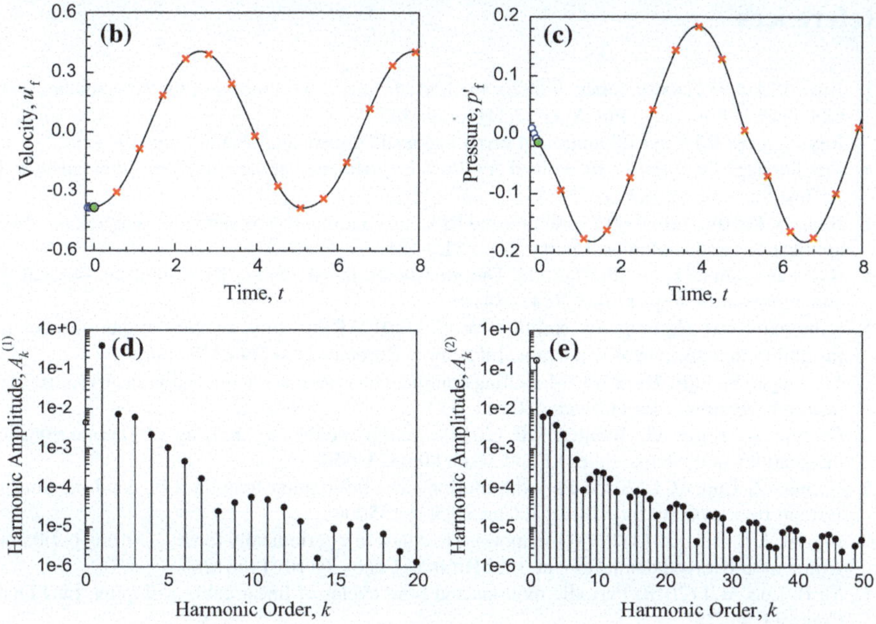


**Fig. 6** (continued)

motion is continuously exists in the range of $Q_1 = [0, 4]$, and no coexisting solution has been found in such a range. But for numerical bifurcation, it is known that period-1 motion disappears when the excitation amplitude continues to decrease at the critical point $Q_1 = 1.736$. By carrying out numerical simulation at $Q_1 = 1.73$ and 2.00, both the analytic solution before and after such a critical point should be exact, since the numerical prediction with initial conditions computed from the analytic solution of period-1 motion correlates with the analytic solution very well for the first many periods. For $Q_1 = 2.00$, the motion always sticks to the period-1 orbit. By observing the motion involving process using the Poincare map, it can be found that the motion stick with the analytic solution of period-1 motion for the first 100 periods, and then it runs to the orbit of period-17 motion after 300 periods. It might be because the analytic solution of period-1 motion for such a horizontal Rijke tube is unstable for $Q_1 < 1.736$. Therefore, one has to continue to find out the way to determining the stability of periodic motion for such a Rijke tube which time-delay terms are included in the model in the future.

**Acknowledgements** This work was sponsored by Shanghai Sailing Program (Grant No. 19YF1421600).

## References

1. Rijke PLLXXI (1859) Notice of a new method of causing a vibration of the air contained in a tube open at both ends. Phil Mag 17(116):419–422
2. Rayleigh (1878) The explanation of certain acoustic phenomena. Nature 18:319–321
3. Friedlander MM, Smith TJB, Powell A (1964) Experiments on the Rijke-tube phenomenon. J Acoust Soc Am 36(9):1737–1738
4. Bisio G, Rubatto G (1999) Sondhauss and Rijke oscillations-thermodynamic analysis, possible applications and analogies. Energy 24(2):117–131
5. Balasubramanian K, Sujith RI (2008) Thermoacoustic instability in a Rijke tube: non-normality and nonlinearity. Phys Fluids 20(4):044103
6. Subramanian P, Mariappan S, Sujith S, Wahi P (2010) Bifurcation analysis of thermoacoustic instability in a horizontal Rijke tube. Int J Spray Combust Dyn 2(4):325–356
7. Mariappan S, Sujith RI (2011) Modelling nonlinear thermoacoustic instability in an electrically heated Rijke tube. J Fluid Mech 680:511–533
8. Orchini A, Rigas G, Juniper MP (2016) Weakly nonlinear analysis of thermoacoustic bifurcations in the Rijke tube. J Fluid Mech 805:523–550
9. Huang JZ, Luo ACJ (2015) Periodic motions and bifurcation trees in a buckled, nonlinear Jeffcott rotor system. Int J Bifurcat Chaos 25(1):1550002
10. Luo ACJ, Jin HX (2014) Period-m motions to chaos in a periodically forced, Duffing oscillator with a time-delayed displacement. Int J Bifurcat Chaos 24(10):1450126
11. Yu B, Luo ACJ (2018) Periodic motions and limit cycles of linear cable galloping. Int J Dyn Control 6:41–78
12. Luo ACJ (2015) Discretization and implicit mapping dynamics. Springer, Berlin
13. Wang DH, Huang JZ (2016) Periodic motions and chaos for a damped mobile piston system in a high pressure gas cylinder with P control. Chaos, Solitons Fractals 95:168–178
14. Guo Y, Luo ACJ (2017) Complete bifurcation trees of a parametrically driving pendulum. J Vib Test Syst Dyn 1(2):93–134
15. Luo ACJ, Xing SY (2017) Time-delay effects on periodic motions in a periodically forced, time-delayed, hardening Duffing oscillator. J Vib Test Syst Dyn 1(1):73–91
16. Matveev KI, Culick FEC (2003) A model for combustion instability involving vortex shedding. Combust Sci Technol 175(6):1059–1083

# Real-Time Integrity Monitoring for Civil Aviation with Improved Navigation Performance

**Jin Chang, Xingqun Zhan and Yawei Zhai**

**Abstract** With the worldwide development in the past decades, multi-constellation Global Navigation Satellite System (GNSS) are able to provide consistent and reliable navigation services today, which are expected to bring significant performance improvement to civil aviation in the future. For the GNSS-based aircraft navigation, meeting the integrity and continuity requirements is of the most importance. In the currently proposed baseline Advanced Receiver Autonomous Integrity Monitoring (ARAIM) user algorithm, the integrity risk is evaluated using a conservative upper bound. Despite its computational efficiency, this bound is not tight enough, which may lead to overly conservative results. Operationally, the system may incorrectly alert the user, which severely impacts navigation continuity. Therefore, in this work, we develop a new method to tightly bound the integrity risk and establish a multi-constellation ARAIM test platform to validate the theory. The new approach takes advantage of the independence between position estimation error and detection test statistics and expresses the integrity risk evaluation as a convex optimization problem. It is shown that the global maximum of the objective function is a tight bound on integrity risk, and it can be efficiently computed using an numerical method. Other than the theoretical derivations, another major contribution of this work is prototyping the ARAIM user segment in the Guidance, Navigation, and Control (GNC) laboratory at Shanghai Jiao Tong University. Both of the ARAIM Multiple Hypothesis Solution Separation (MHSS) algorithm and the new approach are incorporated into the prototype, and the real-time integrity monitoring results are visually displayed in terms of horizontal and vertical protection levels, effective monitoring threshold, integrity risk, etc. As compared to the existing MHSS theory, the results suggest that the navigation service availability can be noticeably improved using the proposed method, especially when the constellations are subject to larger ranging errors.

**Keywords** Multi-constellation GNSS · Integrity monitoring · Integrity risk · ARAIM test platform

---

J. Chang · X. Zhan (✉) · Y. Zhai
Shanghai Jiao Tong University, Shanghai, China
e-mail: xqzhan@sjtu.edu.cn



Z. Jing (ed.), *Proceedings of the International Conference on Aerospace System Science and Engineering 2019*, Lecture Notes in Electrical Engineering 622,
https://doi.org/10.1007/978-981-15-1773-0_9

# 1 Introduction

Nowadays navigation system is indispensable for civil air transportation. Navigation system allows aircraft to determine its location and to fly on a predetermined route. This can avoid many hazardous aviation accidents. For instance, when aircraft drifts off course, pilot is able to correct the error of position with navigation system. To build a robust navigation system, several work has been made in the past century. Before 1970s, the research of aircrafts navigation system focused on Radio Navigation System (RNS) and Inertial Navigation System (INS). But these two systems both have drawbacks such as limited system coverage and increasing navigation errors over time [1]. After 1970s, satellite-based positioning system which provides users with global coverage and lower positioning error began to flourish [2].

From the 1970s up to the present, many countries and regions have established their own satellite-based positioning systems such as Global Navigation System (GPS, USA), GLONASS (Russia), BeiDou Navigation Satellite System (BDS, China), and Galileo (Europe). At the same period, some augmentation systems, like Ground-Based Augmentation System (GBAS) and Satellite-Based Augmentation Systems (SBAS), have been built to improve the navigation performance. The combination of these satellite-based positioning systems (GPS, BDS, GLONASS, and Galileo) is referred to as Global Navigation Satellite System (GNSS), and each individual satellite-based positioning system is termed as constellation [3].

For civil aviation, navigation system which meets stringent requirements is critical to guarantee the safety during the flight. The International Civil Aviation Organization (ICAO) put forward some metrics for satellite-based positioning system [4]. Two of the most challenging requirements for civil aviation are integrity and continuity [5]. Integrity is a measure of trust, which is used to determine whether the positioning solution provided by the navigation system is correct. Continuity measures navigation system's ability to operate without unplanned interruptions [4]. Receiver Autonomous Integrity Monitoring (RAIM) is a system that can provide users with real-time integrity monitoring results [6]. When satellite fault occurs during the flight, timely warning can be sent by RAIM to users. After mid-1990s, by using single frequency signal and single constellation, RAIM has become a backup navigation tool and it can be applied to en-route flight [5]. But previous research has found the drawbacks of RAIM. One of the major drawbacks is occasional lack of availability [7].

With the great development of satellite-based positioning system, single frequency signal and single constellation positioning are gradually replaced by dual-frequency signal and multi-constellation positioning because it possesses two advantages. (a) Dual-frequency signal can cancel the ionosphere delay [8]. (b) Multi-constellation positioning with increased number of satellites improves user's geometry [9].

Due to these two advantages, dual-frequency signal and multi-constellation positioning can significantly improve the accuracy and stability of positioning [10]. The superiority and greater redundancy of dual-frequency multi-constellation positioning lead to consider making up the shortcomings of RAIM. So Advanced

Receive Autonomous Monitoring (ARAIM) has been put forward to overcome these shortcomings. Many researchers have made great contributions to this. For instance, Professor Juan Blanch from Stanford University and Professor Boris Pervan from Illinois Institute of Technology have already built a solid theoretical foundation for ARAIM Multiple Hypothesis Solution Separation (MHSS) algorithm. Some institutions, take Stanford GPS Lab as an example, have devised ARAIM prototype that can be equipped on aircraft. Though dual-frequency signal and multi-constellation positioning have better accuracy and greater redundancy for integrity monitoring, it also brings the higher probability of having faulted satellites which can severely affect the positioning solution.

Since the current augmentation systems, SBAS and GBAS, cannot cover every region of the world [11, 12], it is urgent to develop ARAIM MHSS algorithm for the place with poor navigation performance. ARAIM MHSS algorithm enables aircraft to detect navigation fault and to correct positioning result by excluding some faulty satellites or constellations. Meanwhile, today's ARAIM MHSS algorithm also has some weaknesses, and one of them is that the probability bound of ARAIM algorithm is not tight enough, which can significantly affect navigation performance especially when there exists exclusion function in ARAIM algorithm. To avoid this weakness, we propose a new approach of calculating the probability bound and establish multi-constellation ARAIM test platform to validate the theory.

The second part of this paper will introduce the basic principles of ARAIM. The third part of this paper focuses on introducing the new approach. In this chapter, the probability bound will be rewritten mathematically and new Probability of Hazardous Misleading Information (PHMI) equation will be introduced. The fourth part of this paper explains the working principles of multi-constellation ARAIM test platform. The fifth part of this paper shows the navigation performance under different conditions (MHSS algorithm and new approach).

## 2 ARIAM Overview

### *2.1 ARAIM Background*

Based on linearized pseudorange equations, GNSS receiver can compute the user's location by solving linear algebraic equation, which can be expressed as:

$$\mathbf{y} = \mathbf{H}\mathbf{x} + \boldsymbol{\epsilon} + \mathbf{f} \tag{1}$$

where $\mathbf{y}$ is a $n \times 1$ measurement vector, and this vector represents the corrected pseudorange of satellites. $\mathbf{H}$ is a $n \times (3 + N)$ matrix in which $n$ and $N$ are the number of satellites and constellations, respectively. $\mathbf{x}$ is a $(3 + N) \times 1$ vector, which includes the user's location and the clock error of different constellations. $\boldsymbol{\epsilon}$ is the noise, which comes from clock error, ephemeris error, tropospheric error, multipath, and receiver noise, and it follows normal distribution. **b** and **V** are defined as bias vector and covariance matrix of $\boldsymbol{\epsilon}$, respectively. **f** is a $n \times 1$ vector, which indicates the fault magnitude of navigation system. The solution of equation Eq. (1) is all-in-view estimation positioning solution, and it can be written as:

$$\hat{\mathbf{x}}_0 = (\mathbf{H}^T\mathbf{V}^{-1}\mathbf{H})^{-1}\mathbf{H}^T\mathbf{V}^{-1}\mathbf{y} \tag{2}$$

When one satellite or more satellites that corresponds to the $i$th fault model was excluded, satellite-removed positioning solution $\hat{\mathbf{x}}_i$ can be written as:

$$\hat{\mathbf{x}}_i = (\mathbf{H}^T\mathbf{V}_i^{-1}\mathbf{H})^{-1}\mathbf{H}^T\mathbf{V}_i^{-1}\mathbf{y} \tag{3}$$

$\mathbf{V}_i$ in Eq. (3) is the covariance matrix which excludes the faulty satellite or satellites corresponding to $i$th fault model. For $i$th fault model, we assume that $p$th satellite is in $i$th fault model and $q$th satellite is faultless. Based on this assumption, we can compute covariance matrix $\mathbf{V}_i$ below:

$$\mathbf{V}_i(p, p) = 0, \quad \mathbf{V}_i(q, q) = \mathbf{V}(g, g) \tag{4}$$

$$\hat{\mathbf{x}}_{ex,k} = (\mathbf{H}^T\mathbf{V}_{ex,k}^{-1}\mathbf{H})^{-1}\mathbf{H}^T\mathbf{V}_{ex,k}^{-1}\mathbf{y} \tag{5}$$

When considering MHSS ARAIM exclusion function, we introduce $\mathbf{V}_{ex,k}$, which excludes the $k$th subset fault model, under the $ex$th fault model. For fault model $ex$ and the subset fault model $k$, we assume that the $p$th satellite is in the $ex$th fault model or the $k$th subset fault model and $q$th satellite is faultless. Based on this assumption, we can compute covariance matrix $\mathbf{V}_{ex,k}$ below:

$$\mathbf{V}_{ex,k}(p, p) = 0, \quad \mathbf{V}_{ex,k}(q, q) = \mathbf{V}(q, q) \tag{6}$$

If fault model $ex$ and subset fault mode $k$ have been excluded, the estimation positioning solution can be written as:

$$\hat{\mathbf{x}}_{ex,k} = (\mathbf{H}^T\mathbf{V}_{ex,k}^{-1}\mathbf{H})^{-1}\mathbf{H}^T\mathbf{V}_{ex,k}^{-1}\mathbf{y} \tag{7}$$

By using Eqs. (1), (2) and (4), error test statistic $\epsilon_0$ and detection test statistic $\Delta_i$ can be defined as:

$$\epsilon_0 = \hat{\mathbf{x}}_0 - \mathbf{x}, \quad \epsilon_0 \sim (\mu_{HI,0}, \quad \mathbf{P}_0) \tag{8}$$

where $\mu_{HI,0}$ and $\mathbf{P}_0$ are equal to $\mathbf{S}_0\mathbf{f} + \mathbf{S}_0\mathbf{b}$ and $(\mathbf{H}^T\mathbf{V}^{-1}\mathbf{H})^{-1}$, respectively.

$$\Delta_i = \hat{\mathbf{x}}_0 - \hat{\mathbf{x}}_i, \quad \Delta_i \sim (\mu_{HI,i}, \quad \mathbf{P}_i) \tag{9}$$

where $\mu_{ND,i}$ and $\mathbf{P}_i$ are equal to $\mathbf{S}_0\mathbf{f} + \mathbf{S}_0\mathbf{b} - \mathbf{S}_i\mathbf{b}$ and $(\mathbf{H}^T\mathbf{V}_i^{-1}\mathbf{H})^{-1} - (\mathbf{H}^T\mathbf{V}^{-1}\mathbf{H})^{-1}$, respectively.

$\mathbf{S}_0$ and $\mathbf{S}_i$ are equal to $(\mathbf{H}^T\mathbf{V}^{-1}\mathbf{H})^{-1}\mathbf{H}^T\mathbf{V}^{-1}$ and $(\mathbf{H}^T\mathbf{V}_i^{-1}\mathbf{H})^{-1}\mathbf{H}^T\mathbf{V}_i^{-1}$, respectively. By using Eqs. (1), (3), and (5), for the second layer detection of MHSS ARIAM, error test statistic $\epsilon_{ex}$ and detection test statistic $\Delta_{ex,k}$ can be defined as:

$$\epsilon_{ex} = \hat{\mathbf{x}}_{ex} - \mathbf{x}, \quad \epsilon_{ex} \sim \left(\mu_{HI,ex}, \quad \mathbf{P}_{ex}\right) \tag{10}$$

where $\mu_{HI,ex}$ and $\mathbf{P}_{ex}$ are equal to $\mathbf{S}_{ex}\mathbf{b}$ and $(\mathbf{H}^T\mathbf{V}_{ex}^{-1}\mathbf{H})^{-1}$, respectively.

$$\Delta_{ex,k} = \hat{\mathbf{x}}_{ex,k} - \hat{\mathbf{x}}_{ex}, \ \Delta_{ex,k} \sim (\mu_{ND,ex}, \ \mathbf{P}_{ex,k}) \tag{11}$$

where $\mu_{ND,ex}$ and $\mathbf{P}_{ex,k}$ are equal to $\mathbf{S}_{ex}\mathbf{b}$ and $(\mathbf{H}^T\mathbf{V}_{ex}^{-1}\mathbf{H})^{-1}$, respectively.

$\mathbf{S}_{ex,k}$ are equal to $(\mathbf{H}^T\mathbf{V}_{ex,k}^{-1}\mathbf{H})^{-1}\mathbf{H}^T\mathbf{V}_{ex,k}^{-1}$, respectively.

In the following paper, ARAIM is referred to as MHSS ARAIM.

## 2.2 ARAIM User Algorithm

ARAIM user algorithm consists of five functions:

(1) Analyzing the fault models

In practice, the number of fault models increases sharply with the number of satellites and constellations. To reduce computational burden, ARAIM user algorithm does not need to monitor all the fault models. Based on required threshold $P_{thresh}$ ($8 \times 10^{-8}$) [13–15], the sum of unmonitored fault models should meet the following requirement:

$$\sum_{j=1}^{h} P_j <= P_{thresh} \tag{12}$$

where $P_j$ is the probability of the $j$th unmonitored fault model occurring in navigation system, and $h$ is the number of unmonitored fault models.

(2) Determining detection and exclusion threshold

Detection threshold $T_i$ can be derived from the allocated probability $P_{FA\_NE}$, which represents the probability that test statistic $\Delta_i$ exceeds their threshold under fault-free hypothesis [13, 14].

To calculate the exclusion threshold $T_{ex,k}$, we use the allocated probability $P_{FD\_NE}$, which represents the probability that test statistic $\Delta_{ex,k}$ exceeds its threshold under second layer fault-free hypothesis [13, 14].

(3) Evaluating ARAIM user algorithms availability

Before conducting the fault detection and exclusion function, the availability of ARAIM algorithm needs to be evaluated. If ARAIM user algorithm is not available, the algorithm will skip the fault detection and exclusion functions. The requirement of ARAIM availability depends on different application scenarios. LPV-200 is a stringent requirement for ARAIM in approach phase. LPV-200 requirements of availability are as follows: (a) Vertical protection level (VPL) is lesser than vertical alert limit (VAL) (VPL $\leq 35$ m), (b) Effective monitor threshold (EMT) $\leq 15$ m, (c) 95% of the time, vertical accuracy $\leq 4$ m, (d) 99.99999% of the time, fault-free vertical accuracy $\leq 10$ m. The method of calculating VPL can derive from the equation of PHMI.

(4) Fault detection

When ARAIM algorithm is available, fault detection function can be conducted. The result of fault detection function depends on the equation below:

$$\Delta_{i,d} = |\mathbf{x}_{0,d} - \mathbf{x}_{k,d}| \qquad (d = 1, 2, 3) \tag{13}$$

where the subscript $d$ represents the horizontal ($d = 1, 2$) and vertical ($d = 3$) direction. If $\Delta_{i,d}$ is larger than detection threshold $T_{i,d}$, it shows that the fault is detected by ARIAM user algorithm and fault exclusion function is about to exclude the fault. Otherwise, the algorithm skips the fault exclusion function.

(5) Fault exclusion

If fault has been detected, exclusion function attempts to correct the navigation solution by excluding some satellites or constellations [16]. To determine the correctness of exclusion, we introduce $\Delta_{ex,k,d}$, which can be expressed by the following:

$$\Delta_{ex,k,d} = |\mathbf{x}_{ex,d} - \mathbf{x}_{ex,k,d}| \qquad (d = 1, 2, 3) \tag{14}$$

when $\Delta_{ex,k,d}$ is larger than exclusion threshold $T_{ex,k,d}$, it shows that the fault cannot be eliminated by ARAIM algorithm and the user needs to be alerted.

## 3 Optimized ARAIM Algorithm

### *3.1 Tighter Bound of PHMI*

PHMI evaluation is of great importance in ARAIM user algorithm. This evaluation can monitor the efficiency of algorithm. But the current evaluation of PHMI accumulates errors in summation process. This may lead to safety problems.

To calculate PHMI more precisely, we use the independence of $\epsilon_0$ and $\Delta_i$, $\epsilon_{ex}$, and $\Delta_{ex,k}$ [17] to simplify the current PHMI equation. The simplified form can be given by:

$$
\begin{aligned}
P_{HMI} &\le P(|\epsilon_0| > l|H_0)P_{H0} \\
&+ \sum_{i=1}^{m} \max_{\mathbf{f}_i} \left( P(|\epsilon_0| > l|H_i, \mathbf{f}_i) \cdot P(|\Delta_i| < T_i|H_i, \mathbf{f}_i) \right) P_{H_i} \\
&+ \sum_{ex=1}^{m} (P(|\epsilon_{ex}| > l|H_0)P_{H_0} + P(|\epsilon_{ex}| > l|H_{ex})P_{H_{ex}} \\
&+ \sum_{\substack{k=1 \\ k \ne ex}}^{m_1} \max_{\mathbf{f}_k} \left( P(|\epsilon_{ex}| > l|H_k, \mathbf{f}_k) \cdot \right. \\
&\qquad \left. P(|\Delta_{ex,k}| < T_{ex,k}|H_k, \mathbf{f}_k)) \right) P_{H_k}
\end{aligned}
\tag{15}
$$

where $\max_{\mathbf{f}_i}$ and $\max_{\mathbf{f}_k}$ represent the function of calculating the maximum value under fault model $i$ and subset fault model $k$. $\mathbf{f}_i$ and $\mathbf{f}_k$ are the fault magnitude corresponding to the fault model $i$ and subset fault model $k$. $m$ and $m_1$ are the number of fault models that need to be monitored in detection function and exclusion function.

The first term of Eq. (15) is PHMI of detection function under fault-free condition, so fault magnitude does not exist in this term.

For the second term of Eq. (15), it indicates a condition that the fault has existed in navigation system but detection function fails to detect the fault. We denote $P_{HI,i,d}$ and $P_{ND,i,d}$ as $P(|\epsilon_0| > l|H_i, \mathbf{f}_i)$ and $P(|\Delta_i| < T_i|H_i, \mathbf{f}_i)$, respectively. In Chapter "Ultra-Rapid Direct Satellite Selection Algorithm for Multi-GNSS", we explained the bias and variance of random variable $\epsilon_0$ and $\Delta_i$. These numerical characteristics can rewrite the second term of Eq. (15) mathematically.

As for $P_{HI,i,d}$, it can be rewritten as:

$$
P_{HI,0,d} = 1 - \int_{-l_d}^{l_d} \frac{1}{\sqrt{2\pi \mathbf{P}_0(d,d)}} exp \left( \frac{(x - \mu_{HI,0,d})^2}{2\mathbf{P}_0(d,d)} \right) dx \tag{16}
$$

where $\mathbf{P}_0$ is covariance matrix of $\epsilon_0$ and $\mu_{HI,i,d}$ can be written as:

$$
\mu_{HI,0,d} = \sum_{e=1}^{n} |\mathbf{S}_0(d,e)|\mathbf{b}(e,d) + \sum_{e=1}^{n} |\mathbf{S}_0(d,e)|\mathbf{f}_i(e,1) \tag{17}
$$

For $P_{ND,i,d}$, it can be written as:

$$
P_{ND,i,d} = \int_{-T_{\Delta_d}}^{T_{\Delta_d}} \frac{1}{\sqrt{2\pi \mathbf{P}_i(d,d)}} exp \left( \frac{(y - \mu_{ND,i,d})^2}{2\mathbf{P}_i(d,d)} \right) dy \tag{18}
$$

where $\mathbf{P}_i$ is covariance matrix of $\Delta_i$ and $\mu_{ND,i,d}$ can be written as:

$$\begin{aligned}\mu_{ND,i,d} = &-\sum_{e=1}^{n} |\mathbf{S}_0(d,e) - \mathbf{S}_i(d,e)|\mathbf{b}(e,d) \\ &+\sum_{e=1}^{n} |\mathbf{S}_0(d,e)|\mathbf{f}_i(e,1)\end{aligned} \tag{19}$$

Since the only variable in Eqs. (17) and (19) is $\mathbf{f}_i$, we can rewrite $P_{HI,i,d} \cdot P_{ND,i,d}$ as:

$$\begin{aligned}P_{HMI,2} &= P_{HI,i,d} \cdot P_{ND,i,d} \\ &= k_i \int_{-T_{\Delta_d}}^{T_{\Delta_d}} \exp\left(\frac{(y-M+b_i)^2}{2\mathbf{P}_i(d,d)}\right) dy \\ &- k_i k_0 \int_{-l_d}^{l_d} \int_{-T_{\Delta_d}}^{T_{\Delta_d}} \exp\left(-\frac{(y-M+b_i)^2}{2\mathbf{P}_i(d,d)}\right) \\ &\exp\left(-\frac{(x-M-b_0)^2}{2\mathbf{P}_0(d,d)}\right) dxdy\end{aligned} \tag{20}$$

where $k_i = \frac{1}{\sqrt{2\pi \mathbf{P}_i(d,d)}}$, $k_0 = \frac{1}{\sqrt{2\pi \mathbf{P}_0(d,d)}}$, $y - M + b_i = y - \mu_{ND,i,d}$, and $x - M - b_0 = x - \mu_{HI,0,d}$. The subscript 2 in $P_{HMI,2}$ represents the second term of Eq. (15). Figure 1 shows the image function of Eqs. (15), (17), and (19).

In Eqs. (16) and (18), $b_0$ and $b_i$ are defined as positive and negative values, respectively. This definition takes into account the worst-case scenario. Under this definition, the maximum value of PHMI can be obtained. In Figs. 2 and 3, we discuss four conditions of $b_0$ and $b_i$ ($b_0 > 0$, $b_0 < 0$, $b_i > 0$, $b_i < 0$). Figure 2 shows the function image of $P_{HI,0,d}$ and $P_{ND,0,d}$.

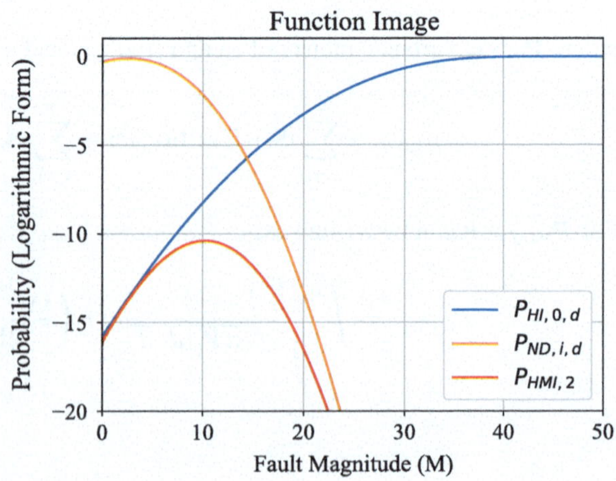


**Fig. 1** Function image (logarithmic form) of $P_{HI,0,d}$ $P_{ND,i,d}$ and $P_{HI,0,d} \cdot P_{ND,i,d}$

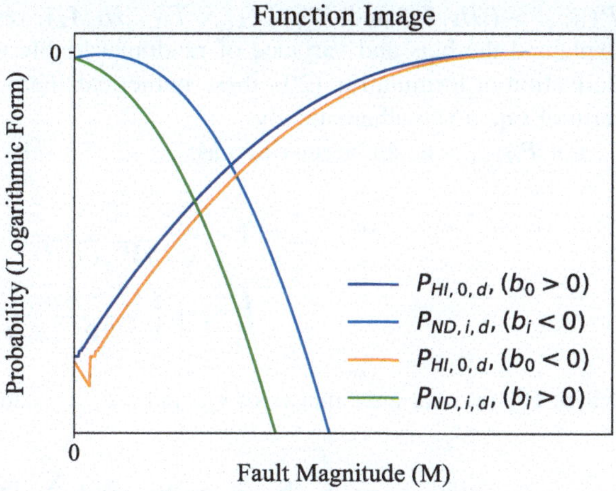


**Fig. 2** Function image (logarithmic form) of $P_{HI,0,d}$ and $P_{ND,i,d}$ with different biases

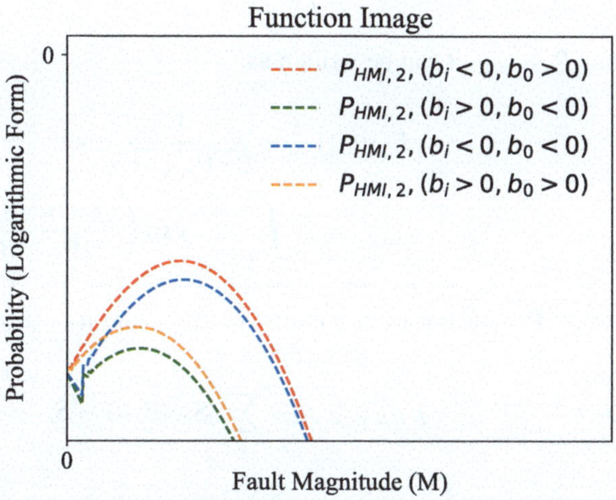


**Fig. 3** Function image (logarithmic form) of $P_{HI,0,d} \cdot P_{ND,i,d}$ with different biases

In Fig. 2 , the function images of $b_0 > 0$ and $b_i < 0$ are above the other function images, so their product is lager than other conditions. Figure 3 is the function image of $P_{HMI,2}$ under different conditions. It shows that the product of $P_{HI,0,d}(b_0 > 0)$ and $P_{ND,i,d}(b_i < 0)$ is above other function images. So we draw the conclusion that the product of $b_0 > 0$ and $b_i < 0$ can be used in calculating the maximum value of PHMI.

The third and the fourth terms in Eq. (16) are the PHMI of exclusion function which are under fault-free condition and right exclusion condition, respectively. So fault magnitude does not exist in these two terms.

For the fifth term of Eq. (15), it indicates a condition that even exclusion function has excluded some satellites but the fault still exists in navigation system and the second layer detection fail to detect the fault. We denote $P_{HI,ex,d}$ and $P_{ND,ex,d}$ as

$P(|\epsilon_{ex}| > l|H_k, \mathbf{f}_k)$ and $P(|\Delta_{ex,k}| < T_{ex,k}|H_k, \mathbf{f}_k)$, respectively. In Chapter 2, we explained the bias and variance of random variable $\epsilon_{ex}$ and $\Delta_{ex,k}$. Similar to the derivation of formula Eq. (20), these numerical characteristics can rewrite the fifth term of Eq. (15) mathematically.

For $P_{HI,ex,d}$, it can be rewritten as:

$$P_{HI,ex,d} = 1 - \frac{1}{\sqrt{2\pi \mathbf{P}_{ex}(d,d)}} \cdot \int_{-l_d}^{l_d} \exp\left(\frac{(x - \mu_{HI,ex,d})^2}{2\mathbf{P}_{ex}(d,d)}\right) dx \tag{21}$$

where $\mathbf{P}_{ex}$ is covariance matrix of $\epsilon_{ex}$ and $\mu_{HI,ex,d}$ can be written as:

$$\mu_{HI,ex,d} = \sum_{e=1}^{n} |\mathbf{S}_{ex}(d,e)|\mathbf{b}(e,d) + \sum_{e=1}^{n} |\mathbf{S}_{ex}(d,e)|\mathbf{f}_k(e,1) \tag{22}$$

For $P_{ND,ex,d}$, it can be written as:

$$P_{ND,ex,d} = \frac{1}{\sqrt{2\pi \mathbf{P}_{ex,k}(d,d)}} \cdot \int_{-T_{\Delta_{ex,k,d}}}^{T_{\Delta_{ex,k,d}}} \exp\left(\frac{(y - \mu_{ND,ex,d})^2}{2\mathbf{P}_{ex,k}(d,d)}\right) dy \tag{23}$$

where $\mathbf{P}_{ex,k}$ is covariance matrix of $\Delta_{ex,k}$ and $\mu_{ND,ex,d}$ can be written as:

$$\mu_{ND,ex,d} = -\sum_{e=1}^{n} |\mathbf{S}_{ex}(d,e) - \mathbf{S}_{ex,k}(d,e)|\mathbf{b}(e,d) + \sum_{e=1}^{n} |\mathbf{S}_{ex}(d,e)|\mathbf{f}_k(e,1) \tag{24}$$

we can rewrite $P_{HI,ex,d} \cdot P_{ND,ex,d}$ as:

$$\begin{aligned} P_{HMI,5} &= P_{HI,ex,d} \cdot P_{ND,ex,d} \\ &= k_{ex,k} \int_{-T_{\Delta_{ex,k,d}}}^{T_{\Delta_{ex,k,d}}} \exp\left(\frac{(y - M + b_{ex,k})^2}{2\mathbf{P}_{ex,k}(d,d)}\right) dy \\ &\quad - k_{ex,k}k_{ex} \int_{-l_d}^{l_d} \int_{-T_{\Delta_{ex,k,d}}}^{T_{\Delta_{ex,k,d}}} \exp\left(-\frac{(y - M + b_{ex,k})^2}{2\mathbf{P}_{ex,k}(d,d)}\right) \\ &\quad \exp\left(-\frac{(x - M + b_{ex})^2}{2\mathbf{P}_{ex}(d,d)}\right) dxdy \end{aligned} \tag{25}$$

where $k_{ex,k} = \frac{1}{\sqrt{2\pi \mathbf{P}_{ex,k}(d,d)}}$, $k_{ex} = \frac{1}{\sqrt{2\pi \mathbf{P}_{ex}(d,d)}}$, $y - M + b_{ex,k} = y - \mu_{ND,ex,d}$, and $x - M + b_{ex} = x - \mu_{HI,ex,d}$. The derivation of $P_{HMI,5}$ is similar to $P_{HMI,2}$, so the function image of $P_{HMI,5}$ is similar to Fig. 1. As for $b_{ex,k}$ and $b_{ex}$, the definition is same as $b_i$ and $b_0$.

In order to evaluate navigation performance, we need to derive the equation of calculating VPL. When the integrity risk requirement $I_{req}$ is specified, VPL can be derived from Eq. (15). Corresponding to Eq. (15), $I_{req}$ is divided into five terms.

$$I_{req} \leq I_1 + I_2 + I_3 + I_4 + I_5 \tag{26}$$

Based on the first term in Eq. (15), the first term of Eq. (26) can be expressed as:

$$I_1 = \left[ \bar{Q}\left(\frac{-VPL - b_{0,3}}{\sigma_{0,3}}\right) + Q\left(\frac{VPL - b_{0,3}}{\sigma_{0,3}}\right)\right] P_{H_0} \tag{27}$$

where function $Q$ and $\bar{Q}$ are defined as:

$$Q(x) = \frac{1}{2\pi}\int_x^{\infty} \exp(-\frac{t^2}{2})dt, \quad \bar{Q}(x) = 1 - Q(x) \tag{28}$$

In Eq. (15), the second term represents the system contains a specific fault $H_i$ and the magnitude $\mathbf{f}_i$ of this fault model. $I_2$ in Eq. (26) is under this condition, and it can be expressed as:

$$\begin{aligned} I_2 &= \sum_{i=1}^{m} \max_{\mathbf{f}_i}\left(P(|\epsilon_0| > l|H_i, \mathbf{f}_i) \cdot P(|\Delta_i| < T_i|H_i, \mathbf{f}_i)\right) P_{H_i} \\ &= \sum_{i=1}^{m}\left[\bar{Q}\left(\frac{-VPL - \mu_{HI,i,3}}{\mathbf{P}_0(3,3)}\right) + Q\left(\frac{VPL - \mu_{HI,i,3}}{\mathbf{P}_0(3,3)}\right)\right] \\ &\quad \cdot \left[Q\left(\frac{-T_{\Delta_{i,3}} - \mu_{ND,i,3}}{\mathbf{P}_i(3,3)}\right) - Q\left(\frac{T_{\Delta_{i,3}} - \mu_{ND,i,3}}{\mathbf{P}_i(3,3)}\right)\right] P_{H_i} \end{aligned} \tag{29}$$

where $\mu_{HI,i,3}$, $\mathbf{P}_0(3,3)$ and $\mu_{ND,i,3}$, $\mathbf{P}_i(3,3)$ in Eq. (29) are bias and covariance corresponding to the maximum value of $P(|\epsilon_0| > l|H_i, \mathbf{f}_i) \cdot P(|\Delta_i| < T_i|H_i, \mathbf{f}_i)$.

As for the third and the fourth terms of Eq. (26), these two terms represent the fault-free model in exclusion function. The third term $I_3$ can be written as:

$$I_3 = \sum_{ex=0}^{m}\left[\bar{Q}\left(\frac{-VPL - b_{ex,3}}{\sigma_{ex,3}}\right) + Q\left(\frac{VPL - b_{ex,3}}{\sigma_{ex,3}}\right)\right] P_{H_0} \tag{30}$$

and the fourth term can be written as:

$$I_4 = \sum_{ex=0}^{m} \left[ \bar{Q}\left(\frac{-VPL - b_{ex,3}}{\sigma_{ex,3}}\right) + Q\left(\frac{VPL - b_{ex,3}}{\sigma_{ex,3}}\right)\right] P_{H_{ex}} \tag{31}$$

The last term $I_5$ of Eq. (26) still exist the fault, and its magnitude can affect the value of PHMI. This term can be calculated as:

$$\begin{aligned} I_5 &= \sum_{ex=0}^{m} \sum_{\substack{k=1 \\ k \neq ex}}^{m_1} \max \mathbf{f}_k \Big( P(|\epsilon_{ex}| > l | H_k, \mathbf{f}_k) \cdot P(|\Delta_{ex,k}| < T_{ex,k} | H_k, \mathbf{f}_k) \Big) P_{H_k} \\ &= \sum_{ex=0}^{m} \sum_{\substack{k=1 \\ k \neq ex}}^{m_1} \left[ \bar{Q}\left(\frac{-VPL - \mu_{HI,ex,3}}{\mathbf{P}_{ex}(3,3)}\right) + Q\left(\frac{VPL - \mu_{HI,ex,3}}{\mathbf{P}_{ex}(3,3)}\right)\right] \\ &\cdot \left[ Q\left(\frac{-T_{\Delta_{ex,k,3}} - \mu_{ND,ex,3}}{\mathbf{P}_{ex,k}(3,3)}\right) - Q\left(\frac{T_{\Delta_{ex,k,3}} - \mu_{ND,ex,3}}{\mathbf{P}_{ex,k}(3,3)}\right)\right] P_{H_k} \end{aligned} \tag{32}$$

where $\mu_{HI,ex,3}$, $\mathbf{P}_{ex}(3,3)$, and $\mu_{ND,ex,3}$, $\mathbf{P}_{ex,k}(3,3)$ in Eq. (32) are bias and covariance corresponding to the maximum value of $P(|\epsilon_{ex}| > l | H_k, \mathbf{f}_k) \cdot P(|\Delta_{ex,k}| < T_{ex,k} | H_k, \mathbf{f}_k) P_{H_k}$.

From Eqs. (27)–(32), the only variable is *VPL*. So we can derive the value of VPL by using the numerical method in APPENDIX E of [14]. In order to compute the maximum value in Eqs. (29) and (32), we introduce a numerical method in the following section and it could seek the maximum value.

## 3.2 Numerical Method and Test Example

To calculate the maximum value in Fig. 3, we introduce a numerical method: optimization method (Fig. 4).

Before we introduce the method in detail, we denote $f(x(k))$ as the maximum value of the function. At the beginning of optimization method, we first need to determine the interval where the extremum is located. Based on the properties of probability density function (PDF) of normal distribution, the initial searching interval is set to [0, T + 37], where T is the threshold for test statistic. If we assume the extremum

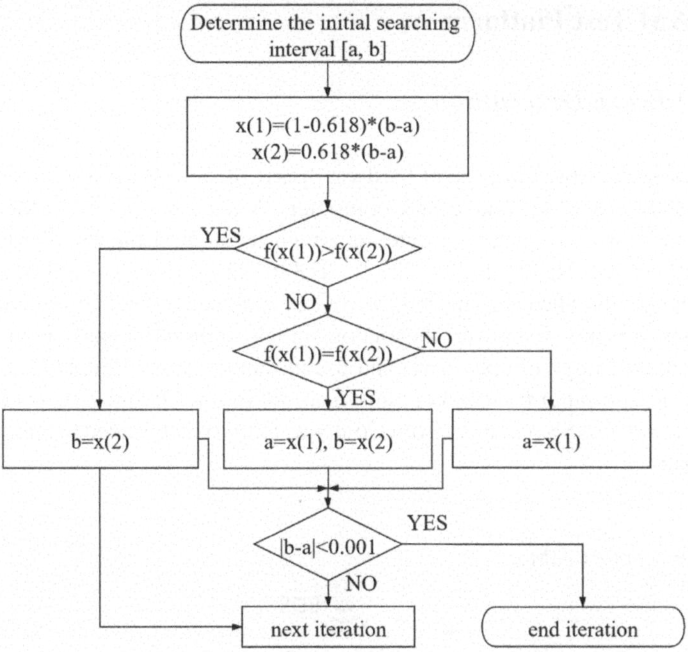


**Fig. 4** Flowchart of optimization method

**Table 1** Test example result

| Termination condition | Error $(x(i) - x(k))$ | Time/s |
|---|---|---|
| 1 | −0.073 | 0.00042 |
| 0.6 | −0.073 | 0.00060 |
| 0.2 | 0.0623 | 0.00099 |
| 0.1 | 0.0106 | 0.00210 |
| 0.05 | −0.0410 | 0.00221 |
| 0.01 | −0.0016 | 0.00284 |

locates within [a, b], then the method chooses two point $x\,(1)$ and $x\,(2)$ in $[a, b]$ where $x\,(1) = (1 - 0.618) \times (b - a)$ and $x\,(2) = 0.618 \times (b - a)$. If $f\,(x(1)) > f\,(x(2))$, the searching interval is changed to $[x\,(1)\,, b]$. If $f\,(x(1)) < f\,(x(2))$, the searching interval is changed to $[a, x\,(2)]$. Especially, when $f\,(x(1)) = f\,(x(2))$, the searching interval is changed to $[x\,(1)\,, x\,(2)]$. When $x\,(1)$ and $x\,(2)$ are close enough ($|x(1) - x(2)| < 0.01$), the search will be terminated.

Take $Z = -(x - 5)^2 + 70$ as an example, the maximum value of $Z$ is 70 and the corresponding point is 5. In order to find this maximum value, the initial interval is set to [4, 7]. Table 1 shows the result of optimization method with different termination conditions.

# 4 ARIAM Test Platform

## *4.1 Hardware Description*

Figure 5 shows the three important hardware parts of ARAIM test platform: GNSS antenna, GNSS receiver, and user's computer. When we operating ARAIM test platform, we first use chock ring antenna which located on the top of SJTU GNC Lab (31.026105436 N, 121.442347327 E, 36.836202 m) to receive GPS, BDS, and GLONASS signals. Then analog signals will be processed by multi-frequency GNSS receiver, which is able to convert the analog signals to digital signals and to provide users with the message of ephemeris and pseudorange. Next, digital signals will be transmitted to computer by serial communication. Figures 6 and 7 shows the chock ring antenna and GNSS receiver (from ComNav, China) that is used to build ARAIM test platform.

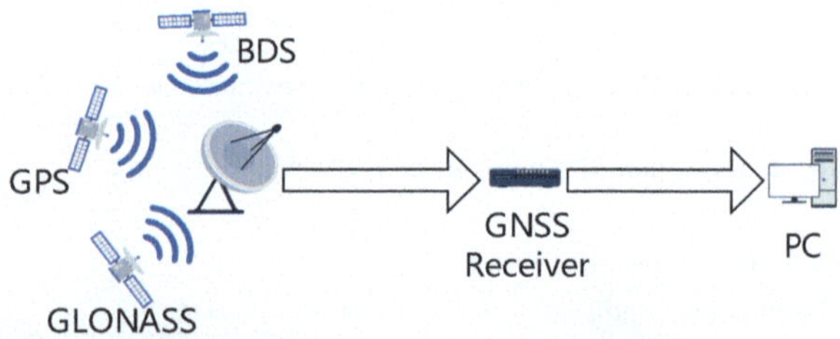


**Fig. 5** Hardware of ARAIM test platform

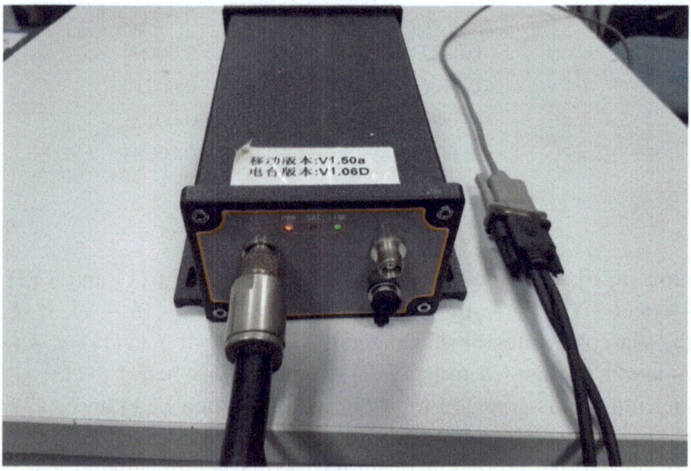

**Fig. 6** ComNav GNSS receiver

**Fig. 7** Chock ring antenna

## 4.2 Software Description

The software of ARAIM test platform consists of two parts: online version and offline version. For online version, when binary input stream is transmitted by serial communication from GNSS receiver to user's computer, the computer first decodes the binary input stream based on IEEE 754 protocol. After we acquiring the message of ephemeris and pseudorange from decoded binary input stream, local computer is able to calculate user's location and implement ARIAM algorithm. Figure 8 shows the structure of online version ARAIM test platform. For offline version, it is designed for some harsh circumstance that local computer is malfunctioning and serial communication is interrupted. Due to GNSS receiver has 100 megabytes of memory, the lost data processed by GNSS receiver can be stored in the receiver as RENIX format. Then, by using RENIX file, we can calculate the user's location and start running the ARAIM algorithm. Figure 9 shows the structure of offline version ARAIM test platform.

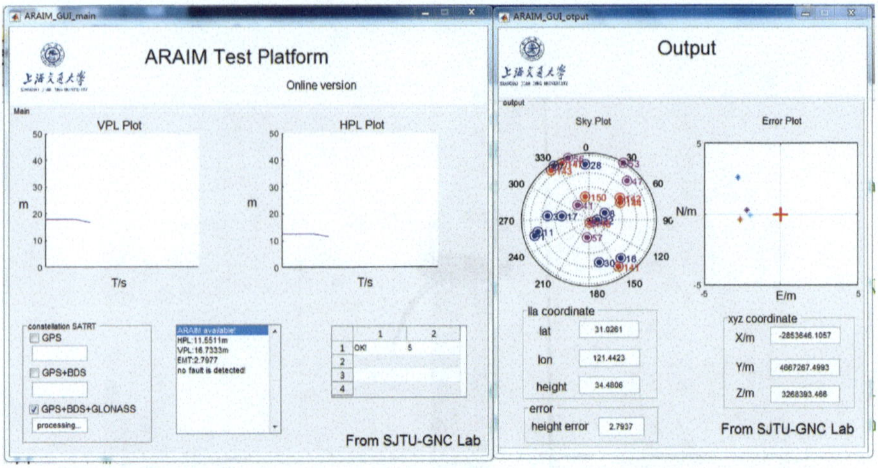


**Fig. 8** Online ARAIM test platform

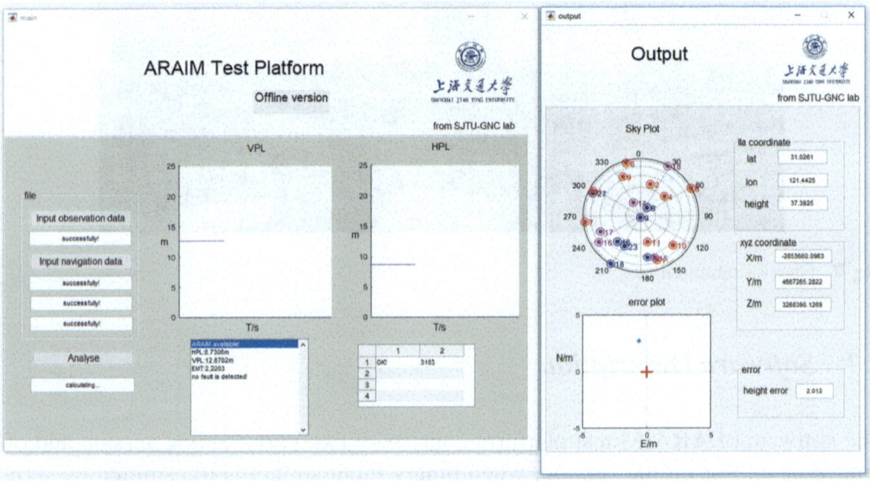


**Fig. 9** Offline ARAIM test platform

# 5 Result

## 5.1 Global Almanac Data Result

We use 6-h almanac data from International GNSS Service (IGS) to calculate the worldwide VPL and 99.5% coverage. For ARAIM Fault Detection (FD) algorithm and ARAIM Fault Detection and Exclusion (FDE) algorithm, based on the require-

**Table 2** Computational burden of different method

| Algorithm | Method | Time (all) (min) | Time (epoch) (s) |
|---|---|---|---|
| FD | MHSS | 3.28 | 0.008 |
| | Tighter bound | 12.87 | 0.033 |
| FDE | MHSS | 40.72 | 0.104 |
| | Tighter bound | 221 | 0.568 |

**Table 3** VPL for ARAIM FD algorithm

| URA | Bias | VPL (MHSS) (m) | VPL (tighter bound) (m) |
|---|---|---|---|
| URA = 0.5 | b = 0.1 | 15.75 | 5.70 |
| | b = 0.3 | 16.87 | 6.60 |
| URA = 1 | b = 0.1 | 20.59 | 8.74 |
| | b = 0.3 | 21.70 | 9.67 |
| URA = 1.6 | b = 0.1 | 27.87 | 13.84 |
| | b = 0.3 | 28.99 | 15.25 |

**Table 4** Coverage for ARAIM FD algorithm

| URA | Bias | Coverage (MHSS) (%) | Coverage (tighter bound) (%) |
|---|---|---|---|
| URA = 0.5 | b = 0.1 | 100 | 100 |
| | b = 0.3 | 100 | 100 |
| URA = 1 | b = 0.1 | 99.89 | 100 |
| | b = 0.3 | 99.89 | 100 |
| URA = 1.6 | b = 0.1 | 94.53 | 99.14 |
| | b = 0.3 | 94.02 | 98.82 |

ment of continuity [5], we analyze the VPL and 99.5% coverage. The computational burden of ARAIM algorithm and our new method is also listed below.

The results which are listed in Tables 2, 3, 4, 5, and 6 show tighter bound ARIAM algorithm can lower the protection level and improve the coverage worldwide without increasing too much computational burden for real-time capability.

## 5.2 *ARAIM Test Platform Result*

Using ARAIM test platform, we collect 6-h, three constellations' (GPS, BDS, GLONASS) real-time data to verify the effect of our new method on navigation performance. The total time of processing 6-h real-time data using MHSS algo-

**Table 5** VPL for ARAIM FDE algorithm

| URA | Bias | VPL (MHSS) (m) | VPL (tighter bound) (m) |
|---|---|---|---|
| URA = 0.5 | b = 0.1 | 33.01 | 19.73 |
| | b = 0.3 | 35.56 | 23.79 |
| URA = 1 | b = 0.1 | 43.92 | 29.27 |
| | b = 0.3 | 46.45 | 32.93 |
| URA = 1.6 | b = 0.1 | 60.38 | 39.13 |
| | b = 0.3 | 62.92 | 40.76 |

**Table 6** Coverage for ARAIM FDE algorithm

| URA | Bias | Coverage (MHSS) (%) | Coverage (tighter bound) (%) |
|---|---|---|---|
| URA = 0.5 | b = 0.1 | 81.59 | 85.39 |
| | b = 0.3 | 76.40 | 79.12 |
| URA = 1 | b = 0.1 | 51.93 | 71.61 |
| | b = 0.3 | 46.13 | 60.76 |
| URA = 1.6 | b = 0.1 | 8.35 | 34.98 |
| | b = 0.3 | 5.70 | 23.21 |

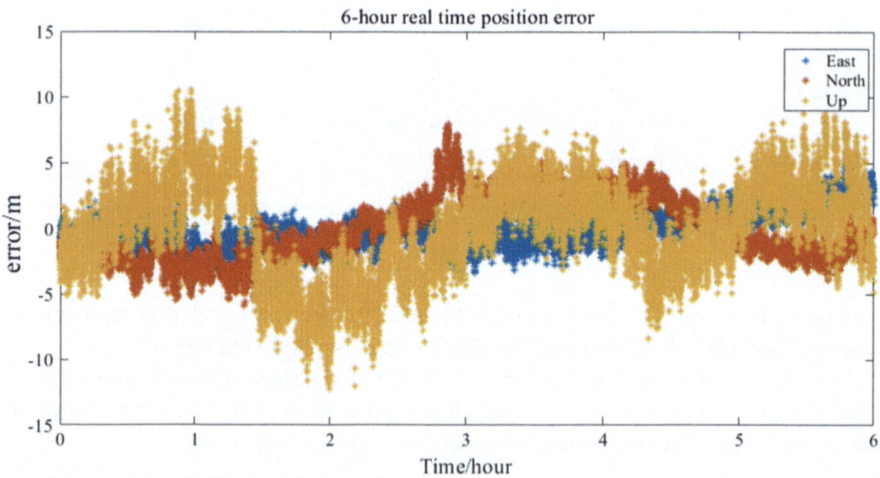


**Fig. 10** 6-h positioning error

rithm and tighter bound algorithm is 392 and 443 min. For each epoch, the time of MHSS algorithm and tighter bound algorithm 1.09 and 1.23 s, respectively. Figure 10 shows the 6-h positioning error. The average errors in east, north, and up direction are 1.12 m, 1.99 m, and 3.00 m, respectively.

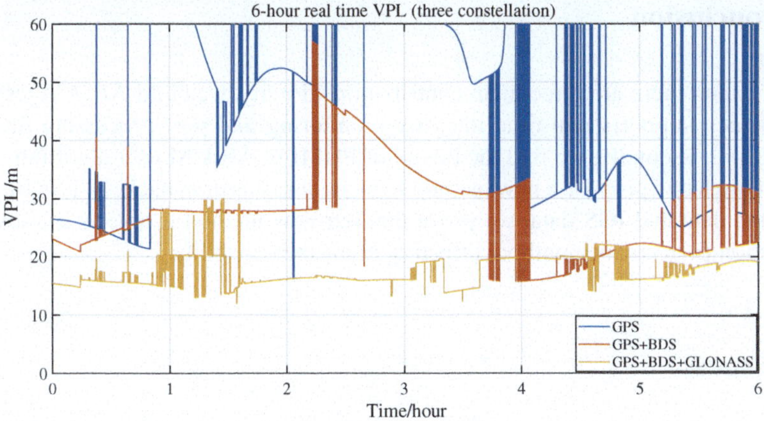


**Fig. 11** VPL using different constellations

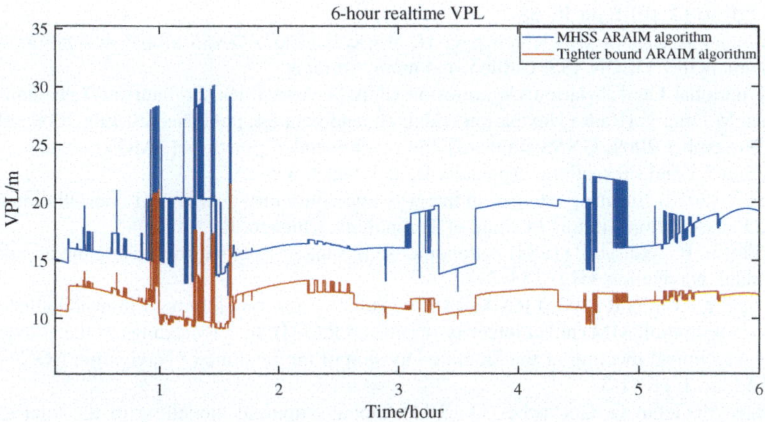


**Fig. 12** VPL of different methods (URA = 1, b = 0.75)

Figure 11 shows the 6-h VPL by using ARAIM FD algorithm. The blue line, red line, and yellow line in Fig. 11 represent the VPL of single constellation, dual constellation, and three constellation. Navigation performance increases with the number of constellations. Three constellation can provide the best navigation performance for user.

Figure 12 shows the 6-h VPL by using MHSS ARAIM FD algorithm and tighter bound ARAIM algorithm. When the test platform applies three constellations, VPL can be declined by our tighter bound ARAIM algorithm. The blue line and red line in Fig. 12 represent the VPL of MHSS algorithm and tighter bound ARAIM algorithm.

## 6 Conclusion

In this paper, there are three main contributions. Firstly, we build ARIAM test platform which supports real-time integrity monitoring and post-processing integrity monitoring. Secondly, we find the bound of PHMI in ARAIM user algorithm is not tight enough and propose a new method to overcome this drawback. At last, by using real-time data and IGS data, we prove that our new method is able to improve the navigation performance without affecting real-time capability.

## References

1. Mohamed AH, Schwarz KP (1999) Adaptive Kalman filtering for INS/GPS. J Geodesy 73(4):193–203
2. Alexander I (1993) Perspective/navigation-the global positioning system. In: IEEE spectrum, vol 30, no 12. IEEE pp 36–38
3. Hofmann-Wellenhof B, Lichtenegger H, Wasle E (2007) GNSS-global navigation satellite systems: GPS, GLONASS, Galileo, and more. Springer
4. International Civil Aviation Organization (ICAO), Annex 10:Aeronautical Telecommunications. Volume 1: (Radio Navigation Aids), Amendment 84, published 20 July 2009, effective 19 November 2009, GNSS standards and recommended practices (SARPs) are contained in Section 3.7 and subsections, Appendix B, and Attachment D
5. Zhai Y (2018) Ensuring navigation integrity and continuity using multi-constellation GNSS. Ph.D. dissertation, Illinois Institute of Technology, Chicago, IL
6. Parkinson B, Axelrad P (1988) Autonomous GPS integrity monitoring using the pseudorange residual. Navigation 35(2):255–274
7. Patrick Y, Collins R (2006) RAIM-FDE revisited: a new breakthrough in availability performance with nioRAIM (novel integrity-optimized RAIM). In: Proceedings of the 18th international technical meeting of the Satellite Division of the Institute of Navigation (ION GNSS), vol 53, no 1, pp 41-51
8. Ivanov V, Gefan G, Gorbachev O (2011) Global empirical modelling of the total electron content of the ionosphere for satellite radio navigation systems. J Atmos Solar-Terrestrial Phys 73(13):1703–1707
9. El-Mowafy A, Yang C (2016) Limited sensitivity analysis of ARAIM availability for LPV-200 over Australia using real data. Adv Space Res 57(2):659–670
10. Li X, Zhang X, Ren X, Fritsche M, Wickert J, Schuh H (2015) Precise positioning with current multi-constellation global navigation satellite systems: GPS, GLONASS, Galileo and BeiDou. Sci Rep 5:8328
11. Aleshkin AP, Myslivtsev TO, Nikiforov SV, Savochkin PV, Sakhno IV, Semenov AA, Troitskii BV (2019) Calculation of navigation corrections for a single-frequency GNSS receiver based on satellite radio occultation data. Gyroscopy Navig 10(1):15–20
12. Pan W, Zhan X, Zhang X, Liu B (2019) GNSS/INS integrity monitoring considering nominal bias for civil aircraft CAT-I approach. In: AIAA Scitech 2019 forum, vol 0361
13. Blanch J, Walker T, Enge P, Lee Y, Pervan B, Rippl M, Spletter A, Kropp V (2015) Baseline advanced RAIM user algorithm and possible improvements. IEEE Trans Aerosp Electron Syst 51(1):713–732
14. Cassel R (2017) Real-time ARAIM using GPS, GLONASS, and Galileo. Master thesis, Illinois Institute of Technology
15. EU-U.S. Cooperation on Satellite Navigation Working Group C-ARAIM Technical Subgroup (2016) Milestone 3 report. Final version

16. Zhai Y, Joerger M, Pervan B (2015) Continuity and availability in dual-frequency multi-constellation ARAIM. In: Proceedings of the 28th international technical meeting of the Satellite Division of the Institute of Navigation (ION GNSS)
17. Joerger M, Chan F, Pervan B (2014) Solution separation versus residual-based RAIM. NAVIGATION: J Inst Navig 61(4):273–291

# Space Vehicle Orbital Determination Performance Analysis Considering GNSS Side Lobe Signals

**Xiaoya Liu, Xingqun Zhan, Jihong Huang, Chao Qin and Cheng Chi**

**Abstract** Global navigation satellite systems (GNSS), which were originally designed for terrestrial location service, are now developed for utility in space, which are called GNSS Space Service Volume (SSV). One major problem that SSV confronts is the poor satellite availability in high orbital vehicles when main beam signals are solely used. Recently, side lobe signals, the signals emitted sideways from transmitter antennas, have been proposed to improve the SSV performance. It is necessary to study the GNSS SSV with the presence of side lobe signals. In this paper, the system improvement owing to the side lobe signals in high orbital vehicles for specific missions is quantitatively evaluated. Different multi-constellation conditions and system setups are taken into account as factors. GPS and BDS III constellations are simulated, and the satellite availability and the maximum outage duration (MOD) are assessed in three scenarios, the GEO, HEO, and lunar trajectory, in different heights. The experimental results indicate that the side lobe signals can effectively improve satellite availability for space users, especially in the upper SSV. Moreover, it shows that the advantages of multi-constellation interoperability lie in improving signal availability and shortening the MOD.

**Keywords** Space service volume · Side lobe signal · Satellite availability · Maximum outage duration · Orbital determination

## 1 Introduction

GNSS initially predicted Position, Navigation, and Timing (PNT) services for terrestrial users, but it is now increasingly used for autonomous navigation in space. Historically, most space users are at low altitudes, where to receive signals just like

X. Liu · X. Zhan (✉) · J. Huang · C. Qin · C. Chi
School of Aeronautics and Astronautics, Shanghai Jiao Tong University, Shanghai, China
e-mail: xqzhan@sjtu.edu.cn

X. Liu
e-mail: liuxy95@sjtu.edu.cn



Z. Jing (ed.), *Proceedings of the International Conference on Aerospace System Science and Engineering 2019*, Lecture Notes in Electrical Engineering 622,
https://doi.org/10.1007/978-981-15-1773-0_10

in the ground [1]. However, it is often difficult to provide reliable PNT services for spacecraft above 3,000 km above sea level. Commonly used navigation systems are the GPS in the USA, the GLONASS in Russia, China's BDS and Europe's Galileo system [2]. The availability and performance of GNSS signals at high altitude are documented as the SSV [1]. And the SSV is developed for the benefit of all high-altitude users, especially in GEO and HEO [3].The environment in which the space receiver operates in the SSV is significantly different from the environment of a conventional terrestrial receiver or a low Earth orbit GNSS receiver. SSV users span a very dynamic and ever-changing environment as they cross the GNSS constellation.

GNSS navigation technology for spacecraft provides real-time, high-precision orbit data for low-orbit satellites and manned spacecraft [4], greatly reducing the burden of terrestrial tracking, telemetry, and command (TT&C) networks.However, since the antenna of the GNSS navigation satellite points to the Earth and the angle of the main lobe signal is limited [1], when the spacecraft orbital altitude exceeds GNSS constellation, only the navigation satellite signal from the other side of the Earth can be received [5]. Therefore, in the high orbit environment, the availability and signal quality of GNSS navigation satellite will be limited by the Earth occlusion and the increase of signal-free space loss [6]. At the same time, because the orbital altitude of the high-orbit spacecraft is very high and the received signal comes from the other side of the Earth, the geometric configuration of the visible stars will be seriously affected, resulting in a sharp increase in the geometric accuracy factor [7]. Therefore, the application of GNSS is still mainly in low- and medium-orbit spacecraft [3]. In recent years, the research on GNSS positioning technology for high orbit aircraft has been gradually propelled, but the extant literature focuses on a research level. The main lobe signal reception of GNSS satellite limits the application scope [8]. Currently, the working group B of the United Nations International Committee on GNSS only takes the main beam signals into consideration, but the nature of the high-altitude positioning environments severely limits the GNSS accuracy and availability. Meanwhile, side lobe signals, the signals emitted sideways from transmitter antennas, have recently been proposed to assist GNSS in space. Therefore, it is necessary to study the GNSS SSV service performance with the presence of side lobe signals. Side lobe signals, the signals emitted sideways from transmitter antennas, have recently been proposed to assist GNSS in space [9, 10]. High-altitude applications of GNSS are more challenging due to reduced signal power levels and availability, potentially reduced pseudorange accuracy, less optimal geometric diversity, and in the case of elliptical orbits, highly dynamic motion [11]. In these environments, an increased number of available GNSS signals of sufficient power and accuracy would substantially improve the potential signal availability, and thus mission navigation performance [10]. In this paper, we study the effect of the side lobe signals to GNSS SSV performance. By taking side lobe signals into account, the performance improvement of the GNSS is quantitatively analyzed in multi-constellation conditions and system setups. We simulate the GPS, BDS III constellation, and evaluate the system performance in three scenarios, the GEO, HEO, and lunar trajectory, with different heights. The experimental results show that the side lobe signal, as a supplement to the main lobe signal, has potential in the future high-altitude positioning.

And we prove that in the multi-constellation GNSS, increasing the receiver sensitivity can greatly shorten MOD, improving the SSV service. Our results demonstrate that the side lobe signal is an important signal source to improve GNSS SSV service performance.

## 2 SSV Performance Assessment

### *2.1 Definition*

The parameters characterizing SSV performance are closely related to the GNSS constellation and navigation satellite design, including the transmit power and gain modes of the antenna. Figures 1 and 2 show the feature models for each parameter of the shaded area [11].

As the height of the SSV increases, the performance of the GNSS decreases. According to the definition given in the SSV booklet [1], the SSV is divided into two distinct areas :

(1) Lower SSV for medium Earth orbits: 3000–8000 km altitude.
(2) Upper SSV for geostationary and high Earth orbits: 8000–36,000 km altitude.

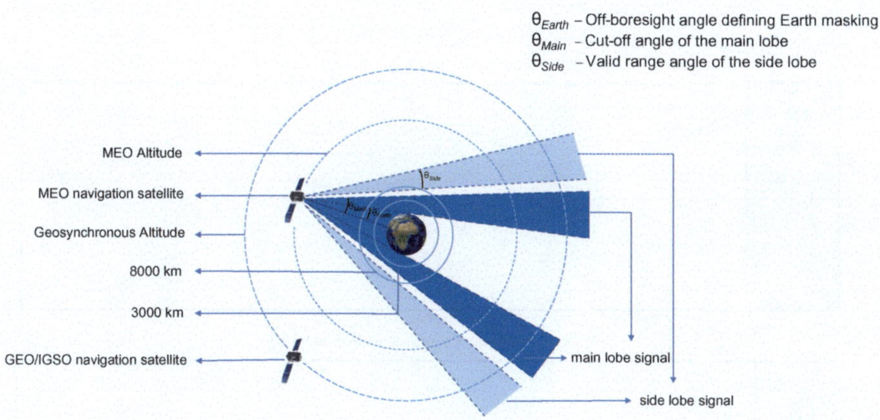


**Fig. 1** SSV availability for BDS MEO satellite

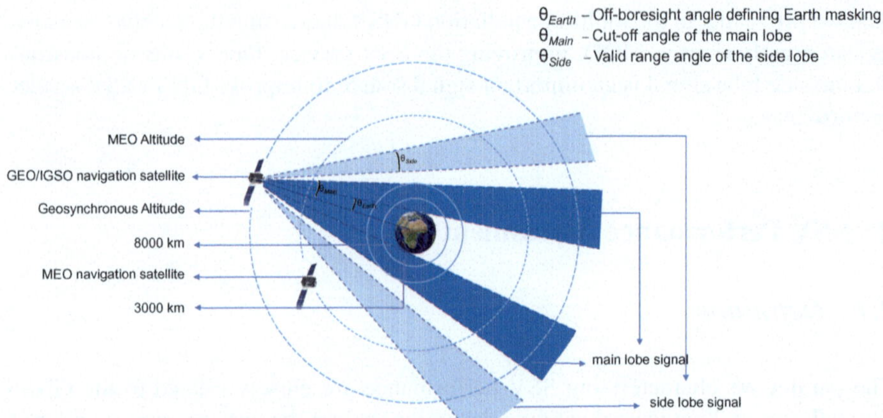


**Fig. 2** SSV availability for BDS GEO/IGSO satellite

## 2.2 *Characterization Metrics*

Evaluation indicators selected here are minimum received power,satellite availability and maximum outage duration. More details can be researched in the SSV booklet [1] (Figs. 3 and 4).

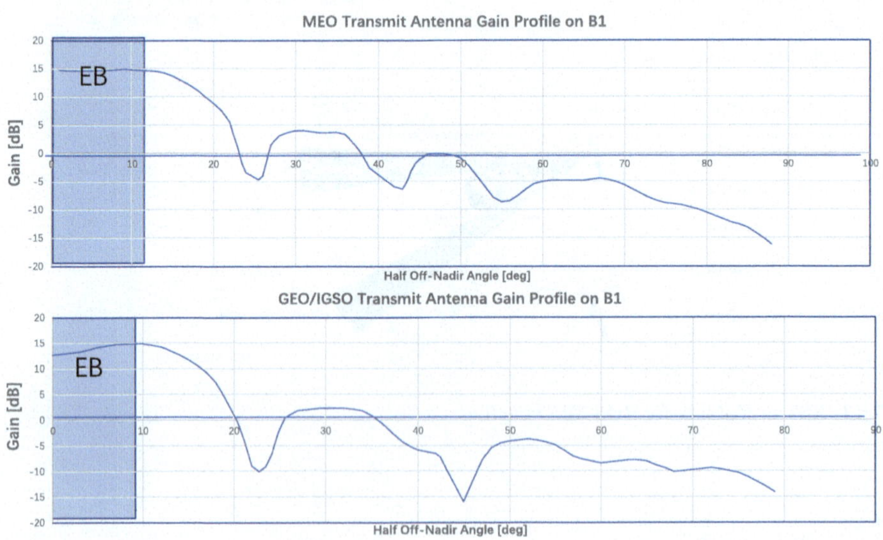


**Fig. 3** Assumed BDS III transmitting gain for B1 versus off-nadir angle (EB represents blocked by Earth)

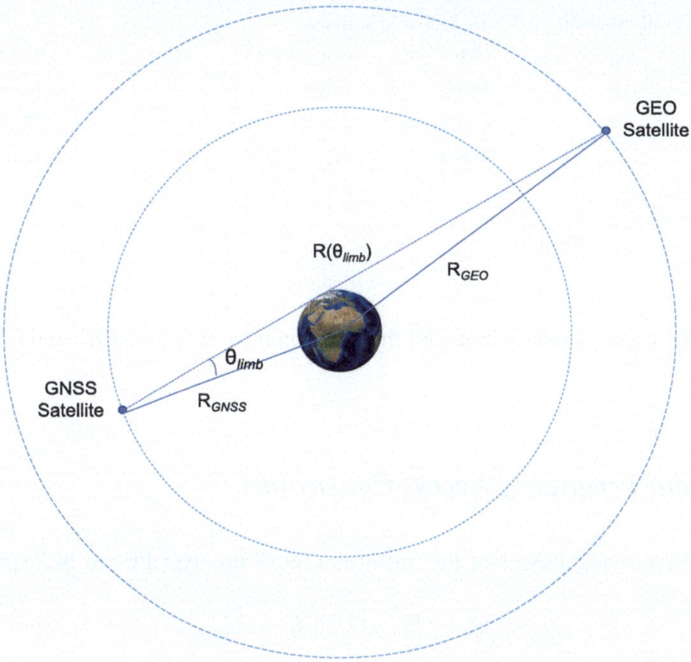


**Fig. 4** Geometry used in MRTP calculation

# 3 Availability Evaluation Index for Global SSV Performance

## *3.1 Geometric Constraints*

The lower the transmit antenna gain threshold, the larger the range of available effective main beam angles.On the contrary, the smaller [12].

The Earth occlusion angle of MEO is 13.2°, and that of GEO and IGSO is 8.7°. If the gain threshold 0 dB transmit antenna is selected, the available angle range of the sub-beam of GPS L1 is 27.5°–33.0° and the sub-beam of L2/L5 is too weak to be used. The sub-beam availability angle of B1 frequency signal of BDS GEO/IGSO satellite is 25°–38° and the sub-beam of B2/B3 frequency signal is too weak to be used, the sub-beam availability angle of B1 frequency signal of BDS MEO satellite is 29°–37°, and the sub-beam of other two frequency signals of MEO satellite is not available [13, 14], as listed in Table 1.

Assume that each SSV user installs two 0 dBic receive antennas, one facing the Earth and the other facing the zenith.This will enable the user to capture all available signals from the satellite up and down.

**Table 1** Overall off-nadir angles of available signals

| Constell. | Sig. | Orb. | EBA[a] (°) | MLA[b] (°) | RSLVA[c] (°) |
|---|---|---|---|---|---|
| BD | B1 | MEO | 13.2 | 22 | 29–37 |
| | | GEO/IGSO | 8.7 | 21 | 25–38 |
| GPS | L1 | MEO | 13.8 | 23.5 | 27.5–33 |

[a]Earth blocked angle
[b]Main lobe cut-off angle
[c]Range of side lobe valid angles

So in this paper, we only consider the performance of BDS on B1 and GPS on L1 signal.

## 3.2 *Radio Frequency Access Constraints*

The signal receiving power of the airborne GNSS receiver Pr can be expressed as [15]:

$$P_r = P_t + G_t + L_s + G_r \quad (1)$$

where $G_t$ is the gain of transmitting antenna, $G_r$ is the gain of the antenna, $P_t$ is the transmission power of signal, and $L_s$ is the loss of space propagation path.

Each constellation MRTP can be obtained by performing reverse link budget calculation using the minimum received power of a given constellation [1], as shown in Fig. 5. MRTP is constant when all well angles are less than the reference well angle.

The inverse link budget is defined as

$$MRTP = P_{\min} + L_s \quad (2)$$

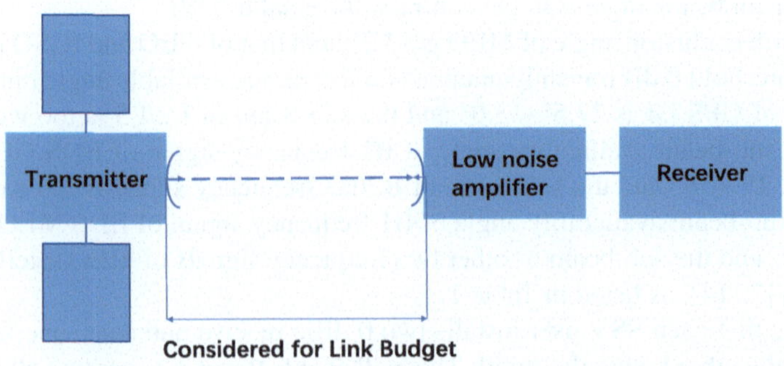


**Fig. 5** Link Budget calculation scenario, where $T_x$ is transmitter onboard the GNSS satellite, LNA is the low-noise amplifier and $R_x$ is the user receiver

**Table 2** GNSS Radio Frequency (RF) parameters

| Constell. | Sig. | Orb. | Freq. | MCP[a] (dBW) | MRTP |
|---|---|---|---|---|---|
| BD | B1 | MEO | 1575.42 | −184.2 | 9 |
| | | GEO/IGSO | 1575.42 | −185.9 | 9 |
| GPS | L1 | MEO | 1575.42 | −184 | 9.1 |

[a]Minimum received power

where $P_{\min}$ is the specified minimum received power at GEO and $L_s$ is the free space path loss at the worst-case Earth-limb distance The resulting MRTPs calculated with this method are shown in Table 2 [1]. More details can be found in the SSV booklet.

# 4 Simulations

In order to fully understand the influence of side lobe signals on SSV service performance, we synthetically evaluated single and multi-GNSS constellations. By considering side lobe signals, the performance improvement of GNSS under multi-constellation conditions and system settings is quantitatively analyzed. The evaluation of multiple constellations is based on BDS III and GPS [16]. Their informations are attached in Table 3.

According to the SSV booklet, two types of performance estimates are provided: globally averaged, and mission-specific. Global performance is estimated by simulating signal availability at a fixed grid of points in space, at both the lower SSV altitude of 8000 km and the upper SSV at 36,000 km. Mission-specific performance estimates are obtained by estimating signal availability for a spacecraft on a particular trajectory within the SSV.

**Table 3** Basic orbit parameters of BDS III and GPS constellation

| Constell. | Orb. | NS[a] | NP[b] | Hgt[c] (km) | Ic.[d] (°) | Ec.[e] |
|---|---|---|---|---|---|---|
| BDS | MEO | 27 | 3 | 21528 | 55 | 0 |
| | GEO | 5 | 1 | 35786 | 0 | 0 |
| | IGSO | 3 | 3 | 35786 | 55 | 0 |
| GPS | | 24 | 6 | 26561.75 | 55 | 0 |

[a]Number of satellites
[b]Number of planes
[c]Height
[d]Inclination
[e]Eccentricity

## 4.1 Global Space Service Volume Performance

This section will cover the globally averaged SSV simulations. These simulations analyze the SSV using both geometrical access constraints alone as well as combined geometrical and radio frequency access constraints.

### 4.1.1 Geometrical Analysis Methodology

Figures 6 and 7 show available performance at 8000 km and 36,000 km altitudes considering a zero-gain user antenna when considering only geometrical access constraints.

| Signal | Constellation | Signal availability (%) | | Max Outage Duration(min) | |
|---|---|---|---|---|---|
| | | At least 1 signal | 4 or more signals | At least 1 signal | 4 or more signals |
| Main lobe | BD | 76.59 | 12.26 | 994 | 6700 |
| | GPS | 57.86 | 0 | 712 | 1432 |
| | Combined | 88.76 | 24.81 | 705 | 783 |
| Main lobe and Side lobe | BD | 97.70 | 60.97 | 51 | 545 |
| | GPS | 83.77 | 0.84 | 106 | 1004 |
| | Combined | 99.64 | 82.02 | 46 | 383 |

**Fig. 6** Global performance estimates of availability and maximum outage duration for each constellation and all constellations together. Results for omni pointing antenna (nadir and zenith) in the lower SSV

| Signal | Constellation | Signal availability (%) | | Max Outage Duration(min) | |
|---|---|---|---|---|---|
| | | At least 1 signal | 4 or more signals | At least 1 signal | 4 or more signals |
| Main lobe | BD | 64.93 | 0.33 | 422 | 1905 |
| | GPS | 57.75 | 0 | 717 | 2433 |
| | Combined | 83.92 | 4.78 | 717 | 935 |
| Main lobe and Side lobe | BD | 96.01 | 26.89 | 64 | 850 |
| | GPS | 83.69 | 0.83 | 106 | 647 |
| | Combined | 99.40 | 62.33 | 46 | 514 |

**Fig. 7** Global performance estimates of availability and maximum outage duration for each constellation and constellations together. Results for nadir-pointing antenna in the upper SSV

No matter in the lower SSV or upper SSV, the participation of the side lobe signals improves the service performance of BDS and GPS dramatically, and one-signal availability significantly exceeds four-signal availability.

Performance in the lower SSV is estimated to be significantly better than that in the upper SSV, due to the improved geometric availability at the lower altitude.

The addition of side lobe signals is especially effective in the upper SSV. The availability of more than one satellite of the reaches 96.01% while considering side lobe signals, and the availability of four or more satellites increases to 62.33% when multiple systems are interoperable.

On the basis of the availability of side lobe signals, MOD can be effectively reduced under interoperability of satellites systems, and the service performance will be better when multiple systems are interoperable.

### 4.1.2 RF Access Analysis Methodology

Figure 8 shows the signal availability and the MOD for a user in the upper SSV as a function of different $C/No$ thresholds for each individual constellation and for all constellations combined.

When side lobe signals are considered, one-signal availability is nearly 100% for 15 dB-Hz threshold. The abundance of signals available in an interoperable multi-GNSS SSV greatly reduces constraints imposed by navigation in the upper SSV. At the highest threshold of 25 dB/Hz, availability is nearly 0.0% in the upper SSV shown.

With the increase of altitude, the availability of satellite decreases obviously. Nevertheless, in an interoperable multi-GNSS SSV, the availability of satellite increases obviously and the MOD becomes shorter. While side lobe signals are considered, the availability of signals increased dramatically, especially for the more than four sig-

| Signal | Constellation | $C/NO_{min}$=15dBHz | | | | $C/NO_{min}$=20dBHz | | | | $C/NO_{min}$=25dBHz | | | |
|---|---|---|---|---|---|---|---|---|---|---|---|---|---|
| | | At least 1 signal | | 4 or more signals | | At least 1 signal | | 4 or more signals | | At least 1 signal | | 4 or more signals | |
| | | Avail(%) | MOD (min) | Avail(%) | MOD (min) | Avail(%) | MOD (min) | Avail(%) | MOD (min) | Avail(%) | MOD (min) | Avail(%) | MOD (min) |
| Main lobe | BD | 91.04 | 109 | 5.09 | 10325 | 14.14 | 133 | 0 | 10051 | * | * | * | * |
| | GPS | 91.04 | 109 | 5.09 | 10325 | 91.04 | 109 | 5.09 | 10325 | * | * | * | * |
| | Combination | 99.50 | 58 | 66.42 | 348 | 92.40 | 92 | 8.20 | 1216 | * | * | * | * |
| Combined | BD | 99.22 | 27 | 45.26 | 309 | 97.31 | 41 | 53.97 | 322 | * | * | * | * |
| | GPS | 99.22 | 27 | 45.26 | 309 | 99.22 | 27 | 45.26 | 309 | * | * | * | * |
| | Combination | 100 | 0 | 99.94 | 23 | 99.97 | 22 | 94.53 | 69 | * | * | * | * |

*No signal observed for the worst-case grid location for maximum simulation

**Fig. 8** Upper SSV performance with RF constraints, for various $C/No$ thresholds nals at the 20 dB/Hz threshold, where the availability of four signals reaches 94.53% from 8.20%, which also reflects the great potential of side lobe signals for deep space orbit determination in the future.

## 4.2 Mission-Specific Performance

Mission-specific simulated usage scenarios are considered as actual use cases for GNSS spatial users. Three representative mission scenarios were selected for simulation, a geostationary orbit mission, a highly elliptical orbit mission, and a lunar mission in the SSV booklet. The contribution of side lobe signals to space service and the influence of Individual constellation to multi-GNSS in SSV are simulated and analyzed.

### 4.2.1 Common Assumptions and Methods

For mission-specific analysis, an antenna beam pattern for the user spacecraft is included in the link power calculation. In particular, two different user antenna gain characteristics were used: a patch antenna with a gain of approximately 2 dBi, and a "high-gain" antenna with a gain of 8 to 9 dBi. The assumed acquisition threshold of the space user receiver is 20 dB-Hz.

### 4.2.2 Geostationary Orbit Mission

For all six GEO receivers at L1 frequencies, the satellite availability is shown from Fig. 9a–d.

Obviously, when only the main lobe signal is available, the availability of satellite is low, but when the side lobe signal is added, the availability increases obviously. Especially when multiple systems are user-operated, the availability of side lobe signals makes the number of visible satellites almost always four or more, and even 19 in some areas.The service performance of BDS is better than that of GPS. At 180 deg, the availability of four satellites of BDS is from 30.0 to 100.0% when side lobe signal is added.

### 4.2.3 Scientific Highly Elliptical Orbit Mission

The onboard GNSS antennas are configured in both nadir- and zenith-facing sides of the spacecraft showed in Fig. 10 [1].

Figure 12 shows the GNSS signal availability of all GNSS constellations for the HEO nadir and zenith-pointing antennas over the time of 1.5 HEO orbital periods.

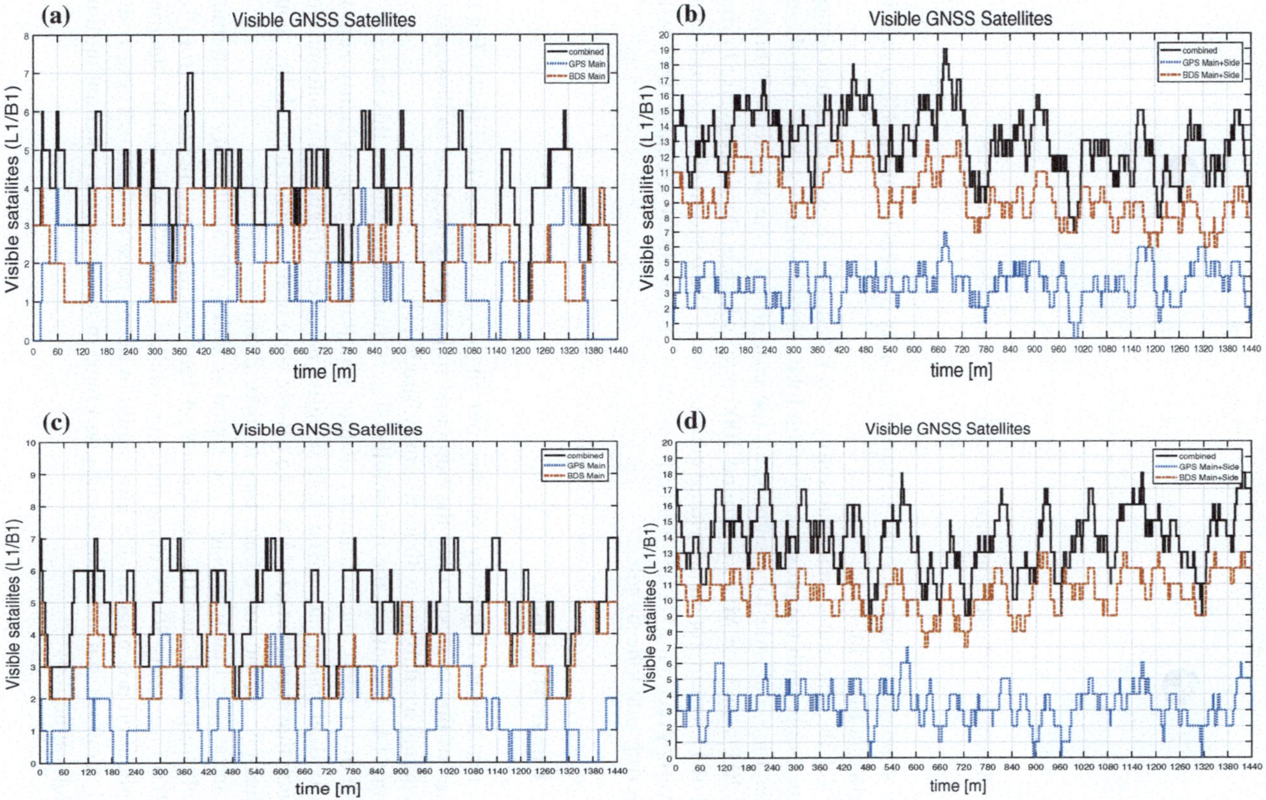


**Fig. 9** **a** L1/B1 availability for GEO at 60° with main lobe signal. **b** L1/B1 availability for GEO at 60° with main lobe signal and side lobe signal. **c** L1/B1 availability for GEO at 180° with main lobe signal. **d** L1/B1 availability for GEO at 180° with main lobe signal and side lobe signal

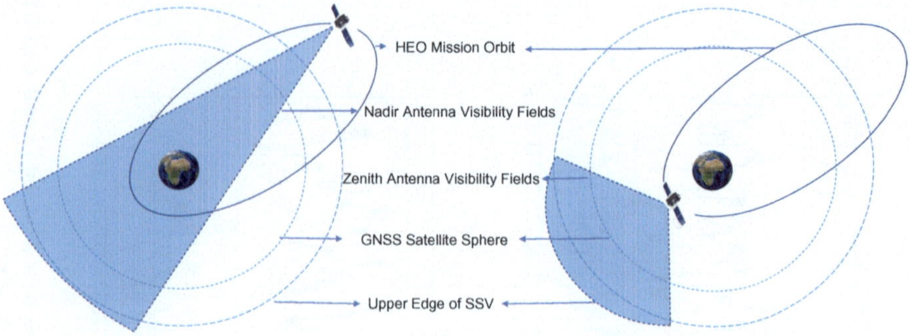


**Fig. 10** Schematic of the HEO mission with nadir and zenith-pointing antennas

The simulated results for the signal availability and MOD of the HEO mission are shown in Fig. 11a, b.

For L1/B1, the one-signal availability can reach 84.81% with all constellations combined only with the presence of main lobe signals, while 99.96% with the use of main lobe and side lobe signals. In the case of L1, four-signal availability is below 20.0% and the MOD is around 1000 min, which is close to the HEO orbital period of 1130 min, for an individual constellation. The performance is significantly improved by receiving signals from all constellations combined to nearly 100% when considering side lobe signals.

### 4.2.4 Lunar Mission

A full lunar mission trajectory contains four phases, as showcased in Fig. 13. For the purposes of this analysis, only the outbound trajectory is modeled to illustrate the GNSS signal availability with increasing altitude [1].

Figure 14a, b contains the full simulated performance results for this mission. In the case of L1 band, the availability of four simultaneous signals is nearly zero for any individual constellation. If a more sensitive receiver or higher-gain antenna were used such that signals at a $C/No$ of 15 dB-Hz were usable, signal availability would be achievable for the entire trajectory to lunar distance.

It can be seen that when side lobe signal is added, the availability of satellite is generally improved, and the availability of four satellites has increased from less than 50.0% to nearly 80.0% in the multi-constellation GNSS, which fully illustrates the application potential of side lobe signal in deep space navigation.These results strongly confirm that side lobe signals will have more prominent potential in future orbit determination.

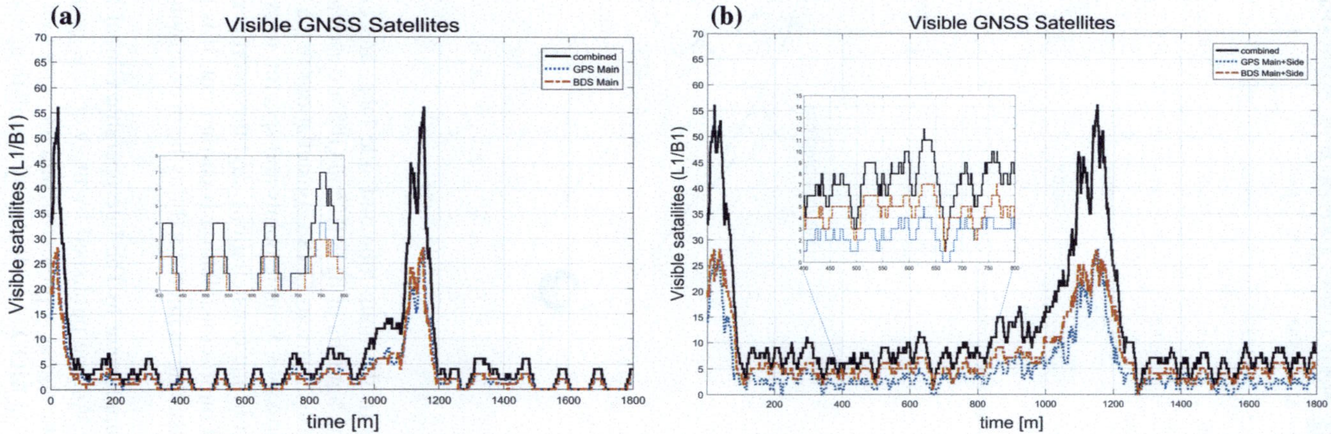


**Fig. 11** **a** Visible GNSS satellites over 1.5 orbital periods of HEO with main lobe signal. **b** Visible GNSS satellites over 1.5 orbital periods of HEO with the main lobe signal and side lobe signal

| Signal | Constellation | Signal availability (%) | | Max Outage Duration(min) | |
|---|---|---|---|---|---|
| | | At least 1 signal | 4 or more signals | At least 1 signal | 4 or more signals |
| Main lobe | BD | 76.84 | 16.89 | 73 | 1020 |
| | GPS | 82.41 | 19.60 | 62 | 1006 |
| | Combination | 84.81 | 51.80 | 62 | 91 |
| Combined | BD | 99.89 | 84.56 | 11 | 74 |
| | GPS | 98.14 | 43.84 | 15 | 276 |
| | Combination | 99.96 | 96.67 | 5 | 19 |

**Fig. 12** HEO mission simulated performance result

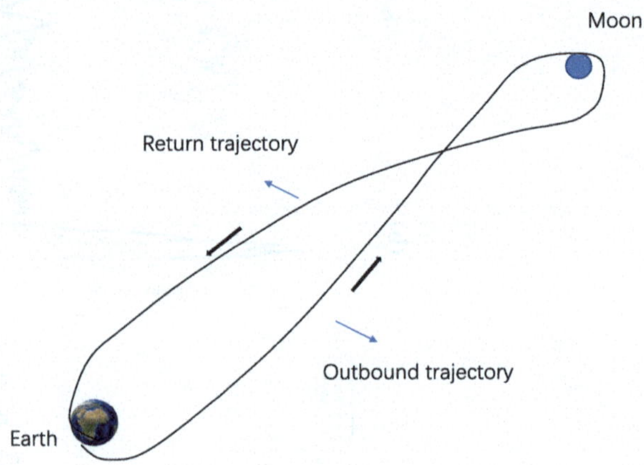


**Fig. 13** Lunar trajectory phases

## 5 Conclusion

In this paper, the range of the main lobe and side lobe, and carrier-to-noise ratio threshold of the downlink antenna of BDS III and GPS are used as evaluation indicators to simulate and establish various GNSS constellations in the way of approaching the real state as far as possible. Sample points and specific missions at different orbital altitudes are added to the constellations, and the interoperability between the single constellation and multi-constellation is carried out. It is pointed out that the advantages of multi-constellation interoperability lie in improving signal availability and shortening MOD.

On this basis, as a supplement to The Interoperable Global Navigation Satellite Systems Space Service Volume published by the working group B of ICG in October 2018. The side lobe signal is presented, and its impact on improving satellite availability and shortening MOD is evaluated by simulation, which fully proves the potential of side lobe signals in orbit determination.

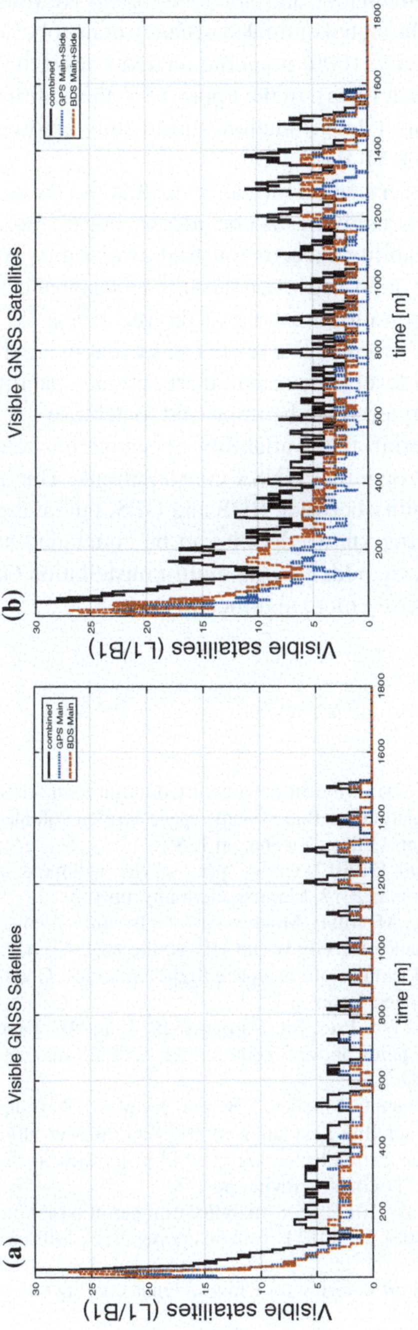


**Fig. 14** **a** Signal availability by trajectory altitude, to the limit of available signals at 36 RE with main lobe signal. **b** Signal availability by trajectory altitude, to the limit of available signals at 36 RE with the main lobe signal and side lobe signal

Taking BDS III, GPS single constellation and their interoperability as examples, the performances of the main lobe and side lobe signals are simulated and compared. Dual-frequency main lobe signal or dual-frequency main lobe signal enhanced by the side lobe signal can provide 100% potential navigation service for MEO spacecraft (more than four visible satellites). In the upper SSV, the addition of side lobe signals reflect its potential for orbit determination, significantly improves the availability of satellites and shortens the MOD.

In the global average service performance evaluation, the availability of satellites under different receiver sensitivities is considered. For $C/No_{\text{min}} = 15$ dB-Hz or 20 dB-Hz, one-signal availability exceeds fourfold availability. And the availability of the satellite is obviously improved when the side lobe signal is introduced. However, when the carrier-to-noise ratio threshold of the receiver is 25 dB/Hz, performance drops to 0.0% availability for all. The results show that in order to make use of these very weak GNSS signal levels, the required reception capability is quite harsh, that is, the receiver sensitivity needs to be improved to achieve.

In the case of high orbit, the availability of navigation satellites can be greatly improved by using interoperable GNSS constellations. This paper only discusses the status of interoperability between BDS and GPS, but on the basis of experimental simulation and existing knowledge, it can be concluded by analogy that when Galileo and GLONASS are added to the multi-constellation GNSS, the availability of satellites can be improved more significantly.

## References

1. Working Group B (WG-B) of the International Committee on GNSS (ICG) (2018) The interoperable global navigation satellite systems space service volume. Publishing and Library Section of United Nations Office , Vienna, pp 1–35
2. Enderle W, Schoenemann E (2017) GNSS space service volume & user perspective. Munich Satellite Navigation Summit 2017, Munich, Germany, pp 1–5
3. Enderle W, Schmidhuber M, Gill E, Montenbruck O, Braun A, Lemke N, Balbach O, Eisfeller B (1998) GPS performance for GEOs and HEOs: the EQUATOR-S spacecraft mission. In: Thirteenth international symposium on space flight dynamics. Goddard Space Flight Center, Greenbelt Maryland, United States, p 1
4. Winternitz LB, Bamford WA, Price SR, Carpenter JR, Long AC, Farahmand M (2017) Global positioning system navigation above 76,000 km for NASA'S magnetospheric multiscale mission, vol 64, pp 289–300
5. Filippi H, Gottzein E, Kuehl C, Mueller C, Barrios-Montalvo A, Dauphin H (2010) Feasibility of GNSS receivers for satellite navigation in GEO and higher altitudes. In: 2010 5th ESA workshop on satellite navigation technologies and European workshop on GNSS signals and signal processing (NAVITEC), Noordwijk, pp 1–8
6. Zhan CX, Liu B, Yuan W (2017) SSV visibility evaluation based on different GPS transmitting antenna characteristics. In: 2017 Forum on cooperative positioning and service (CPGPS), Harbin, pp 2–3
7. Yang W (2010) Phasing orbit design for Chinese lunar satellite CE-1. Chinese Space Science and Technology
8. Chen L (2016) Research of high orbit spacecraft positioning technology based on GNSS. Ph.D. thesis, National University of Defense Technology

9. GPS World Staff (2013) GIOVE—a uses GPS side lobe signals for far-out space navigation. GPS World, pp 1–2
10. Balbach O, Eissfeller B, Hein GW, Enderle W, Schmidhuber M, Lemke N (1998) Tracking GPS above GPS satellite altitude: first results of the GPS experiment on the HEO mission Equator-S. In: IEEE 1998 position location and navigation symposium
11. Jing S, Zhan X, Lu J, Feng S, Ochieng WY (2015) Characterisation of GNSS space service volume. J Navig 68:107–125
12. ICD, GPS (2011) Global positioning systems directorate system engineering & integration interface specification IS-GPS-200F. Navstar GPS Space Segment Navigation User Interfaces
13. ICD-BDS (2012) BeiDou navigation satellite system signal in space interface control document open service signal B1I (version 1.0)
14. Stanton BJ, Parker Temple LIII, Edgar CE (2006) Analysis of signal availability in the GPS space service volume. In: Proceedings of the ION GNSS 2006
15. Hogg DC (1993) Fun with the Friis free-space transmission formula. Antennas and Propagation Magazine, IEEE
16. Jing S (2017) Study on GNSS performance evaluation and related auxiliary in the space service volume. Ph.D. thesis, Shanghai Jiao Tong University

# Formation Control and Obstacle/Collision Avoidance with Dynamic Constraints

**Shaoyang Mu and Pingfang Zhou**

**Abstract** This paper studies the leader-follower formation control problem of unmanned air vehicles(UAVs) flying with dynamic constraints in an obstacle-laden environment. Firstly, formation protocols are presented for UAV swarm systems. Necessary and sufficient conditions for UAV swarm systems to achieve formations are presented based on graph theory. A formation tracking protocol is designed to drive the followers to track the leader. Then, an improved obstacle avoidance method based on potential field method is designed by combining with local rules considering dynamic constraints. The improved method can avoid potential field falling into local minimum and have good global searching ability. The stability of the presented approach is proved by using Lyapunov stability theory. Finally, numerical simulations are presented to demonstrate the effectiveness of the designed approach.

**Keywords** Leader-follower formation · Consensus algorithm · Potential filed · Collision avoidance

## 1 Introduction

Cooperative control problems of multiagent systems have attracted a lot of attention in recent years [1, 2]. This is partly due to its broad applications in flocking, formation control and robot target tracking [3–8].

Typical approaches for formation control include leader-follower, behavioral and virtual structure/virtual leader approaches. Some papers [9, 10] discuss the leader-follower approach. In these papers, the first vehicle is regarded as the leader, and others are regarded as followers. The leader vehicle flies according to the pre-planned trajectory, the follower vehicles follow the leader's motion, maintaining a desired ge-

S. Mu (✉) · P. Zhou
School of Aeronautics and Astronautics, Shanghai Jiao Tong University, Shanghai, China
e-mail: shaoyangmu@sjtu.edu.cn

P. Zhou
e-mail: zhoupf@sjtu.edu.cn

 
Z. Jing (ed.), *Proceedings of the International Conference on Aerospace System Science and Engineering 2019*, Lecture Notes in Electrical Engineering 622,
https://doi.org/10.1007/978-981-15-1773-0_11

ometric structure. The leader-follower approach simplifies the control of multiagent systems. However, if the leader is out of control, the formation will not be maintained. The behavioral approach is introduced by papers [11, 12]. The main content of this approach is to design various basic behaviors, obstacle/collision avoidance, target searching and formation maintaining, i.e., the behavioral approach is a fully distributed control structure with real-time feedback characteristics and has better flexibility. However, it is difficult to carry out mathematical analysis and corresponding stability analysis of the system. These papers [13–15] discuss the approach of virtual structure/virtual leader. The entire formation is designed as a single rigid virtual structure, and each vehicle is a fixed point. During the formation flying, each vehicle only needs to track the fixed point motion corresponding to the rigid structure, the virtual structure/virtual leader approach simplifies task description and assignment. However, it is a centralized control approach with poor reliability.

Consensus problems are important and challenging research topics in multiagent coordination. The basic idea of consensus is that a group of agents reach an agreement on their common states via local interaction. In [16], the author has proved that most approaches of leader-follower, behavioral and virtual structure/virtual leader can be regarded as special cases of consensus-based approaches. And the weakness of the previous approaches can be overcome. In the past decade, consensus algorithms have been studied in various multiagent systems [17–25]. In [17], the formation control protocol is designed to guarantee the consistency of vehicles in altitude. In [18], considering the input constraints, a non-linear centrally free consistency algorithm is used to achieve altitude maintaining of multi-UAV systems. In [18–20], consensus algorithm takes the form of first-order dynamics is used, and in [21–23], the authors extend it to second-order dynamics under undirected information flow. In [24], the consensus problem of multiagent systems with switching topologies is discussed, a formation protocol is designed to solve the average consensus problem. In [25], a formation control problem with time-varying is discussed under the case of a spanning tree in the network topology. The stability conditions are obtained, and the experimental verification is carried out.

Most of the aforementioned works have not taken collision/obstacle avoidance into consideration, but collision would occur while the UAVs are flying in formation. Note that in [26] and [27], the artificial potential field approach is used to achieve collision/obstacle avoidance. Attractive potential fields are assigned to target points and repulsive potential are assigned to obstacles. The vehicles move along the direction under resultant forces and toward the attractive target point. However, the weakness of the artificial potential field approach is that it is easy to fall into local minimum. What is more, vehicles have dynamic constraints, maximum turning angle and maximum climb angle, etc.

In this paper, we study the consensus problem of second-order continuous-time UAVs system. Graph theory is used to describe network topology of the system. A consensus algorithm to achieve formation control is presented. The stability conditions of the consensus algorithm are analyzed by using a Lyapunov function and LaSalle's principle. Contributions of this paper include using a consensus algorithm to achieve formation control and taking collision/obstacle avoidance into consideration compared with the second-order consensus algorithm studied in [22]. Compared with [26], considering dynamic constraints, an improved artificial potential field-based approach with dynamic constraints is designed to achieve collision/obstacle avoidance. Our method can avoid potential field function falling into a local minimum and improve the global searching ability.

In Sect. 2, some basic concepts and results about graph theory and the system model are introduced. In Sect. 3, a formation protocol is presented and stability conditions are analyzed. In Sect. 4, simulation results of formation control, collision/obstacle avoidance in 2-D are shown. Conclusions and future work are discussed in Sect. 5.

## 2 Preliminaries and System Model

### 2.1 Notations

$R_n$ denotes the set of all $n$ dimensional real column vectors.
$1_n$ and $0_n$ denote the $n \times 1$ column vector of all-one and all-zero, respectively.
$\boldsymbol{I}_n$ and $\boldsymbol{0}_{(n\times n)}$ denote the identity matrix and zero matrix with dimension $n$, respectively.
For a given $\lambda \in C$, $Re(\lambda)$ and $Im(\lambda)$ denote the real part and the imaginary part of $\lambda$, respectively.
Let $\otimes$ denote the Kronecker product.
$\| \boldsymbol{A} \|$ denotes the norm of the matrix/vector $\boldsymbol{A}$.

### 2.2 Graph Theory

Here, we introduce some definitions and results about graph theory (for more details, please refer to [28]).

An undirected graph $G = (V, E, \boldsymbol{A})$ is used to describe the communication among $n$ agents. $V = s_1, s_2, \ldots, s_n$ is the set of nodes. $E = (s_i, s_j) \in s \times s, i \neq j$ is the set of edges. An edge of $G$ is denoted by $s_{ij} = (s_i, s_j)$, if $s_i \in E \longleftrightarrow s_{ji} \in E$, the graph is said to be undirected graph. $\boldsymbol{A} = [a_{ij}]$ is a weighted adjacency matrix with $a_{ii} = 0, a_{ij} = a_{ji} \geq 0$, where $a_{ij} > 0$ if and only if $s_{ij} \in E$. $N_i = s_j \in V : (s_i, s_j) \in E$ denotes the set of neighbors of node $s_i$. The Laplacian matrix of undirected graph is defined as $\boldsymbol{L} = [l_{ij}]$, where $l_{ij} = \sum_{j=1}^{n} a_{ij}$ and $l_{ij} = -a_{ji}, i \neq j$. If there is a path from any node to each other node, the graph is considered as connected. Assuming that a collection of graphs $\overline{G}_1, \overline{G}_2, \ldots, \overline{G}_m$ with the same node set $V$ is defined as the graph $\overline{G}_{1\sim m}$. The node set is $V$ and the edge set $\overline{G}_{1\sim m}$ is the union of the edge sets of all graphs in the collection. Besides, if the graph $\overline{G}_{1\sim m}$ is connected , the collection of graphs $\overline{G}_1, \overline{G}_2, \ldots, \overline{G}_m$ will be considered as jointly connected.

**Lemma 1** *[21]: $\mathbf{L}$ is the matrix of an undirected graph, then $\mathbf{L}$ has a simple zero eigenvalue and $\boldsymbol{I}_n$ is the associated eigenvector, that means, $\boldsymbol{L}\boldsymbol{I}_n = 0$.*

### 2.3 System Dynamics and Control Protocol

In order to focus on studying formation control law, an individual UAV is modeled as a point-mass system. The consensus problem for second-order multiagent systems in 2-D plane is considered. The dynamics of each UAV can be described as a double integrator as [16, 21]

$$\begin{cases} \dot{\xi}_i(t) = \zeta_i(t) \\ \dot{\zeta}_i(t) = \boldsymbol{u}_i(t) \end{cases} \tag{1}$$

where $i = 1, 2, \ldots, N$, $\xi_i(t) = x = [x_i, y_i]^T \in R^n$ and $\zeta_i(t) = v = [v_{xi} \cos\theta_i, v_{yi} \sin\theta_i]^T \in R^n$ denote the position state and velocity state of UAV $i$, respectively. $\theta_i$ and $u_i(t) \in R^n$ denote the heading angle and control input, respectively. If the control protocol can guarantee that the state of all UAVs reach $[\xi_j(t) \to \xi_i(t)] \to \boldsymbol{r}_{ij}$ and $[\zeta_j(t) \to \zeta_i(t)] \to \zeta^*$ as $t \to \infty$ for all UAVs from $1, 2, \ldots, N$, then we can say the formation is achieved with desired speed $\zeta^*$, where $\boldsymbol{r}_{ij}$ denotes the desired distance difference between UAV $i$ and $j$, $\zeta^*$ denotes the desired speed.

In [22], the consensus protocol is proposed to solve the formation control without considering dynamic constraints and collision/obstacle avoidance. In this paper, our main objective is to extend the work of [22] to solve the consensus problem with jointly connected topologies and collision/obstacle avoidance. Here, we presented the following linear consensus protocol

$$\boldsymbol{u}_i(t) = \sum_{V_j \in N_i(t)} a_{ij}\{k_1[\xi_j(t) - \xi_i(t) - \boldsymbol{r}_{ij}] + k_2[\zeta_j(t) - \zeta_i(t)]\} - k_3(\zeta_i - \zeta^*) \tag{2}$$

where $a_{ij}$ denotes the element of the weighted adjacency matrix, $N_i(t)$ denotes the set of neighbors of node $s_i$ and the value of $k_1, k_2, k_3$ all are positive.

## 3 Formation Control Protocol Design and Analysis

### 3.1 Formation Control Protocol Design

In order to analyze the performance of formation control protocol, we give the concept of formation center, which is the center of formation structure. For example, as is shown in Fig. 1, assume that the formation structure is a regular triangle. Where $O$ denotes the origin of Cartesian coordinate system, $O_C$ denotes the formation center,

**Fig. 1** Formation plane

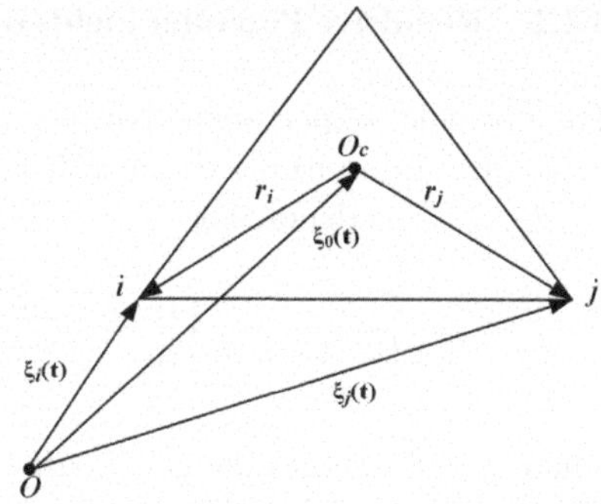


$\xi_i(t), \xi_j(t), \xi_0(t)$denote the position state of UAV and formation center, respectively, $\boldsymbol{r}_i, \boldsymbol{r}_j$ denote the distance from formation center to UAV $i, j$, respectively.

Then, consensus protocol (2) can be transformed into

$$\boldsymbol{u}_i(t) = \sum_{V_j \in N_i(t)} a_{ij}\{k_1[(\xi_j(t) - \boldsymbol{r}_j) - (\xi_i(t) - \boldsymbol{r}_i)] + k_2[\zeta_j(t) - \zeta_i(t)]\} - k_3(\zeta_i - \zeta^*) \tag{3}$$

where $\boldsymbol{r}_{ji} = \boldsymbol{r}_j - \boldsymbol{r}_i$. In the next, the consensus protocol (3) will be extended to achieve formation control.

## 3.2 *Potential Forces*

According to the potential field method in [22], the potential forces will be designed to achieve collision/obstacle avoidance as following.

### 3.2.1 Attractive Potential Fields of Target

The position state of UAV $i$ is denoted as $\xi_i = [x_i, y_i]^T \in R^2$. The target applies attractive force to UAV and the value of the force is in reverse proportion to the distance from UAV to target. The attractive potential field $U_{att}(\xi_i, \xi_G)$ is denoted as

$$U_{att}(\xi_i, \xi_G) = 0.5\varepsilon d^2(\xi_i, \xi_G) \tag{4}$$

where $\varepsilon > 0$, denotes the gain coefficient of attractive potential field, $\xi_G = [x_g, y_g]^T \in R^2$ denotes the position state of target $G$, $d(\xi_i, \xi_G)$ denotes the distance from UAV $i$ to target $G$ and the attractive force $\boldsymbol{F}_{att}(\xi_i, \xi_G)$ is denoted as

$$\boldsymbol{F}_{att}(\xi_i, \xi_j) = -\nabla U_{att}(\xi_i, \xi_j) = -\varepsilon(\xi_i - \xi_G) \tag{5}$$

### 3.2.2 Repulsive Potential Fields of Obstacle

The obstacles expert repulsive force to UAV and the value of the force are in reverse proportion to the distance from UAV to target. The repulsive potential field $U_{rep}(\xi_i, \xi_{obs})$ is denoted as

$$U_{rep}(\xi_i, \xi_{obs}) = \begin{cases} 0.5\eta(\frac{1}{d(\xi_i,\xi_{obs})} - \frac{1}{d_0})^2 d^n(\xi_i, \xi_G) & , d(\xi_i, \xi_{obs}) \le 0 \\ 0 & , d(\xi_i, \xi_{obs}) > d_0 \end{cases} \tag{6}$$

where $\eta > 0$ denotes the gain coefficient of repulsive potential field, $d(\xi_i, \xi_{obs})$ denotes the distance from UAV $i$ to obstacles. $d_0$ denotes the affects distance of obstacle. $d_0$ is generally chosen as $d_0 \le min(d_1, d_2)$ in order to avoid repulsive potential field falling into local minimum, where $d_1$ represents half of the smallest distance between obstacles, $d_2$ represents the minimum distance from target to each obstacle. In order to operate the UAV more flexible, $d_0$ can be chosen dynamically according to the speed of UAV.

$$d_0 = d_0^{min} + k_0 v \tag{7}$$

where $d_0^{min}$ denotes the minimum distance to avoid obstacle, $k_0$ is associated with the performance of UAV.

The repulsive force is denoted as $\boldsymbol{F}_{rep}(\xi_i, \xi_{obs}) = -\nabla_{\xi_i} U_{rep}(\xi_i, \xi_{obs})$

$$\boldsymbol{F}_{rep}(\xi_i, \xi_{obs}) = \begin{cases} \boldsymbol{F}_{rep1}\boldsymbol{n}_{ov} + \boldsymbol{F}_{rep2}\boldsymbol{n}_{vG} & , d(\xi_i, \xi_{obs}) \le d_0 \\ 0 & , d(\xi_i, \xi_{obs}) > d_0 \end{cases} \tag{8}$$

where $\boldsymbol{n}_{ov}$ and $\boldsymbol{n}_{vG}$ denote unit vectors with the direction from obstacle to UAVs and from UAVs to target.

$$\boldsymbol{F}_{rep1} = \eta(\frac{1}{d(\xi_i, \xi_{obs})} - \frac{1}{d_0})\frac{d^n(\xi_i, \xi_G)}{d^2(\xi_i, \xi_{obs})} \tag{9}$$

$$\boldsymbol{F}_{rep2} = 0.5n\eta(\frac{1}{d(\xi_i, \xi_{obs})} - \frac{1}{d_0})d^{n-1}(\xi_i, \xi_G) \tag{10}$$

### 3.2.3 Collision Avoidance Potential Field of UAVs

In this paper, obstacle avoidance method is used to achieve collision avoidance. For UAV $i$, the other UAVs all are obstacles. The other UAVs expert repulsive forces to UAV $i$. The collision avoidance potential field is denoted as

$$U_{rep}(\xi_i, \xi_j) = \begin{cases} 0.5k(\frac{1}{d(\xi_i,\xi_j)} - \frac{1}{d_{safe}})^2 & , d(\xi_i, \xi_j) \le d_{safe} \\ 0, & , d(\xi_i, \xi_j) > d_{safe} \end{cases} \tag{11}$$

where $k > 0$ denotes the gain coefficient of collision avoidance potential field, $d(\xi_i, \xi_j)$ denotes the distance between UAV $i$ and $j$, $d_{safe}$ denotes the safety distance of each UAV. The repulsive force between UAV $i$ and $j$ is denoted as $\boldsymbol{F}_{ij} = -\nabla U_{rep}(\xi_i, \xi_j)$.

$$\boldsymbol{F}_{ij} = \begin{cases} k(\frac{1}{d(\xi_i,\xi_j)} - \frac{1}{d_{safe}})^2 \frac{1}{d^2(\xi_i,\xi_j)} \boldsymbol{n}_{ij} & , d(\xi_i, \xi_j) \leq d_{safe} \\ 0 & , d(\xi_i, \xi_j) > d_{safe} \end{cases} \tag{12}$$

where $\boldsymbol{n}_{ij}$ is a unit vector with the direction from UAV $j$ to $i$. Noted that the value of $\boldsymbol{F}_{ij}$ is associated with safety distance $d_{safe}$. Then, the total forces $\boldsymbol{F}_{ij}^{all}$ of UAV $i$ is denoted as

$$\boldsymbol{F}_{ij}^{all} = \begin{cases} \sum_{j=1}^{n} \boldsymbol{F}_{ij} & , d(\xi_i, \xi_j) \leq d_{safe} \\ 0 & , d(\xi_i, \xi_j) > d_{safe} \end{cases} \tag{13}$$

## 3.3 Dynamic Constraints

As in [29], we take velocity constraints into consideration. The velocity $v$ is associated with heading angle, $v = [v_{xi}\cos(\theta_i), v_{yi}\sin(\theta_i)]^T \in R^n$, and the heading angle is related to resultant force. However, due to the performance constraints of UAVs, the motion direction of UAVs cannot be completely determined by the calculated resultant force in each step. Heading angle should be corrected after calculating resultant force in each step. As shown in Fig. 2, assume that node $s$ and flight direction of UAV $i$ at current are known. $\angle \boldsymbol{COB}$ denotes the maximum heading angle $\theta$. Therefore, the next motion direction of UAV $i$ is limited to $\angle \boldsymbol{AOB} = 2\theta$.

If the direction of resultant force is under the range of $2\theta$, motion direction of UAV $i$ will be the direction of resultant force. Otherwise, it should be corrected. The corrected heading angle is represented as

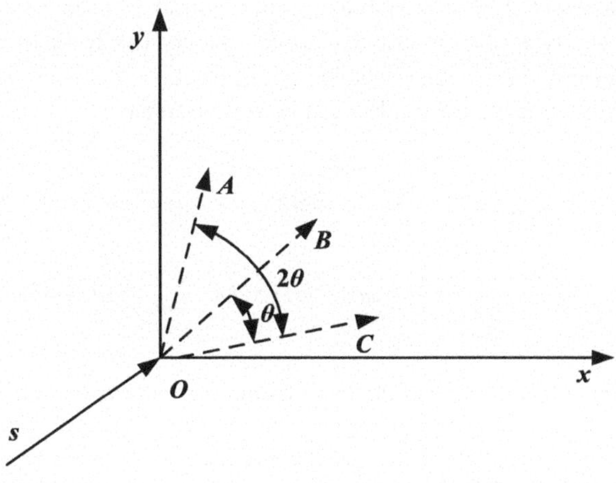


**Fig. 2** The range of motion

$$\theta_0 = \begin{cases} \theta & , \theta_{real} > \theta \\ \theta_{real} & , |\theta_{real}| < \theta \\ -\theta & , \theta_{real} < -\theta \end{cases} \tag{14}$$

where $\theta_{real}$ denotes the direction angle of resultant force, and $\theta_0$ denotes the heading angle of UAV in next step after being corrected. The positive heading angle is counterclockwise. By limiting heading angle, the velocity constraints is achieved.

***Remark 1*** Agents are subject to bounded maximum linear velocity because of dynamic limitations, and most of the agents are subject to the positive-minimum velocity in case of stall conditions. In 3-D plane, it should be pointed out that pitch angle also should be taken into account.

## 3.4 Collision/Obstacle Avoidance Protocol Design

Based on the theory introduced above and the protocol (3), here, collision/obstacle avoidance protocol for UAV $i$ is designed as

$$\begin{aligned} \boldsymbol{u}_i(t) = k_1 \sum_{s_j \in N_i(t)} a_{ij}(t)[\xi_j(t) - \xi_i(t) - \boldsymbol{r}_{ij}] + k_2 \sum_{obs=1}^{l} \boldsymbol{F}_{rep}(\xi_i, \xi_{obs} + k_3 \boldsymbol{F}_{att}(\xi_i, \xi_G) \\ + k_4 \sum_{s_j \in \overline{N}_i(t)} \boldsymbol{F}_{ij} - \zeta_i(t) + v_0 \end{aligned} \tag{15}$$

where $k_1, k_2, k_3, k_4$ all are positive number, $a_{ij}(t)$ denotes the weighted adjacency of communication topology $G$, $N_i(t)$ denotes the set of neighbors of UAV $i$, $\overline{N}_i(t)$ denotes the set of collision neighbors of UAV $i$, $l$ denotes the number of obstacles currently detected by UAV $i$.

***Remark 2*** The variable-weight method is used to make the protocol (15) more widely applied in practice, which means, (a): If the obstacle is not detected, $k_1, k_3 > k_2 = 0$, (b): if the obstacle is detected,$k_2 > k_1, k_3 > 0$. Substituting (4),(8),(12) into (15), then protocol (15) can be rewritten as

$$\begin{aligned} \boldsymbol{u}_i(t) = k_1 \sum_{s_j \in N_i(t)} a_{ij}(t)[\xi_j(t) - \xi_i(t) - \boldsymbol{r}_{ij}] - k_2 \sum_{obs=1}^{l} \nabla_{\xi_i} U_{rep}(\xi_i, \xi_{obs}) \\ - k_3 \nabla_{\xi_i} U_{att}(\xi_i, \xi_G) - k_4 \sum_{s_j \in \overline{N}_i(t)} \nabla_{\xi_i} U_{rep}(\xi_i, \xi_j) - \zeta_i(t) + \boldsymbol{v}_0 \end{aligned} \tag{16}$$

According to the concept of formation center in Sect. 2, (16) can be rewritten as

$$\begin{aligned} \boldsymbol{u}_i(t) = k_1 \sum_{s_j \in N_i(t)} a_{ij}(t)[(\xi_j(t) - \boldsymbol{r}_j) - (\xi_i(t) - \boldsymbol{r}_i)] - k_2 \sum_{obs=1}^{l} \nabla_{\xi_i} U_{rep}(\xi_i, \xi_{obs}) \\ - k_3 \nabla_{\xi_i} U_{att}(\xi_i, \xi_G) - k_4 \sum_{s_j \in \overline{N}_i(t)} \nabla_{\xi_i} U_{rep}(\xi_i, \xi_j) - \zeta_i(t) + \boldsymbol{v}_0 \end{aligned} \tag{17}$$

where $\boldsymbol{r}_{ji} = \boldsymbol{r}_j - \boldsymbol{r}_i$. Let $\hat{\xi}_i(t) = \xi_i(t) - \xi_0(t) - \boldsymbol{r}_i$, $\hat{\zeta}_i(t)\zeta_i(t) - \boldsymbol{v}_0(t)$, then, (17) can be transformed into

$$\begin{aligned} \boldsymbol{u}_i(t) = k_1 \sum_{s_j \in N_i(t)} a_{ij}(t)[\hat{\xi}_j(t) - \hat{\xi}_i(t)] - k_2 \sum_{obs=1}^{l} \nabla_{\xi_i} U_{rep}(\hat{\xi}_i + \xi_0 + \boldsymbol{r}_i, \xi_{obs}) \\ - k_3 \nabla_{\xi_i} U_{att}(\hat{\xi}_i + \xi_0 + \boldsymbol{r}_i, \xi_G) - k_4 \sum_{s_j \in \overline{N}_i(t)} \nabla_{\xi_i} U_{rep}(\hat{\xi}_i, \hat{\xi}_j) - \hat{\zeta}_i \end{aligned} \tag{18}$$

By protocol (18), then, swarm system (1) can be rewritten as

$$\begin{cases} \dot{\hat{\xi}}_i(t) = \hat{\zeta}_i(t) \\ \dot{\hat{\xi}}_i(t) = k_1 \sum_{s_j \in N_i(t)} a_{ij}(t)[\hat{\xi}_j(t) - \hat{\xi}_i(t)] - k_2 \sum_{obs=1}^{l} \nabla_{\xi_i} U_{rep}(\hat{\xi}_i + \xi_0 + \boldsymbol{r}_i, \xi_{obs}) \\ \quad -k_3 \nabla_{\xi_i} U_{att}(\hat{\xi}_i + \xi_0 + \boldsymbol{r}_i, \xi_G) - k_4 \sum_{s_j \in \overline{N}_i(t)} \nabla_{\xi_i} U_{rep}(\hat{\xi}_i, \hat{\xi}_j) - \hat{\zeta}_i \end{cases} \tag{19}$$

In the next, the stability of the system will be analyzed.

### 3.5 Formation Analysis

**Lemma 2** *Considering a UAV system with n agents and the communication topology is jointly connected. Assume that the number of obstacles $\xi_{obs}$ is $l$ ($obs = 1, 2, \ldots, l$) and the total energy $V_t$ of UAV system is a finite value during the flight, that is, $V_t \le V_0$. Then, under protocol* (16), *the desired formation is achieved, that is, $\xi_j(t) - \xi_i(t) = \boldsymbol{r}_{ij}$.*

***Proof*** Here, we present the following Lyapunov candidate function $V$.

$$\begin{aligned} V = \sum_{i=1}^{n} [\frac{1}{2} k_4 \sum_{s_j \in \overline{N}_i(t)} U_{rep}^{ij} + k_2 \sum_{obs=1}^{l} U_{rep}^{i}(\hat{\xi}_i + \xi_0 + \boldsymbol{r}_i, \xi_{obs}) + \frac{1}{2} \hat{\zeta}_i^T \hat{\zeta}_i \\ + \frac{k_1}{4} \sum_{s_j \in N_i(t)} a_{ij}(\hat{\xi}_i - \hat{\xi}_j)^2 + k_3 U_{att}^{i}(\hat{\xi}_i + \xi_0 + \boldsymbol{r}_i, \xi_G)] \end{aligned} \tag{20}$$

Clearly, we can see that $V$ is a semidefinite function. Substituting (19) into (20), the time derivative of function $V$ is given by

$$\begin{aligned}\dot{V} = \sum_{i=1}^{n}[k_4\hat{\zeta}_i^T \sum_{s_j \in \overline{N}_i(t)} \nabla_{\xi_i} U_{rep}(\hat{\xi}_i, \hat{\xi}_j) + k_2 \sum_{obs=1}^{l} \nabla_{\xi_i} U_{rep}(\hat{\xi}_i + \xi_0 + \mathbf{r}_i, \xi_{obs})\\ + k_3 \nabla_{\xi_i} U_{att}(\hat{\xi}_i + \xi_0 + \boldsymbol{r}_i, \xi_G) + \hat{\zeta}_i^T k_1 \sum_{s_j \in N_i(t)} a_{ij}(t)[\hat{\xi}_j(t) - \hat{\xi}_i(t)]\\ - k_2 \hat{\zeta}_i^T k_2 \sum_{obs=1}^{l} \nabla_{\xi_i} U_{rep}(\hat{\xi}_i + \xi_0 + \boldsymbol{r}_i, \xi_{obs}) - k_3 \hat{\zeta}_i^T \nabla_{\xi_i} U_{att}(\hat{\xi}_i + \xi_0 + \boldsymbol{r}_i, \xi_G)\\ - k_4 \hat{\zeta}_i^T \sum_{s_j \in \overline{N}_i(t)} \nabla_{\xi_i} U_{rep}(\hat{\xi}_i, \hat{\xi}_j) + \hat{\zeta}_i^T k_1 \sum_{s_j \in N_i(t)} a_{ij}(t)[\hat{\xi}_j(t) - \hat{\xi}_i(t)]\\ - k_3 \nabla_{\xi_i} U_{att}(\hat{\xi}_i + \xi_0 + \boldsymbol{r}_i, \xi_G)]\end{aligned} \tag{21}$$

By simplifying (21), we can obtain that

$$\dot{V} = \sum_{i=1}^{n}[-\hat{\zeta}_i^T \hat{\xi}_i] \tag{22}$$

From (22), we can know that $\dot{V} \leq 0$. According to Lyapunov's second method for stability, the system described by (19) is asymptotically. Considering a set $\Omega = \{(\hat{\xi}_{ij}, \hat{\zeta}_i) | V(t) \leq V_0\}$, where $\hat{\xi}_{ij} = \hat{\xi}_i - \hat{\xi}_j$ is bounded and $\hat{\zeta}_i^T \hat{\zeta}_i \leq \sum_{i=1}^{n} \hat{\zeta}_i^T \hat{\zeta}_i \leq 2V \leq 2V_0$. Therefore, $\| \hat{\zeta} \| \leq \sqrt{2V_0}$ , that means $\hat{\zeta}_i$ is bounded. Then, we can obtain that $\Omega$ is a compact set. Based on LaSalle's principle in [10], if the initial solution of the system is in $\Omega$, the trajectories will converge to the maximum invariant set in $\overline{\Omega} = \{(\hat{\xi}_{ij}, \hat{\zeta}_i) \in \Omega | \dot{V} = 0\}$.

From (22), we can know that if $\dot{V} = 0$, then $\hat{\xi}_i = 0 (i = 1, 2, \ldots, n)$ and $\zeta_i(t) = v_0(t)$ in $\overline{\Omega}$. Noted that $\frac{d(\hat{\xi}_i(t))}{dt}$, then $\hat{\xi}_i(t) \equiv c$. Therefore $\hat{\xi}_i - \hat{\xi}_j = 0$, that is $\xi_j - \xi_i = \boldsymbol{r}_{ij}$ and the desired formation is achieved under protocol (16).

**Lemma 3** *Considering a UAV system with n agents and the communication topology is jointly connected. Under protocol* (16)*, the desired formation is achieved with collision avoidance.*

***Proof*** Here, reduction to absurdity is used to prove that collision avoidance is realized between UAVs. Suppose UAV $i$ and $j$ collide at $t_1 > 0$ and their position states satisfy

$$\xi_i(t_1) = \xi_j(t_1), (i \neq j \ and \ i, j \in (1, 2, \ldots, n)) \tag{23}$$

By (12), we can know

$$\frac{1}{2}\sum_{i=1}^{n} k_4[\sum_{s_j \in \overline{N}_i(t)} U_{rep}^{ij}(\hat{\xi}_i, \hat{\xi}_j)] = \frac{1}{2}\sum_{j=1}^{n} k_4[k \sum_{s_j \in \overline{N}_i(t)} 0.5k(\frac{1}{d(\xi_i, \xi_j)} - \frac{1}{d_{safe}})^2] \tag{24}$$

By (23) and(24), then

$$\lim_{t\to t_1}\frac{1}{2}\sum_{i=1}^{n}k_4[\sum_{s_j\in\overline{N}_i(t)}U_{rep}^{ij}(\hat{\xi}_i,\hat{\xi}_j)]=+\infty \tag{25}$$

However, by (20), we can know

$$\begin{aligned}\frac{1}{2}\sum_{i=1}^{n}k_4[\sum_{s_j\in\overline{N}_i(t)}U_{rep}^{ij}(\hat{\xi}_i,\hat{\xi}_j)] &= V-\sum_{i=1}^{n}[k_2\sum_{obs=1}^{l}U_{rep}^{i}(\hat{\xi}_i+\xi_0+\boldsymbol{r}_i,\xi_{obs})-\frac{1}{2}\hat{\zeta}_i^T\hat{\zeta}_i\\ &\quad -\frac{1}{4}\sum_{s_j\in\overline{N}_i(t)}a_{ij}(t)(\hat{\xi}_i-\hat{\xi}_j)^2-k_3U_{att}^{i}(\hat{\xi}_i+\xi_0+\boldsymbol{r}_i,\xi_G)]\\ &\le V\le V_0\end{aligned} \tag{26}$$

That is

$$\lim_{t\to t_1}\frac{1}{2}\sum_{i=1}^{n}k_4[\sum_{s_j\in\overline{N}_i(t)}U_{rep}^{ij}(\hat{\xi}_i,\hat{\xi}_j)]\le V_0 \tag{27}$$

Clearly, (25) and (26) are contradictory. We can obtain that the collision avoidance is achieved during the flight. Also, the obstacle avoidance can be proved with the same method. In order to save space, the proof process is omitted here.

***Remark 3*** Based on the results, we can obtain that the formation can be achieved with collision/obstacle avoidance under the designed control protocol (16). It should be pointed out that the total energy $V_t$ of UAV system is a finite value and it will change constantly during the flight in practice.

## 4 Simulations

In this section, two cases are provided to demonstrate the approach presented in this paper. A practical example adopted in [22] is considered. Different from the consensus protocol presented in [22], an improved protocol with collision/obstacle avoidance is designed. For simplicity, it is assumed that the elements of weighted adjacency matrix are 0 or 1, and all UAVs move in *XY* plane.

*Case 1*: Consider a system with five UAVs in an obstacle-free environment. Their desired formation structure of regular pentagon and undirected interaction topology are shown in Fig. 3, respectively.

The initial conditions of UAVs are given by Table 1.

The parameters of control protocol (20) are given by $k_1=7.5, k_2=0$, $k_3=2.5, k_4=1.0$. The dynamic constraint of velocity is given by $v_0\in[0,15]$, and the safe distance of UAVs is 10 m. The gain coefficient of attractive potential field and

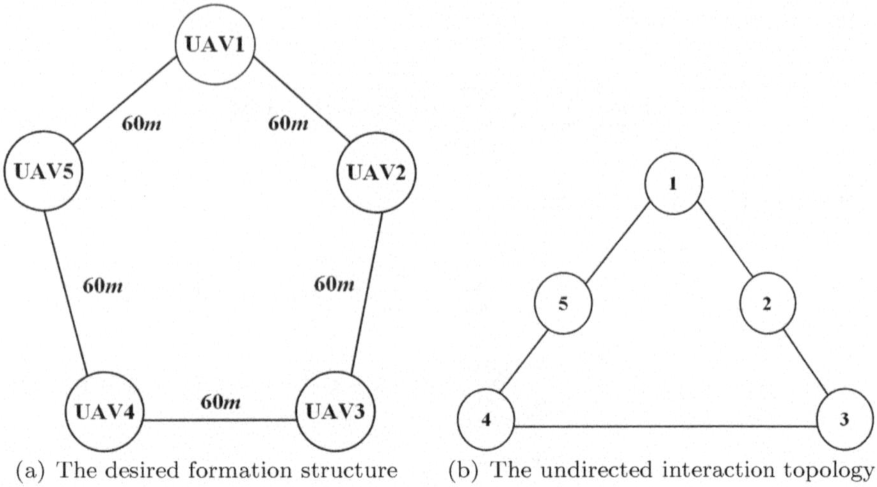


**Fig. 3** The desired formation structure and undirected interaction topology

**Table 1** Initial states of all UAVs

| UAV | $x_i$ (m) | $y_i$ (m) | $v_{xi}$ (m/s) | $v_{yi}$ (m/s) |
|---|---|---|---|---|
| UAV1 | −0.15 | 0.02 | −0.08 | −0.02 |
| UAV2 | −12.25 | −0.06 | 6.35 | −0.06 |
| UAV3 | −12.25 | −0.24 | −0.08 | −0.01 |
| UAV4 | −12.17 | 0.05 | −4.46 | −0.02 |
| UAV5 | −4.71 | −0.08 | −7.1 | 0.04 |

repulsive potential field are chosen as $\varepsilon = 0.03$ and $\eta = 10.2$, respectively. Figure 4a shows the formation trajectories in simulation within 300s, where the initial states of UAVs are denoted by circles. We can see from the figure that all UAVs move along a circle, and at $time = 100$ s, the formation structure of regular pentagon is achieved and maintained with five UAVs. Figure 4b shows the velocity of all UAVs. We can see that when the formation is achieved, the velocities of UAVs are identical. Compared with [22], the presented control protocol (20) has a better consistency. Figure 5a and Fig. 5b show the velocity and position trajectories errors between UAVs and desired formation. It is clear that the velocity error and position error converge to zero finally.

*Case 2*: Consider a UAV system with one leader UAV1 and three followers in an environment with obstacles. Their desired formation structure of square and undirected interaction topology are shown in Fig. 6, respectively.

The initial conditions of the system are given by Table 2

The parameters of control protocol (20) are given by $k_1 = 7.6, k_3 = 2.0, k_2 = 0, k_4 = 1.0$, when the obstacle is detected by UAV, $k_2 = 18$, else $k_2 = 0$. The dynamic constraint of velocity is given by $v_0 \in [0, 15]$, and the safe distance of UAVs is 10 m.

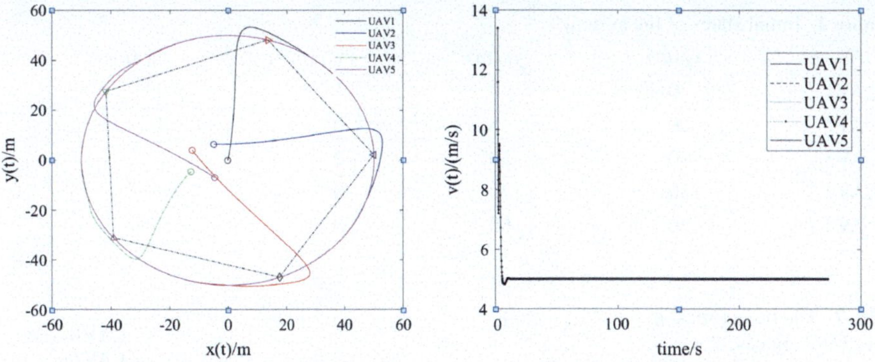

(a) Position trajectories of five UAVs (b) Velocity trajectories of five UAVs

**Fig. 4** Position trajectories and velocity trajectories of five UAVs in case 1

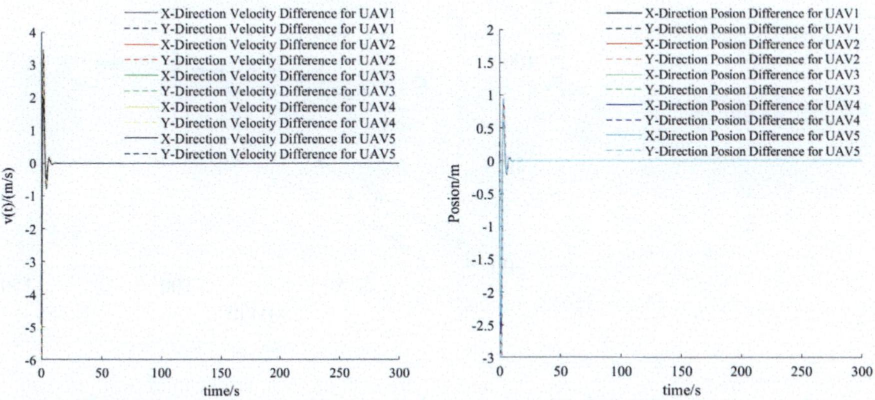

(a) Velocity trajectories errors (b) Position trajectories errors

**Fig. 5** Position trajectories and velocity trajectories of five UAVs in case 1

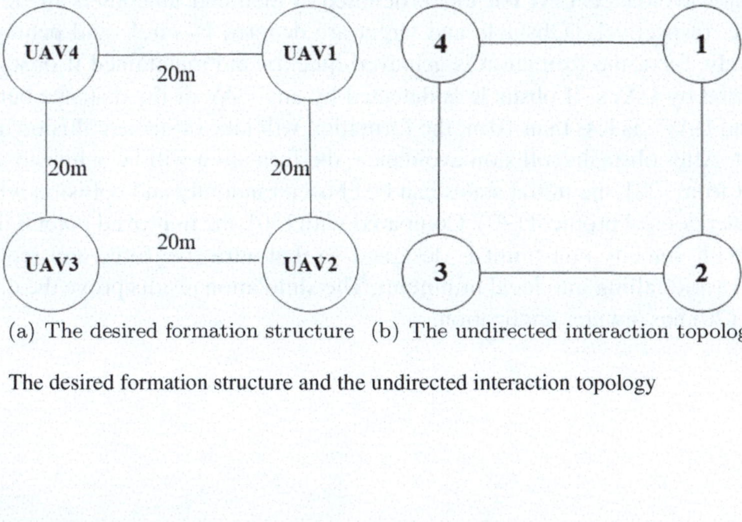

(a) The desired formation structure (b) The undirected interaction topology

**Fig. 6** The desired formation structure and the undirected interaction topology

**Table 2** Initial states of the system

| UAV | $x_i$ (m) | $y_i$ (m) | $v_{xi}$ (m/s) | $v_{yi}$ (m/s) |
|---|---|---|---|---|
| UAV1 | 0 | 0 | 5 | 0 |
| UAV2 | 60 | 5 | 0 | 3 |
| UAV3 | 30 | 50 | 2 | 2 |
| UAV4 | 130 | 120 | 0 | 0 |
| UAV5 | 90 | 80 | 0 | 0 |

**Fig. 7** The trajectories of five UAVs in case 2

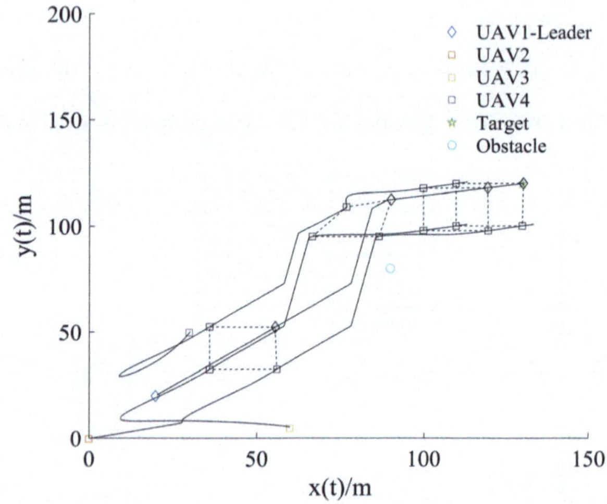


The gain coefficient of attractive potential field and repulsive potential field are chosen as $\varepsilon = 0.03$ and $\eta = 10.2$, respectively.

Figure 7 shows the position trajectories of four UAVs during the flight with obstacle/collision avoidance. UAV1-leader is denoted by diamond, and others are denoted by square, respectively. Obstacle and target are denoted by circle and pentagram, respectively. First, the formation is achieved quickly and maintained if obstacle is not detected by UAVs. If obstacle is detected by any UAV or the distance between UAV $i$ and UAV $j$ is less than 10 m, the formation will take obstacle/collision avoidance first. After obstacle/collision avoidance, the formation will be achieved again. Different from [22], the initial states can be chosen randomly and collision will not occur under control protocol (20). Compared with [30], the improved potential field method with velocity constraint is designed so that attractive force and repulsive force can avoid falling into local minimum. The simulation results prove the control protocol (20) has a wider application.

## 5 Conclusion

In this paper, the leader-follower formation control problem of unmanned air vehicles(UAVs) flying with velocity constraint in an obstacle-laden environment has been studied. The linear control protocol is designed to achieve formation with obstacle/collision avoidance. The stability conditions of the consensus algorithm are analyzed by using a Lyapunov theory and LaSalle's principle. Finally, two applications have been given to show that the presented control protocol is validated and effective for formation flying with obstacle/collision avoidance and velocity constraint.

Future work will include realizing obstacle/collision avoidance in 3-D plane with time delay, and the directed communication topology graph will be taken into consideration.

## References


1. Olfati-Saber R, Fax JA, Murray RM (2007) Consensus and cooperation in networked multi-agent systems. Proc IEEE 95(1):215–233
2. Ma L, Wang Z, Han Q-L (2017) Consensus control of stochastic multi-agent systems: a survey. Sci China Inf Sci 60(12):1
3. Fax JA, Murray RM (2004) Information flow and cooperative control of vehicle formations. IEEE Trans Autom Control 49(9):1465–1476
4. Chen H, Fei J (2018) UAV path planning based on particle swarm optimization with global best path competition. Int J Pattern Recognit Artif Intell **32**
5. Casbeer DW, Beard RW, McLain TW, Li SM, Mehra RK (2005) Forest fire monitoring with multiple small UAVs. In: Proceedings of the American Control Conference. IEEE, pp 3530–3535
6. Kim J, Hespanha JP (2004) Cooperative radar jamming for groups of unmanned air vehicles. In Proc CDC pp 632–637
7. Jadbabaie A, Lin J, Morse AS (2003) Coordination of groups of mobile autonomous agents using nearest neighbor rules. IEEE Trans Autom Control 48(6):988–1001
8. Olfati-saber R (2006) Flocking for multi-agent dynamics systems: algorithms and theory. IEEE Trans Autom Control 51(3), 401–420
9. Wang J, Wu H (2012) Leader-following formation control of multi-agent systems under fixed and switching topologies. Int J Control 85:695–705
10. Edwards DB, Bean TA, Odell DL, Anderson MJ (2004) Leader-follower algorithm for multiple AUV formations. In: Proceeding IEEE/OES Auton Underwater Vehicles, pp 40–46
11. Balch T, Arkin RC (1998) Behavior-based formation control for multi-robot teams. IEEE Trans Robot Autom 14:1–15
12. Brunete A, Hernando M, Gambao E, Torres JE (2012) A behaviour-based control architecture for heterogeneous modular, multi-configurable, chained microrobots. Robot Auton Syst 60:1607–1624
13. Seuret A, Dimarogonas DV, Johansson KH (2008) Consensus under communication delay. In: Proceedings of the 47th IEEE Conference on Decision Control, pp 4922–4927
14. Tian Y-P, Liu C-L (2008) Consensus of multi-agent systems with diverse input and communication delays. IEEE Trans Autom Control 53(9):2122–2128
15. Guzey HM, Dierks T, Jagannathan S, Acar L (2019) Modified consensus-based output feedback control of quadrotor UAV formations using neural networks. J Intell Robot Syst 94(1):283–300

16. Ren W (2007) Consensus strategies for cooperative control of vehicle formations. IET Control Theory Appl 1(2):505–512
17. Lawton JR, Beard RW (2002) Synchronized multiple spacecraft rotations. Automatica 38(8):1359–1364
18. Bauso D, Giarre L, Pesenii R (2003) Attitude alignment of a Team of UAVs under decentralized information structure. In: IEEE Conference on Control Applications, pp 486–491
19. Li Z, Wen G, Duan Z, Ren W (2015) Designing fully distributed consensus protocols for linear multi-agent systems with directed graphs. IEEE Trans Autom Control 60(4):1152–1157
20. Feng T, Fan W, Tang J, Zeng W (2019) Consensus-based robust clustering and leader election algorithm for homogeneous UAV clusters. J Phys: Conf Ser 93(1–2):213–226
21. Ren W, Beard RW (2005) Consensus seeking in multiagent systems under dynamically changing interaction topologies. IEEE Trans Autom Control 50(5):655–661
22. Dong X, Yu B, Shi Z, Zhong Y (2015) Time-varying formation control for unmanned aerial vehicles: theories and applications. IEEE Trans Control Syst Technol 23(1):340–348
23. Tanner HG, Jadbabaie A, Pappas GJ (2003) Stable flocking of mobile agents, Part i: fixed topology. In: Proceedings of the IEEE Conference on Decision and Control, pp 2010–2015
24. Kuriki Y, Namerikawa T (2013) Formation control of UAVs with a fourth-order flight dynamics. Decision and Control (CDC). In: IEEE 52nd Annual Conference, pp 6706–6711
25. Olfati-Saber R, Murray RM (2004) Consensus problems in networks of agents with switching topology and time-delays. IEEE Trans Autom Control 49(9):1520–1533
26. Wang X, Yadav V, Balakrishnan SN (2007) Cooperative UAV formation flying with obstacle/collision avoidance. IEEE Trans Control Syst Technol 15(4):672–679
27. Kuriki Y, Namerikawa T (2014) Consensus-based cooperative formation control with collision avoidance for a multi-UAV system. In: 2014 American Control Conference (ACC)
28. Hassan K (2001) Khalil. Prentice Hall, Nonlinear Systems
29. Yu X, Liu L (2016) Senior Member, IEEE, Distributed formation control of nonholonomic vehicles subject to velocity constraints. IEEE Trans, Industrial Electronisc 63(2)
30. Koren Y, Borenstein J (1991) Potential field methods and their inherent limitations for mobile robot navigation Proceedings. In: 1991 IEEE international conference on robotics and automation, vol 2, pp 1398–1404

# New Model and Analytical Review of Approaches to Buckling Problem Investigation of Structurally Anisotropic Aircraft Panels Made from Composite Materials

**L. M. Gavva**

**Abstract** The different approaches were analyzed to investigate the buckling problems of structurally anisotropic panels made from composite materials. One considered the studies of scientific schools from 2000 year to present time, mainly. The classification of mathematical models, analytical methods of calculations, numerical methods of calculations, and testing results is presented in this review. Aircraft composite structure design in the field of production technology is the outlook research trend. New mathematical model relations for the buckling investigation of structurally anisotropic panels comprising composite materials are presented in this study. The primary scientific novelty of this research is the further development of the theory of thin-walled elastic ribs related to the contact problem for the skin and the rib with an improved rib model. One considers the residual thermal stresses and the preliminary tension of the reinforcing fibers with respect to panel production technology. The mathematical model relations for the pre-critical stress state investigation of structurally anisotropic panels made of composite materials are presented. Furthermore, the mathematical model relations for the buckling problem investigation of structurally anisotropic panels made of composite materials are presented in view of the pre-critical stress state. The critical force definition of the general bending mode of the thin-walled system buckling and the critical force definition of the multi-wave torsion buckling are of the most interest in accordance with traditional design practices. In both cases, bending is integral with the plane stress state. Thus, the buckling problem results in the boundary value problem when solving for the eighth-order partial derivative equation in the rectangular field. The schematization of the panel as structurally anisotropic has been proposed as a design model when the critical forces of total bending mode of buckling are determined. For a multi-wave torsion buckling study, one should use the generalized function set. The solution is designed by a double trigonometric series and by a unitary trigonometric series. A computer program package is developed using the MATLAB operating environment. The computer program package has been utilized for multi-criteria optimization of the design

L. M. Gavva (✉)
Department of Aircraft Design, Department of Machine Components,
Moscow Aviation Institute (NRU), Moscow, Russia
e-mail: rva101@mail.ru

 
Z. Jing (ed.), *Proceedings of the International Conference on Aerospace System Science and Engineering 2019*, Lecture Notes in Electrical Engineering 622,
https://doi.org/10.1007/978-981-15-1773-0_12

of structurally anisotropic aircraft composite panels. The influence of the structure parameters on the level of critical buckling forces for bending and for torsion modes has been analyzed. The results of testing series are presented. The results of new calculations are presented.

**Keywords** Review · Panels made of composite materials · Eccentric longitudinal and lateral set · Thin-walled rib · Non-symmetric package structure · Force and technology temperature action · Pre-critical stress state · Buckling · Bending mode · Torsion mode

## 1 Introduction

A new design problem—the design to cost—is possible to be solved in combining the high-precision models and modern computer technologies and decreasing the test amount. A computer program package in MATLAB was performed for the investigation and multi-criteria optimization of the buckling of structurally anisotropic composite panels of FA.

The buckling problems of a flat rectangular multilayer panel made from polymer fiber composite materials with the eccentric longitudinal and lateral stiffening set are considered. The buckling problems of a flat rectangular composite panel being anisotropic due to non-symmetric package structure over the thickness are also discussed. The panels are subjected to the distributed constant compressive loading applied to the edges in the casing plane in the stationary temperature field. The boundary conditions at the contour are assumed to be the particular case with conformable boundary restrictions for the plane problem and problem of bending, and the solution in closed form is designed by a trigonometric series.

One should take into consideration the technological factors occurring in the fabrication of composites, namely residual thermal stresses arising during cooling after hardening and pre-stressed tension of reinforcing fibers that is performed in order to increase the bearing strength of the structure [1].

The refined statement of the buckling problems has been formulated subject to the pre-critical stress state in the compression of flat rectangular multiplied panels made of polymer fiber composite materials, the casing of which is eccentrically supported by the longitudinal lateral stiffening set.

The buckling problem investigation of flat rectangular structurally anisotropic composite panels is relevant for the design of the bearing surfaces of FA. The majority of the authors of the theoretical studies during the latest twenty years are paying due attention to the features of the composite panel deformation behavior. The classification and survey of the main directions of the buckling theory development of the structurally anisotropic composite panels, the proposed mathematical models, and equations are of the most interest.

**Buckling Problem Statement**:
Setoodeh and Karami [2], Gangadhara [3], Mittelstedt and Schroder [4], Yshii et al. [5], Ragb and Matbuly [6], Castro and Donadon [7].

**Buckling Problem Statement with Thermal and Force Loading**:
Shukla and Nath [8], Chen et al. [9], Matsunaga [10], Cetkovic [11], Cetkovic and Gyorgy [12], Kettaf et al. [13], Naik and Sayyad [14].

**Buckling Problem Statement subject to Production Technology**:
Chen et al. [15].

**Analytical Methods to solve Buckling Problems**:
Pandey et al. [16], Vescovini and Dozio [17], Kazemi [18], Yeter et al. [19], Abramovich et al. [20].

**Numerical Methods to solve Buckling Problems**:
Huang et al. [21], Guo et al. [22], Thankam et al. [23], Tenek [24], Tran et al. [25], Kumar et al. [26], Kumar et al. [27], Zarei and Khosravifard [28], Castro et al. [29].

**Test Investigations**:
Falzon et al. [30], Park et al. [31], Rouse and Assadi [32], Ungbyfkorn and Singhatanadgid [33], Baker [34], Zhao et al. [35], Bai et al. [36], Kumar et al. [37], Sanches et al. [38].

The critical force calculations for the general bending mode of the thin-walled system buckling and the critical force calculations for the multi-wave torsion buckling are of the most actual interest in accordance with traditional design practices [1]. General bending buckling is characterized by the less number of semi-waves than the number of stringers. Multi-wave torsion buckling is characterized by skin bending between cross nodes and rib rotation without deformation of the profile cross section. In both cases, bending is integral with the plane stress state. The schematization of the panel as structurally anisotropic has been proposed as a design model when the stress–strain state and critical forces of total bending mode of buckling are determined. For a multi-wave torsion buckling study, one should use the generalized function techniques.

New mathematical model relations for the pre-critical stress–strain state and for the buckling problem investigation of structurally anisotropic panels comprising composite materials are presented in this study. The mathematical model of a stiffening rib being torsioned under one-sided contact with the skin is refined. The scientific novelty of this research reflects the further development of the theory of thin-walled elastic ribs related to the contact problem for the skin and rib with an improved rib torsion model. The aim of this study is the buckling problem statement and the approach to solve this problem in view of the non-uniform pre-critical stress state and production technology. The buckling problem results in the boundary value problem when solving for the eighth-order partial derivative equation in the rectangular field. The buckling problem statement and the proposed approach to solve this problem are new and are of the interest from the design and manufacture of aircraft outlook specimen made from modern composite materials.

## 2 Buckling of Structurally Anisotropic Composite Panels—Problem Statement

Both cases, the buckling problem as the problem of the pre-critical stress state, are integral, and they cannot be divided into a plane part and a plate bending.

The eighth-order differential equation is resolved for the buckling problems (e.g., [39]). This equation is designed as an equilibrium differential equation with the effect of the reduced loading connected with the normal forces $N_x$, $N_y$ and tangential forces $N_{xy}$, $N_{yx}$

$$\sum_{i=0,1,2\ldots}^{8} K_{8-i,i}\frac{\partial^8\Phi}{\partial x^{8-i}\partial y^i} = N_x\frac{\partial^2 w}{\partial x^2} + \left(N_{xy} + N_{yx}\right)\frac{\partial^2 w}{\partial x\partial y} + N_y\frac{\partial^2 w}{\partial y^2} \quad (1)$$

The deflection function $w(x, y)$ is coupled with the potential function $\Phi(x, y)$ as

$$w = \left(R_{40}\frac{\partial^4}{\partial x^4} + R_{31}\frac{\partial^4}{\partial x^3\partial y} + R_{22}\frac{\partial^4}{\partial x^2\partial y^2} + R_{13}\frac{\partial^4}{\partial x\partial y^3} + R_{04}\frac{\partial^4}{\partial y^4}\right)\Phi \quad (2)$$

The coefficients $R_{ij}$, $i = 4,3,\ldots,0$, $j = 0,1,\ldots, 4$ in the relation formulas (2) and the coefficients $K_{ij}$, $i = 8,7,\ldots,0$, $j = 0,1,\ldots, 8$ in the resolving Eq. (1) are the constant values, which depend on the elastic characteristics of the material and geometrical structure parameters.

For the composite panel of orthotropic structure, the left-hand part of the eighth-order differential equation is limited by the even-numbered partial derivatives of $\Phi(x, y)$ but the odd-numbered partial derivatives in the right-hand part are connected with the shear

$$\frac{K_{80}}{a^8}\frac{\partial^8\Phi}{\partial x^8} + \frac{K_{62}}{a^6b^2}\frac{\partial^8\Phi}{\partial x^6\partial y^2} + \frac{K_{44}}{a^4b^4}\frac{\partial^8\Phi}{\partial x^4\partial y^4} + \frac{K_{26}}{a^2b^6}\frac{\partial^8\Phi}{\partial x^2\partial y^6} + \frac{K_{08}}{b^8}\frac{\partial^8\Phi}{\partial y^8}$$
$$= \begin{bmatrix} \frac{N_x R_{40}}{a^6}\frac{\partial^6\Phi}{\partial x^6} & + \frac{(N_{xy}+N_{yx})R_{40}}{a^5 b}\frac{\partial^6\Phi}{\partial x^5\partial y} + \\ + \frac{(N_x R_{22}+N_y R_{40})}{a^4b^2}\frac{\partial^6\Phi}{\partial x^4\partial y^2} & + \frac{(N_{xy}+N_{yx})R_{22}}{a^3b^3}\frac{\partial^6\Phi}{\partial x^3\partial y^3} + \\ + \frac{(N_x R_{04}+N_y R_{22})}{a^2b^4}\frac{\partial^6\Phi}{\partial x^2\partial y^4} & + \frac{(N_{xy}+N_{yx})R_{04}}{ab^5}\frac{\partial^6\Phi}{\partial x\partial y^5} + \\ + \frac{N_y R_{04}}{b^6}\frac{\partial^6\Phi}{\partial y^6} & \end{bmatrix}, \quad (3)$$

$x = x/a$, $y = y/b$ are the dimensionless coordinates related to the panel length $a$ and to its width $b$.

All components of the stress–strain state including the inner force factors are related with the potential function $\Phi(x, y)$ as

$$\left.\begin{array}{ll} N_x = L_{Nx}\Phi - N_x^T - N_x^H, & N_y = L_{Ny}\Phi - N_y^T - N_y^H \\ N_{xy} = L_{Nxy}\Phi - N_{xy}^T - N_{xy}^H, & N_{yx} = L_{Nyx}\Phi - N_{yx}^T - N_{yx}^H \end{array}\right\} \tag{4}$$

The linear differential operator for the orthotropic structure, for example

$$L_{Nx} = P_{60}^x \frac{\partial^6}{\partial x^6} + P_{42}^x \frac{\partial^6}{\partial x^4 \partial y^2} + P_{24}^x \frac{\partial^6}{\partial x^2 \partial y^4} + P_{06}^x \frac{\partial^6}{\partial y^6},$$

The thermal forces and moments are $N_x^T$, $N_y^T$, $N_{xy}^T$, $N_{yx}^T$, and the tension forces and moments of the composite fibers are $N_x^H$, $N_y^H$, $N_{xy}^H$, $N_{yx}^H$. The coefficients $P_{ij}^x$, $i = 6, 4, 2, 0$, $j = 0, 2, 4, 6$ as the coefficients in the relation formulas (2) depend on the elastic characteristics of the material and geometrical structure parameters.

The buckling problem for a structurally anisotropic composite panel is a nonlinear one according to Eq. (3) and connection formulas (4). The linearization method is used to determine the critical forces for the approximate solution. First, one considers the stress–strain state of the structure at compression, namely the pre-critical main stress state being complicated as it is not divided into the plane problem and bending problem according to the mathematical model proposed. It is necessary to determine the distribution law of the normal and shear inner forces caused by the external loading. Then, one considers the buckling problem as the proper value problem to obtain the additional displacement of the basis surface.

## 3 Pre-critical Stress–Strain State of Structurally Anisotropic Composite Panels in Compression

We consider the pre-buckling stress state of a flat rectangular composite panel with the eccentric stiffening set being orthotropic one. The panel is subjected to the uniform distributed normal compressive loading with $P$ intensity applied to the lateral opposite sides in the skin plane. Boundary restrictions satisfy the hinging condition in respect to bending and the sliding constraint condition in the tangential direction for the plane problem when the panel is loaded by the shear force flows along its longitudinal edges.

First, it is necessary to determine the distribution of the normal forces $N_x$ caused by the external loading with the coordinates $x$ and $y$. The compressive loading is applied at two opposite sides in the longitudinal direction. It is of interest to compare the critical parameter results with the uniform pre-critical stress state and with the variable pre-critical stress state. Then, one does not consider the distribution of inner forces $N_y$, $N_{xy}$, $N_{yx}$.

Equation (3) is uniform

$$\frac{K_{80}}{a^8}\frac{\partial^8\Phi}{\partial x^8}+\frac{K_{62}}{a^6b^2}\frac{\partial^8\Phi}{\partial x^6\partial y^2}+\frac{K_{44}}{a^4b^4}\frac{\partial^8\Phi}{\partial x^4\partial y^4}+\frac{K_{26}}{a^2b^6}\frac{\partial^8\Phi}{\partial x^2\partial y^6}+\frac{K_{08}}{b^8}\frac{\partial^8\Phi}{\partial y^8}=0. \quad (5)$$

The boundary restrictions are

$$\left.\begin{array}{l} x=\pm 1 \quad w=M_x=v_0=0, \quad N_x=-P \\ y=0\ y=1 \quad w=M_y=u_0=N_y=0 \end{array}\right\} \quad (6)$$

The integral of Eq. (5) that satisfies the edge conditions (6) is presented by the single trigonometric series. The boundary value problem is symmetric with respect to the coordinate $x$, the solution contains only even-numbered functions, and the generalized displacement function $\Phi(x, y)$ is determined by four unknown constants

$$\Phi(x,y)=\sum_{i=1,3,5}^{\infty}\left[\sum_{L=1}^{4}B_{iL}ch(\lambda_{iL}x)\right]\sin(i\pi y) \quad (7)$$

Here, $\lambda_{iL}=z_L\lambda_{iy}a, \lambda_{iy}=\frac{i\pi}{b}$, $z_L$ are the roots of the corresponding characteristic polynomial and are calculated using the MATLAB operating environment.

The vector displacement components, the strains and the curves of the basis surface, the rotation angels and the inner force factors are derived by single trigonometric series accurate within the constants $B_{iL}$ (7). The thermal and tension forces and moments are also derived by single trigonometric series. The symmetric SSS components contain only even-numbered functions with respect to the coordinate $x$; the oblique symmetric components consist of odd-numbered functions. The linear differential operators of the relations between the SSS kinematic and static components and the resolving potential function $\Phi(x, y)$ are known.

It is necessary to satisfy the face boundary conditions of the structure with $x=+1$ to determine the unknown constants $B_{iL}$ ($i = 1, 3, 5,\ldots, L = 1, 2, 3, 4,$) and SSS components.

The normal forces $N_x$, corresponding to the pre-critical stress state of the stiffened composite panel compressed along the $x$-axis, are distributed as

$$N_x=P\sum_{i=1,3,5\ldots}^{\infty}\left\{\left[\sum_{L=1}^{4}(N_x)_{iL}ch(\lambda_{iL}x)\right]-\frac{\left(N_x^T\right)_i}{P}-\frac{\left(N_x^H\right)_i}{P}\right\}\sin(i\pi y) \quad (8)$$

$(N_x)_{iL}$ are the coefficients of single trigonometric series for the normal forces $N_x$, and known after the determination of the constants $B_{iL}$, $\left(N_x^T\right)_i$ and $\left(N_x^H\right)_i$ are the coefficients of single trigonometric series for the thermal and tension forces.

The variations of the transverse forces $N_y$ and shear forces $N_{xy}$, $N_{yx}$ are obtained analog to (8).

One can estimate the influence of production technology factors on the bearing strength of structurally anisotropic composite panels if the non-uniform pre-critical stressed state is considered, boundary conditions are non-conformable as (6), and the solution is formed by a unitary trigonometric series.

## 4 Pre-critical Stress–Strain State of Structurally Anisotropic Composite Panels in Compression—Numerical Results

A computer software package was developed in the MATLAB operating environment according to the algorithm presented. Computer programs are intended for the pre-critical stress state analysis and the computer multi-criteria optimization of the design for structurally anisotropic aircraft composite panels.

As an example, Fig. 1 presents the epure of the normal forces $N_x$ of a flat rectangular panels made from carbon plastic eccentrically stiffened in the longitudinal

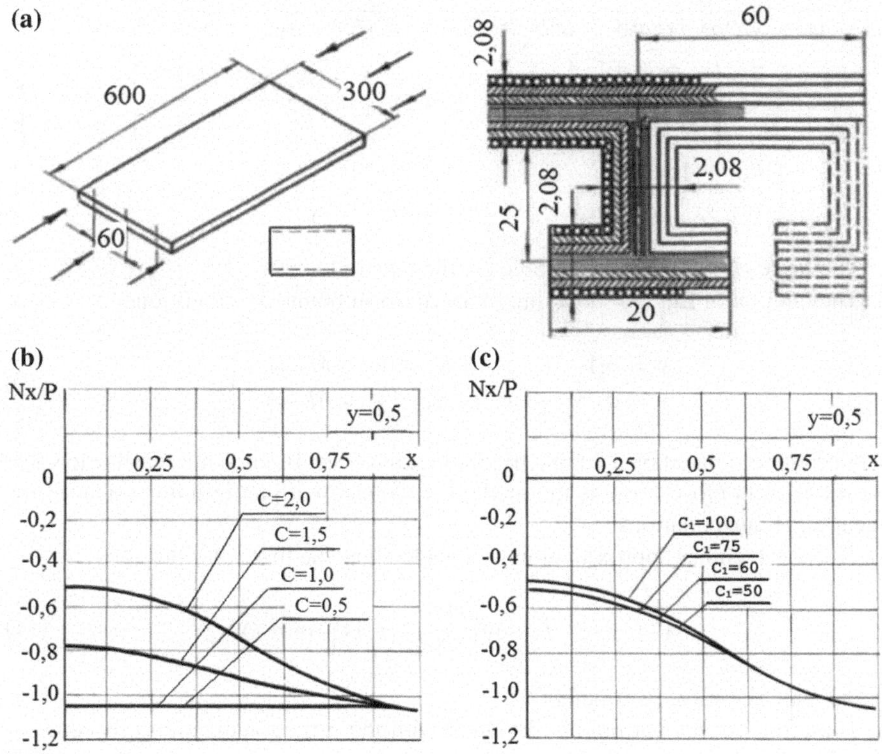


**Fig. 1** Stringer panel compressed in the longitudinal direction: **a** panel sizes, layout of the package and stringer geometry; **b** dependence of normal forces on panel side ratio under in-plane bending; **c** dependence of normal forces on stringer distance under in-plane bending

direction. The panels are under the action of uniform compressive loading per unit length.

The shorter skin panel and the higher stringer stiffness, the pre-critical stress state is closer to uniform. If the panel side ratio c < 1, 0, the inner normal forces are distributed along the panel length almost uniformly.

If the panel side ratio c = 2, 0, the distribution of the normal forces per panel length under in-plane bending does not practically depend on the stringer distance.

## 5 Buckling of Structurally Anisotropic Composite Panels Subject to Pre-critical Stress State

The linear buckling problem is formulated as the proper value problem.

The general differential equation of the curved surface (3) determines the additional equilibrium state of the structure in view of the initial pre-buckling state. With the relation formulas (2) and normal force distribution (8), it is written as

$$\frac{K_{80}}{a^8}\frac{\partial^8\Phi}{\partial x^8}+\frac{K_{62}}{a^6b^2}\frac{\partial^8\Phi}{\partial x^6\partial y^2}+\frac{K_{44}}{a^4b^4}\frac{\partial^8\Phi}{\partial x^4\partial y^4}+\frac{K_{26}}{a^2b^6}\frac{\partial^8\Phi}{\partial x^2\partial y^6}+\frac{K_{08}}{b^8}\frac{\partial^8\Phi}{\partial y^8}$$
$$=P\left\{\sum_{i=1,3,5...}^{\infty}\left[\left[\sum_{L=1}^{4}(N_x)_{iL}ch(\lambda_{iL}x)\right]-\frac{(N_x^T)_i}{P}-\frac{(N_x^H)_i}{P}\right]\sin(i\pi y)\right\}$$
$$\left[\frac{R_{40}}{a^6}\frac{\partial^6\Phi}{\partial x^6}+\frac{R_{22}}{a^4b^2}\frac{\partial^6\Phi}{\partial x^4\partial y^2}+\frac{R_{04}}{a^2b^4}\frac{\partial^6\Phi}{\partial x^2\partial y^4}\right] \quad (9)$$

The forces $N_x$ (8) are the variable coefficients in Eq. (9).

The integral of Eq. (9) satisfying to the uniform boundary conditions

$$\left.\begin{array}{ll} x=\pm 1 & v_0=N_x=w=M_x=0 \\ y=0 \;\; y=1 & u_0=N_y=w=M_y=0 \end{array}\right\}, \quad (10)$$

may be approximated by a double trigonometric series. But it is not feasible to solve the problem in close form as the method is reduced with an infinite system of the linear algebraic equations.

The one-member approximation is considered as the first solution

$$\Phi(x,y)=f_{mn}\sin\left[\frac{m\pi}{2}(x+1)\right]\sin(n\pi y), \quad (11)$$

$m$ and $n$ are the wave parameters.

The critical force formula is designed with the orthogonalization procedure of Eq. (9) subject to the solution (11). The expression

$$P = P^* \left\{ \frac{\pi}{\sum\limits_{i=1,3,5...}^{\infty} \left( -\frac{2}{i} - \frac{1}{2n-i} + \frac{1}{2n+i} \right) \left\{ \left[ \sum\limits_{L=1}^{4} (N_x)_{iL} \frac{\left[(m\pi)^2 - \lambda_{iL}^2\right]}{\left[(m\pi)^2 + \lambda_{iL}^2\right]} \frac{sh\lambda_{iL}}{\lambda_{iL}} \right] - \frac{\left(N_x^T\right)_i}{P} - \frac{\left(N_x^H\right)_i}{P} \right\}} \right\} \tag{12}$$

provides the range of values $P$ for additional possible deformation of the base surface at $m = 1, 2, 3,\ldots, n = 1, 2, 3,\ldots$

The $P$ with «*» is the critical force calculated with the main uniform pre-buckling stress state (e.g., [39])

$$P^* = \frac{\pi^2}{b^2} \frac{K_{80}\left(\frac{m}{c}\right)^8 + K_{62}\left(\frac{m}{c}\right)^6 n^2 + K_{44}\left(\frac{m}{c}\right)^4 n^4 + K_{26}\left(\frac{m}{c}\right)^2 n^6 + K_{08} n^8}{\left[R_{40}\left(\frac{m}{c}\right)^4 + R_{22}\left(\frac{m}{c}\right)^2 n^2 + R_{04} n^4\right]\left(\frac{m}{c}\right)^2},$$

and for the general bending mode of buckling, the panel side ratio $c = 2a/b$, where $a$ and $b$ are the panel half-length and width, correspondingly.

The formula for the critical loading $P$ of the multi-wave torsion buckling problem coincides to the formula (12) within the coefficients $K_{ij}$ and $R_{ij}$. The coefficients $\hat{K}_{ij}, i, j = 0, 2, 4, 6, 8$ and $\hat{R}_{ij}, i, j = 0, 2, 4$ are determined by generalized stiffness characteristics while stiffness averaging for the elements of longitudinal setup to the skin is replaced with the discrete characteristics

$$\frac{1}{c_1} \to \left\{ \begin{array}{l} \frac{2}{b} \sum\limits_{i=1}^{N} \sin^2(n\pi y_i) \\ \frac{2}{b} \sum\limits_{i=1}^{N} \cos^2(n\pi y_i) \end{array} \right. \to \hat{K}_{ij},\ \hat{R}_{ij},$$

$c_1$ is the stringer distance; $y_i$ is the coordinate $y$ of the discrete stringer.

It is allowable to estimate the influence of production technology factors on the bearing strength of the structurally anisotropic composite panels if the complicated pre-critical stress state is considered and the solution is derived in a single approximation of the trigonometric series (11). One considers the residual thermal stresses arising during cooling after hardening and the pre-stressed tension of the reinforcing fibers with respect to panel production technology.

The step-by-step method is used to determine the critical forces. The critical force $P$ with «*» calculated with the uniform pre-buckling stress state is proposed as an initial first approach.

## 6 Buckling of Structurally Anisotropic Composite Panels Subject to Pre-critical Stress State—Results and Discussion

A computer program package is developed using the MATLAB operating environment. The computer program package has been utilized for the determination of the critical forces subject to the uniform and non-uniform pre-buckling stress state and for multi-criteria optimization of the design of structurally anisotropic aircraft composite panels.

The results of determining the critical parameters for rectangular carbon plastic panels eccentrically stiffened and compressed in the longitudinal direction (Fig. 1a) are presented in Fig. 2 as an example in view of the uniform pre-buckling stress state.

For a short panel, if $c < 2, 0$, the multi-wave torsion buckling occurs: $n = 6$, $m = 2, 4, 5, 7, 11$. If $c = 2, 0$, the panel is equally buckled. The buckling for panels with $c > 2{,}0$ has the general bending mode $n = 1$, $m = 1$.

The shorter stringer distance $c_1 < 60$ mm, the higher is probability for the general bending buckling mode $n = 1$, $m = 1$. If $c_1 = 60$ mm, the panel is equally buckled. While the stringer distance $c_1 > 60$ mm, the panel is characterized by torsion buckling mode $n = 4, 5, 6$, $m = 4, 6, 7$.

The lower bending stiffness of the stringer, if the height of stringer cross section wall $c_{4x} < 25$ mm, the higher is probability for the general bending buckling mode $n = 1$, $m = 1$. If $c_{4x} = 25$ mm, the panel is equally buckled. While the height of stringer cross section wall $c_{4x} > 25$ mm, the panel is characterized by torsion buckling mode $n = 6$, $m = 7$.

The testing series of uniform compressed stiffened composite panels for carrying the objects to the moment of stability loss (Fig. 3) has been made using the special fixture for the proposed mathematical model verification [1].

The refined theoretical results and the experimental data are in agreement qualitatively with respect to buckling modes and quantitatively with respect to critical

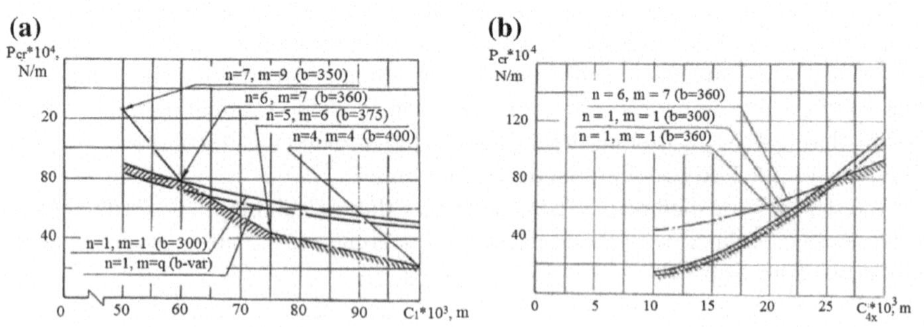


**Fig. 2** Stringer panel compressed in the longitudinal direction: dependence of critical forces of bending mode and torsion mode of buckling **a** on stringer distance; **b** on height of stringer cross-sectional wall (———structurally anisotropic model; – · – · – · discrete ribs)

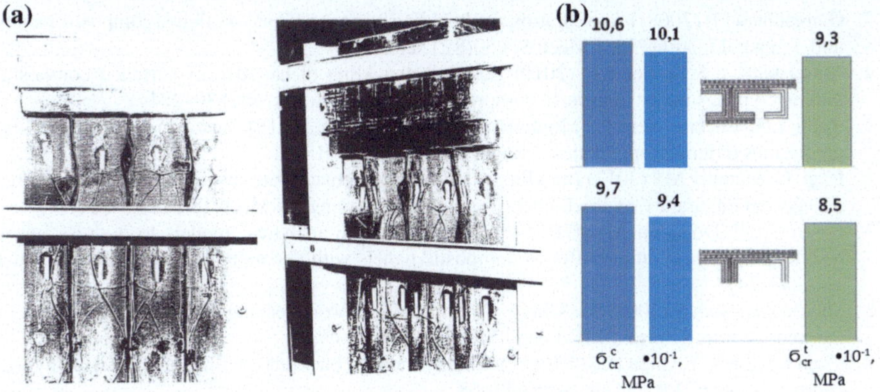


**Fig. 3** Stringer panels compressed in the longitudinal direction: **a** experimental specimen; **b** comparison of the test and calculation results for critical stresses

stresses within 12–13% if pre-buckling stress state is considered uniform. And they reveal a higher precision of 8–10% in view of non-uniform main pre-buckling stress state. Thus, it confirms the authenticity of the presented mathematical model.

## 7 Conclusion

Since the solution is obtained by analytical methods, the calculation time is minimal. This is of interest from the perspective of practical design using parametric analysis. To obtain more accurate data for the critical buckling forces, it is necessary to consider buckling of structurally anisotropic composite panels subject to the non-uniform pre-critical stress state. The results of the stress analysis calculations, as well as the results of the buckling analysis calculations, offer opportunities for reducing and optimizing the weight characteristics of aircraft elements.

**Acknowledgements** The study was performed in the framework of the RFFI (*the project № 17-08-00849/17*).

## References

1. Gavva LM, Endogur AI (2018) Statics and buckling problems of aircraft structurally-anisotropic composite panels with the influence of production technology. IOP Conf Ser: Mater Sci Eng 312(1):012009
2. Setoodeh AR, Karami G (2003) A solution for the vibration and buckling of composite laminates with elastically restrained edges. Compos Struct 60(3):245–253

3. Gangadhara PB (2008) Free vibration and buckling response of hat—stiffened composite panels under general loading. Int J Mech Sci 50(8):1326–1333
4. Mittelstedt C, Schroder KU (2010) Local postbuckling of hat-stringer stiffened composite laminated plates under transverse compression. Compos Struct 92:2830–2844
5. Yshii LN, Lucena Neto E, Monteiro FAC, Santana RC (2018) Accuracy of the buckling predictions of anisotropic plates. J Eng Mech 144(8):04018061
6. Ragb O, Matbuly MS (2017) Buckling analysis of composite plates using moving least squares differential quadrature method. Int J Comput Methods Eng Sci Mech 18(6):292–301
7. Castro SGP, Donadon MV (2017) Assembly of semi-analytical models to address linear buckling and vibration of stiffened composite panels with debonding defect. Compos Struct 160:232–247
8. Shukla KK, Nath Y (2002) Bucklingg of laminated composite rectangular plates under transient thermal loading. Trans ASME J Appl Mech 69(5):684–692
9. Chen CS, Lin CY, Chen RD (2011) Thermally induced buckling of functionally grated hybrid composite plates. Int J Mech Sci 53:51–58
10. Matsunaga H (2005) Thermal buckling of cross-ply laminated composite and sandwich plates according to a global higher-order deformation theory. Compos Struct 68(4):439–454
11. Cetkovic M (2016) Thermal buckling of laminated composite plates using layerwise displacement model. Compos Struct 142:238–253
12. Cetkovic M, Gyorgy L (2016) Thermo-elastic stability of angle-ply laminates-application of layerwise finite elements. Struct Integrity Life 16(1):43–48
13. Kettaf FZ, Benguediab M, Tounsi A (2015) Analytical study of buckling of hybrid multilayer plates. Periodica Polytech Mech Eng 59(4):164–168
14. Naik NS, Sayyad AS (2019) An accurate computational model for thermal analysis of laminated composite and sandwich plates. J Therm Stresses 42(5):559–579
15. Chen X, Dai S, Xu K (2001) Qinghua daxue xuebao. Ziran kexue ban 41(2):77–79, 83
16. Pandey R, Shukla KK, Jain A (2009) Thermoelastic stability analysis of laminated composite plates: an analytical approach. Commun Nonlinear Sci Numer Simul 14(4):1679–1699
17. Vescovini R, Dozio L (2015) Exact refined buckling solutions for laminated plates under uniaxial and biaxial loads. Compos Struct 127:356–368
18. Kazemi M (2015) A new semi-analytical solution for buckling analysis of laminated plates under biaxial compression. Arch Appl Mech 85:1667–1677
19. Yeter E, Erklig A, Bulut M (2014) Hybridization effects on the bucking behavior of laminated composite plates. Compos Struct 118:19–27
20. Abramovich H, Weller T, Bisagni C (2008) Buckling behavior of composite laminated stiffened panels under combined shear and axial compression. J Aircr 45(2):402–413
21. Huang L, Sheikh AH, Ng CT, Griffith MC (2015) An efficient finite element model for buckling analysis of grid stiffened laminated composite plates. Compos Struct 122:41–50
22. Guo MW, Harik IE, Ren WX (2002) Buckling behavior of stiffened laminated plates. Int J Solid Struct 39(11):3039–3055
23. Thankam VS, Singh G, Rao GV, Rath AK (2003) Thermal post-buckling behavior of laminated plates using a shear-flexible element based on coupled-displacement field. Compos Struct 59(3):351–359
24. Tenek LT (2001) Postbuckling of thermally stressed composite plates. AIAA J 39(3):546–548
25. Tran LV, Wahab MA, Kim SE (2017) An isogeometric finite element approach for thermal bending and buckling analyses of laminated composite plates. Compos Struct 179:35–39
26. Kumar S, Kumar R, Mandal S, Ranjan A (2018) Numerical studies on thin wall laminated composite panels under compressive loading. Int J Civ Eng Technol 9(6):586–594
27. Kumar S, Kumar R, Mandal S, Rahul AK (2018) The prediction of buckling load of laminated composite hat-stiffened panels under compressive loading by using of neural networks. Open Civ Eng J 12(1):468–480
28. Zarei A, Khosravifard A (2019) A meshfree method for static and buckling analysis of shear deformable composite laminates considering continuity of interlaminar transverse shearing stresses. Compos Struct 2019:206–218

29. Castro SGP, Donadon MV, Guimaraes TAM (2019) ES-PIM applied to buckling of variable angle tow laminates. Compos Struct 209:67–78
30. Falzon B G, Stevens K, Davies GO (2000) Postbuckling behavior of a blade-stiffened composite panel loaded in uniaxial compression. Compos A 31(5):459–468
31. Park O, Haftka RT, Sakar BV, Starnes JH, Nagendra S (2001) Analytical-experimental correlation for a stiffened composite panel loader in axial compression. J Aircr 38(2):379–387
32. Rouse M, Assadi M (2001) Evalutional of scaling approach for stiffened composite flat panels loaded in compression. J Aircr 38(5):950–955
33. Ungbhakorn V, Singhatanagdid P (2003) Similitude invariants and scaling laws for buckling experiments on anti-symmetrically laminated plates subjected to biaxial loading. Compos Struct 59(4):455–465
34. Baker DJ (2000) Evaluation of thin Kevlar-epoxy fabric panels subjected to shear loading. J Aircr 1(37):138–143
35. Zhao W, Xie Z, Wang X, Li X, Hao J (2019) Buckling behavior of stiffened composite panels with variable thickness skin under compression. Mech Adv Mater Struct
36. Bai R, Bao S, Lei Z, Liu D, Yan C (2018) Experimental study on compressive behavior of I-stiffened CFRP panel using fringe projection profilometry. Ocean Eng 160:382–388
37. Kumar S, Kumar R, Mandal S (2018) Behavior of FRP composite panel subjected to inplane loading. Int J Civ Eng Technol 9(6):1324–1332
38. Sanchez ML, De Almeida SFM, Carrillo J (2017) Evaluation of the effect of thermal residual stress on buckling and post-buckling of composite plates with lateral reinforcement. Revista Latinoamericana de Metalurgia y Materiales 37(1):45–49
39. Firsanov VV, Gavva LM (2017) The investigation of the bending form of buckling for structurally-anisotropic panels made of composite materials in operating MATLAB system. Struct Mech Eng Constr Build 4:66–76 (in Russian)

# Acoustic Fatigue Research for Honeycomb Sandwich Structure with Impact Damage Based on Vibro-Acoustic Coupling Analysis

**Ruowei Li, Haitao Zhao, Mingqing Yuan and Ji'an Chen**

**Abstract** Composite honeycomb sandwich structure is widely used in aircraft wing leading edge, rudder surface, engine fairing, etc. It is susceptible to strong aerodynamic noise and impact loads from birds, hails, and stones. Therefore, it is essential to evaluate the life of structure with sound and collision load. In this paper, an acoustic fatigue life evaluation method based on vibro-acoustic coupling with impact damage is proposed. A representative composite honeycomb sandwich structure is built up to implement the proposed method. Firstly, the low-velocity impact process is simulated by the finite element (FE) method. This analysis case is used to get the structure with damage and material degradation. Secondly, the FE model is linked with the acoustic indirect boundary element (BE) model. A white Gaussian noise load is applied to the coupled FE/indirect BE model, and the power spectral density (PSD) curve of the structural dangerous point is obtained by modal-based vibro-acoustic coupling response case and random post-processing analysis case. At last, the fatigue life of the honeycomb structure is computed by the PSD theory. The result shows that with the increase of the impact energy, the structural life under the same acoustic load decreases. This method reveals high computational efficiency and excellent feasibility. The analytical result has reference value for the acoustic and mechanical properties design of composite honeycomb sandwich structure.

**Keywords** Vibro-acoustic coupling · Low-velocity impact damage · Acoustic fatigue

---

R. Li · H. Zhao (✉) · M. Yuan · J. Chen
School of Aeronautics and Astronautics, Shanghai Jiao Tong University, Shanghai, China
e-mail: zht@sjtu.edu.cn

R. Li
e-mail: liruowei@sjtu.edu.cn

M. Yuan
e-mail: yuanmq@sjtu.edu.cn

J. Chen
e-mail: ja-chen@sjtu.edu.cn



Z. Jing (ed.), *Proceedings of the International Conference on Aerospace System Science and Engineering 2019*, Lecture Notes in Electrical Engineering 622,
https://doi.org/10.1007/978-981-15-1773-0_13

## 1 Introduction

Honeycomb structure [1] has advantages as a special composite material, such as lightweight, fatigue resistance, high strength, high rigidity, and strong heat resistance. So, it has been widely used in aircraft rudder surface, engine fairing, wing tip, floor and interior, etc. The aircraft has been in a strong noise environment during take-off, landing, and flight. Long-term high-intensity noise excitation leads to dynamic response, and then, acoustic fatigue will occur in the structure [2, 3].

During the flight, the aircraft is vulnerable to impact loads such as bird strikes, hail, and gravel impact. The honeycomb sandwich panel made of composite material is very sensitive to impact loads, and its mechanical performance will change a lot under the impact. After the low-velocity impact, it is difficult to see the damage defects only on the surface of the laminate without the aid of the inspection tool. However, complex damages are generated in the interior of the laminate such as fiber breakage, matrix fracture, and delamination. These internal damages seriously degrade the structural properties of the laminate, and the strength can be reduced by 35–40%. The reduction in strength will definitely affect the acoustic fatigue life of the structure.

At present, the methods of calculating acoustic fatigue life are mainly divided into two methods: time domain method and frequency domain method. The main method in the time domain is the rain-flow counting method [4–6]. This method requires loop counting, and the amount of data processing is very large. Therefore, the current mainstream method is the random vibration fatigue life assessment based on the stress probability density function and the power spectral density method [7, 8] in the frequency domain [9]. The frequency domain method can overcome this shortcoming. Therefore, random vibration fatigue life assessment method in frequency domain is popular which is based on the stress probability density function and the power spectral density method. In the meanwhile, research on the low-velocity impact of composite materials has gradually matured. Through the simulation by finite element software and the experiment verification, the damage results of the structures with different materials and different layers after low-velocity impact can be obtained. However, no research has been conducted on the acoustic fatigue life of structures subjected to impact loads.

Aiming at this problem, this paper firstly obtains the stiffness degradation of the structure and the deformed model by performing low-velocity impact simulation [10, 11] on the honeycomb sandwich structure. Then, the acoustic-vibration coupling [12–14] analysis of the honeycomb sandwich panel is conducted under strong noise load. At last, the acoustic fatigue life can be predicted by the power spectral density method.

## 2 Theoretical Considerations

### 2.1 *Hashin Criterion and Stiffness Degradation Criterion*

If the influence of the inter-layer stress is neglected, the damage of the monolayer is determined by the two-dimensional Hashin criterion, which can be described as followed:

Fiber tensile failure:

$$\left(\frac{\sigma_1}{X_t}\right)^2 + \left(\frac{\tau_{12}}{S_{12}}\right)^2 \geq 1, \quad \sigma_1 \geq 0 \tag{1}$$

Fiber compression failure:

$$\left(\frac{\sigma_1}{X_c}\right)^2 \geq 1, \quad \sigma_1 \leq 0 \tag{2}$$

Matrix tensile or shear failure:

$$\left(\frac{\sigma_2}{Y_t}\right)^2 + \left(\frac{\tau_{12}}{S_{12}}\right)^2 \geq 1, \quad \sigma_2 \geq 0 \tag{3}$$

Matrix compression or shear failure:

$$\left(\frac{\sigma_2}{Y_c}\right)^2 + \left(\frac{\tau_{12}}{S_{12}}\right)^2 \geq 1, \quad \sigma_2 \leq 0 \tag{4}$$

where $X_t$ is the longitudinal tensile strength of a monolayer, $X_c$ is the longitudinal compressive strength of a monolayer, and $S_{12}$ is the shear strength of the monolayer 1–2 direction.

The material stiffness will be reduced when the failure mode of the above equation is satisfied. According to Reifsnider [15], if a point fails, the material stiffness reduction is only limited to the vicinity of this point. The degradation model in this paper uses the Camanho degradation model.

Assuming that a kind of damage can occur in an element, it is necessary to calculate cumulative damage, such as matrix and fiber tensile failure. However, if the matrix or fiber is subjected to tensile failure, the compression failure cannot occur at same elements. If elements have multiple failure modes at the same time, the corresponding degradation parameters are accumulated. The specific methods for material degradation are as follows:

Matrix tensile or shear cracking:

$$E_2' = 0.2E_2, G_{12}' = 0.22G_{12}, \nu_{12}' = 0.15\nu_{12} \tag{5}$$

Matrix compression or shear cracking:

$$E_2' = 0.4E_2, G_{12}' = 0.4G_{12}, \nu_{12}' = 0.15\nu_{12} \tag{6}$$

Fiber tensile fracture:

$$E_1' = 0.07E_1, E_2' = 0.07E_2, G_{12}' = 0.07G_{12}, \nu_{12}' = 0.07\nu_{12} \tag{7}$$

Fiber compression fracture:

$$E_1' = 0.14E_1, E_2' = 0.14E_2, G_{12}' = 0.14G_{12}, \nu_{12}' = 0.14\nu_{12} \tag{8}$$

Matrix stretching or shear cracking and fiber tensile fracture occur:

$$E_1' = 0.07E_1, E_2' = 0.2E_2, G_{12}' = 0.22G_{12}, \nu_{12}' = 0.15\nu_{12} \tag{9}$$

Matrix compression or shear cracking and fiber compression cracking occur:

$$E_1' = 0.14E_1, E_2' = 0.4E_2, G_{12}' = 0.4G_{12}, \nu_{12}' = 0.15\nu_{12} \tag{10}$$

where $E_i$, $\upsilon_{ij}$, and $G_{ij}$ are the current material parameters of the monolayer. $E'_i$, $\upsilon'_{ij}$, and $G'_{ij}$ are the material parameter of the degradation of the monolayer after failure.

## 2.2 Metal Material Constitutive Relationship and Failure Mode

The plastic properties of the metal material are defined by the Johnson-Cook constitutive. The description of J-C constitutive parameter has been shown in Table 1.

$$\sigma = [A + B(\bar{\varepsilon}^{\mathrm{pl}})^n][1 + C\ln(\dot{\bar{\varepsilon}}^{\mathrm{pl}}/\dot{\varepsilon}_0)]\left[1 - \left(\frac{T - T_{\mathrm{r}}}{T_{\mathrm{m}} - T_{\mathrm{r}}}\right)^m\right] \tag{11}$$

Damage determination uses maximum failure plastic strain:

$$\varepsilon_f = \left[d_1 + d_2\exp\left(d_3\frac{p}{\sigma_{\mathrm{e}}}\right)\right]\left[1 + d_4\ln\left(\frac{\dot{\varepsilon}}{\dot{\varepsilon}_0}\right)\right]\left(1 + d_5\frac{T - T_{\mathrm{r}}}{T_{\mathrm{m}} - T_{\mathrm{r}}}\right) \tag{12}$$

where $d_1 - d_5$ are failure parameters, $p$ is the three-dimensional average stress, and $\sigma_{\mathrm{e}}$ is the von Mises equivalent stress.

For the Eq. (12) the first bracket indicates that the strain at break decreases as the hydrostatic tension increases, the second bracket indicates the effect of strain rate, and the third bracket indicates the effect of temperature.

**Table 1** Nomenclature and unit

| | Name | Unit |
|---|---|---|
| $\sigma$ | Stress | MPa |
| $A$ | Initial yield stress at reference temperature | Mpa |
| $B$ | Strain hardening coefficient at reference temperature | Mpa |
| $C$ | Strain rate sensitivity index | – |
| $n$ | Strain hardening index at reference temperature | – |
| $\bar{\varepsilon}^{\mathrm{pl}}$ | Plastic strain | – |
| $\dot{\bar{\varepsilon}}^{\mathrm{pl}}$ | Plastic strain rate | – |
| $\dot{\varepsilon}_0$ | Reference strain rate | – |
| $m$ | Thermal softening index at reference temperature | – |
| $T$ | Current temperature | K |
| $T_{\mathrm{r}}$ | Reference temperature | K |
| $T_{\mathrm{m}}$ | Melting temperature | K |

## 2.3 *Vibro-Acoustic Coupling*

A coupling coefficient $\lambda_c$ can be used to determine whether coupling is needed:

$$\lambda_c = \frac{\rho_0 c}{\rho_t T \omega} \tag{13}$$

where $\rho_0$ is the density of the fluid, $c$ is the flow velocity of the sound in the fluid, $\rho_t$ is the density of the structure, $T$ is the equivalent thickness of the structure, and $\omega$ is the angular frequency.

While $\lambda_c > 1$, vibro-acoustic coupling analysis needs to be done. While $\lambda_c \ll 1$, coupling condition could be neglected.

The coupling equation is:

$$\begin{bmatrix} K_s + j\omega C_s - \omega^2 M & L_c & 0 \\ L_c^{T} & \frac{D_{11}}{\rho_0\omega^2} & \frac{D_{12}}{\rho_0\omega^2} \\ 0 & \frac{D_{21}}{\rho_0\omega^2} & \frac{D_{22}}{\rho_0\omega^2} \end{bmatrix} \begin{bmatrix} u_i \\ \mu_{i1} \\ \mu_{i2} \end{bmatrix} = \begin{bmatrix} F_s \\ F_{a1} \\ F_{a2} \end{bmatrix} \tag{14}$$

where $K_s$, $C_s$, $M_s$, and $D_{ij}$ are the stiffness matrix, damping matrix, mass matrix, and coefficient matrix of the structure, respectively. $u_i$ is the displacement of each node. $\mu_{i1}$ is the sound pressure difference between two sides of the mesh which sound field and structure are coupling, and $\mu_{i2}$ is the sound pressure difference between two sides of the mesh which sound field and structure are not coupling. $F_s$ is the external force loading on the structure (excluding the sound pressure load). $F_{a1}$ is the sound pressure loading on the sound field-structure coupled mesh, and $F_{a2}$ is the sound pressure loading on the uncoupled mesh of the sound field. $L_c$ is the coupling matrix.

## 2.4 Random Acoustic

As for a random process, the PSD matrix $\mathbf{S}_{\mathrm{XX}}$ of the input load and the PSD matrix $\mathbf{S}_{\mathrm{YY}}$ of the output response can be described by the transfer function matrix **H**:

$$\mathbf{S}_{\mathrm{YY}} = \mathbf{H}^{*}\mathbf{S}_{\mathrm{XX}}\mathbf{H}^{\mathrm{T}} \tag{15}$$

The input load $\mathbf{S}_{\mathrm{XX}}$ can be decomposed while principal component analysis via SVD decomposition or Schur decomposition:

$$\mathbf{S}_{\mathrm{XX}} = \tilde{\mathbf{X}}^{*}\mathbf{S}_{\mathrm{qq}}\tilde{\mathbf{X}}^{\mathrm{T}} \tag{16}$$

where $\tilde{\mathbf{X}}$ is a matrix determined by each element named principal component. $\mathbf{S}_{\mathrm{qq}}$ is a cross-spectral matrix of a random participation factor **q**. It can be obtained by Eqs. (15, 16).

$$\mathbf{S}_{\mathrm{YY}} = \mathbf{H}^{*}\tilde{\mathbf{X}}^{*}\mathbf{S}_{\mathrm{qq}}\tilde{\mathbf{X}}^{\mathrm{T}}\mathbf{H}^{\mathrm{T}} = \mathbf{Y}^{*}\mathbf{S}_{\mathrm{qq}}\mathbf{Y}^{\mathrm{T}} \tag{17}$$

In general, structural responses can be represented by modal superposition:

$$\tilde{\mathbf{Y}} = \boldsymbol{\Phi}\,\mathbf{p} \tag{18}$$

where $\boldsymbol{\Phi}$ is the modal stress vector. **p** is the modal random participation factor.

Therefore, it can be written as:

$$\mathbf{S}_{\mathrm{YY}} = \boldsymbol{\Phi}^{*}\mathbf{p}^{*}\mathbf{S}_{\mathrm{qq}}\mathbf{p}^{\mathrm{T}}\boldsymbol{\Phi}^{\mathrm{T}} = \boldsymbol{\Phi}^{*}\mathbf{S}_{\mathrm{pp}}\boldsymbol{\Phi}^{\mathrm{T}} \tag{19}$$

$$\mathbf{S}_{\mathrm{pp}} = \mathbf{p}^{*}\mathbf{S}_{\mathrm{qq}}\mathbf{p}^{\mathrm{T}} \tag{20}$$

## 2.5 Power Spectral Density Method

The cross-power spectrum curves of each node of the structure are obtained by numerical calculation. The PSD curve of the dangerous point can be obtained by the superposition of stationary stochastic processes. It can be used as an input to calculate the life of acoustic fatigue by power spectral density method.

Acoustic fatigue life can be expressed by Eq. (21).

$$T = \frac{C}{E(M_{\mathrm{T}})\int_{0}^{\infty} P(s)s^{b}\mathrm{d}s} \tag{21}$$

where $T$ is the life of structural-acoustic fatigue, $b$ and $C$ are the material constants of S-N curve, $E(M_T)$ is the average incidence of stress cycles per unit time, $P(s)$ is the probability density function of stress, and $s$ is stress.

S-N curve can be fitted as a power function:

$$C = Ns^b \tag{22}$$

S-N curve used here has a range of $10^2$–$10^6$ cycles.

If the random process is a narrowband random process, $E(M_T)$ is zero-crossing rate $E(0)$. If it is a wideband random process, $E(M_T)$ is the peak occurrence rate $E(p)$, as shown in Eq. (23).

$$E(0) = \sqrt{\frac{m_2}{m_0}},\ E(p) = \sqrt{\frac{m_4}{m_2}} \tag{23}$$

$m_n$ is the $n$th-order inertia moment of the stress power spectral density.

$$m_n = \int f^n G(f) \mathrm{d}f \tag{24}$$

where $f$ is the frequency and $G(f)$ is the stress power spectral density function of the dangerous point. The relationship between the acceleration power spectral density function $G_g(f)$ and the stress power spectral density function can be expressed as:

$$G(f) = \frac{H^2}{\omega^4} G_g(f) \tag{25}$$

where $H$ is related to the material and structure form and the position of the action reference and the response points. $H$ can be expressed approximately by Eq. (26).

$$H = \frac{\sigma}{A} \tag{26}$$

where $\sigma$ is the stress at the response point, and $A$ is the displacement at the point of action.

In order to determine whether a stationary random process belongs to a wideband process or a narrowband process, an irregular factor is introduced:

$$\gamma = \frac{m_2}{\sqrt{m_0 m_4}} \tag{27}$$

While the irregularity factor is close to 0, the stationary stochastic process can be described as a broadband random process. While the irregular factor is close to 1, it can be described as a narrowband process.

If the random stress process is a narrowband process, the probability density function satisfies Rayleigh distribution which is:

$$P_p(s) = \frac{S}{\sigma_s^2} e^{\frac{-s^2}{2\sigma_s^2}} = \frac{S}{m_0} e^{\frac{-s^2}{2m_0}} \tag{28}$$

The probability density function of a broadband random process is a kind of distribution between Gauss distribution and Rayleigh distribution. It can be described as Dirlik [8] equation which is obtained by Monte Carlo method. It is a semi-empirical equation with good precision.

# 3 Numerical Simulation

## *3.1 Technical Route*

The technical route of random vibration acoustic fatigue life assessment used in this paper is shown in Fig. 1.

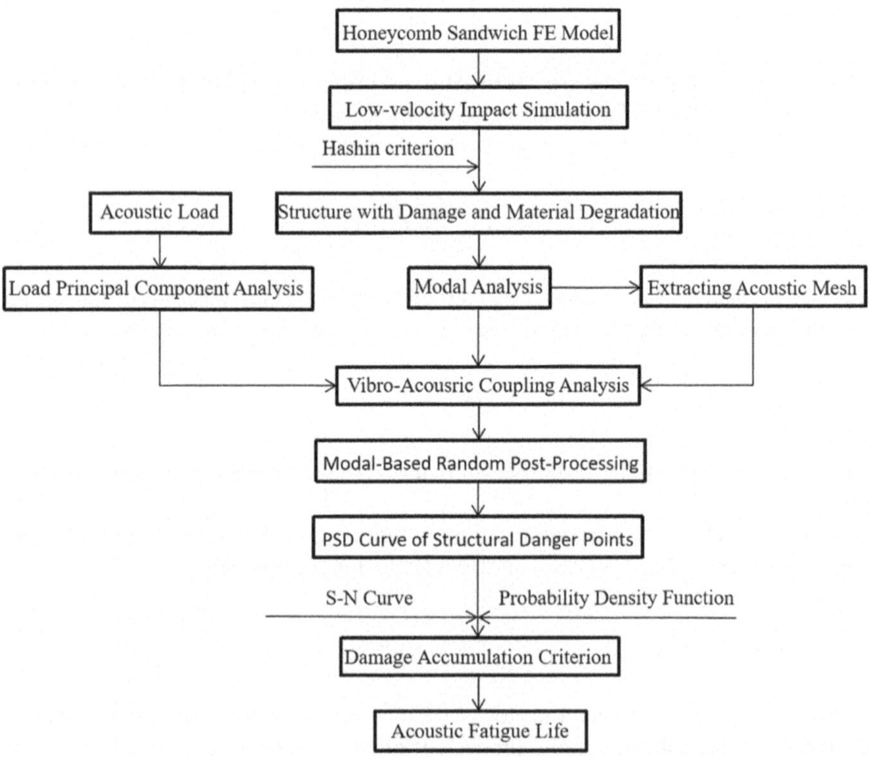


**Fig. 1** Technical road map

Firstly, the low-velocity impact simulation is carried out by finite element analysis. The stiffness degradation model of the composite material and the geometric structure after deformation of the material are obtained by Hashin failure criterion. Secondly, the structure should be used the modal analysis and extracted sound field on the surface of the structure in order to get boundary element mesh. Thirdly, the principal component analysis [16] is applied to the input acoustic load in order to reduce the amount of calculation in the subsequent calculation process. Fourthly, use vibro-acoustic coupling analysis [17]. Apply modal superposition method for modal-based random sound field post-processing. So that the PSD curve of the dangerous point can be obtained. Fifthly, select the appropriate probability density function by the value of the irregularity factor. Combine S-N curve of the material to get the acoustic fatigue life of the structure.

In this paper, the damaged structures under different impact energy are obtained by different initial velocity. Then, the acoustic fatigue life of the structures under the same sound pressure level is calculated.

## 3.2 *Modeling*

The overall size of the honeycomb sandwich structure is 300 mm × 207.846 mm × 20.6 mm. Upper and lower panels are 0.3 mm thick. The honeycomb core height is 20 mm, regular hexagonal honeycomb structure side length L = 15 mm, wall thickness t = 0.06 mm. Punch is modeled by analytical rigid body with a radius of 15 mm and a mass of 100 g. The geometric model is shown in Fig. 2. The detail of the honeycomb structure is shown in Fig. 3.

While acoustic software Virtual.lab Acoustic is used to simulate, the size of the mesh should be consistent. Encryption of the local mesh cannot improve the accuracy because the calculation accuracy of the fluid model is controlled by most elements.

**Fig. 2** Geometric model

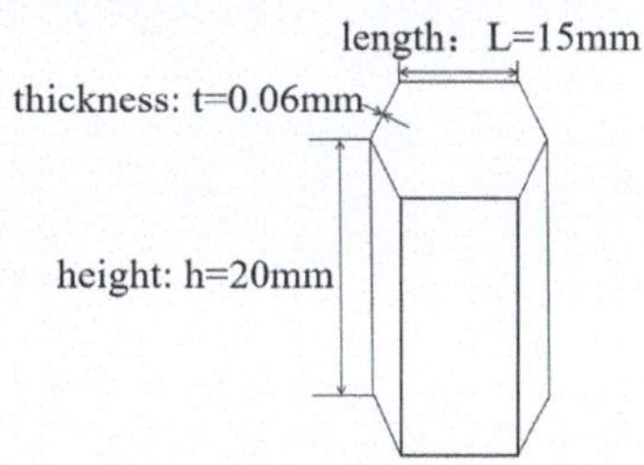


**Fig. 3** Detail of the honeycomb

If the highest frequency $f_{max}$ is known and assume the speed of sound in a fluid medium is $c$, all elements length $L$ should satisfy:

$$L \leq \frac{c}{6 f_{max}} \tag{29}$$

However, in the structural analysis, the mesh should be refined in places where the stress and displacement gradients are large, or in the vicinity of the sound source, the sharp corners, the holes, etc.

The honeycomb sandwich structure is divided into the upper and lower panels and an intermediate honeycomb core. The upper and lower panels are made of E-Glass/epoxy composite material. The honeycomb cores are all made of LY12CZ aluminum alloy. The composite properties and ply information are shown in Table 2. The properties of LY12CZ are shown in Table 3. The properties of the air material are shown in Table 4. Due to the low-velocity impact, the influence of temperature on the material can be neglected for the Johnson-Cook constitutive used in aluminum alloy materials.

The load is 163 dB Gaussian white noise applied on the band range of 10–2500 Hz in which step size is 10 Hz. The PSD curve can be obtained by Eq. (30) and applied as a load on the acoustic mesh surface of the entire model. The acoustic mesh is a two-dimensional boundary element mesh extracted from the surface of the structure. The honeycomb sandwich structure is modeled by the shell element, so the mesh of the acoustic and the structure is consistent. Reference sound pressure $p_0 = 2 \times 10^{-5}$ Pa.

**Table 2** E-glass/epoxy composite material

| | Name | Value |
|---|---|---|
| | Density: $\rho$ | $1.5 \times 10^3$ kg/m$^3$ |
| Ply information | Monolayer thickness | 0.125 mm |
| | Lap angle | $[0/90]_s$ |
| Monolayer material properties | $E_{11}$ | 42 GPa |
| | $E_{22}$ | 11.3 GPa |
| | $\upsilon_{12}$ | 0.3 |
| | $G_{12}$ | 4.5 GPa |
| Monolayer strength parameter | $X_T$ | 900 MPa |
| | $X_C$ | 900 MPa |
| | $Y_T$ | 50 MPa |
| | $Y_C$ | 140 MPa |
| | $S_{12}$ | 72 MPa |
| Fatigue parameter | $C$ | $1.87732 \times 10^{10}$ |
| | $b$ | 1.98216 |

**Table 3** LY12CZ material

| | Name | Value |
|---|---|---|
| | Density: $\rho$ | $2.7 \times 10^3$ kg/m$^3$ |
| Elastic parameter | Young's modulus: $E$ | 70GPa |
| | Poisson's ratio: $\upsilon$ | 0.3 |
| Plasticity parameter | Initial yield stress at reference temperature: $A$ | 324 MPa |
| | Strain hardening coefficient: $B$ | 114 MPa |
| | Strain rate sensitivity index: $C$ | 0.0026 |
| | Strain hardening index: $n$ | 0.42 |
| | Quasi-static reference strain rate $\dot{\varepsilon}_0$ | 1 |
| Plastic failure parameter | No. 1 fracture parameter: $d_1$ | −0.77 |
| | No. 2 fracture parameter: $d_2$ | 1.45 |
| | No. 3 fracture parameter: $d_3$ | 0.47 |
| | No. 4 fracture parameter: $d_4$ | 0 |
| | No. 5 fracture parameter: $d_5$ | 0 |
| Fatigue parameter | Fatigue parameter: $C$ | $3.0218 \times 10^{19}$ |
| | Fatigue parameter: $b$ | 6.447 |

**Table 4** Air material

| Name | Value |
|---|---|
| Sound velocity: $c$ | 340 m/s |
| Density: $\rho$ | 1.225 kg/m$^3$ |

$$dB = 10 \times \log_{10}\left(\frac{p}{p_0}\right)^2 \quad (30)$$

# 4 Results and Discussion

## *4.1 Impact Simulated Results*

With low-velocity impact of different velocities, the stress and deformation cloud diagram is shown in Fig. 4. In order to display the stress level and deformation of the impact point well, the resulting cloud image only shows 1/4 of the structure.

It can be clearly seen that as the impact energy increases, the deformation becomes more and more serious. Honeycomb sandwich made of aluminum alloy has been damaged at an impact velocity of 20 m/s.

The damage of the panel is determined by Hashin criterion which can be divided into six damage modes as shown in Sect. 2.1. Figure 6 shows the damaged cloud

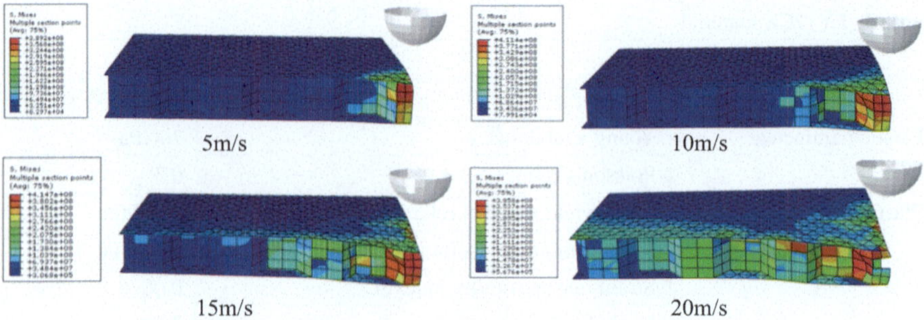


**Fig. 4** Stress cloud diagram of structures at different impact velocities

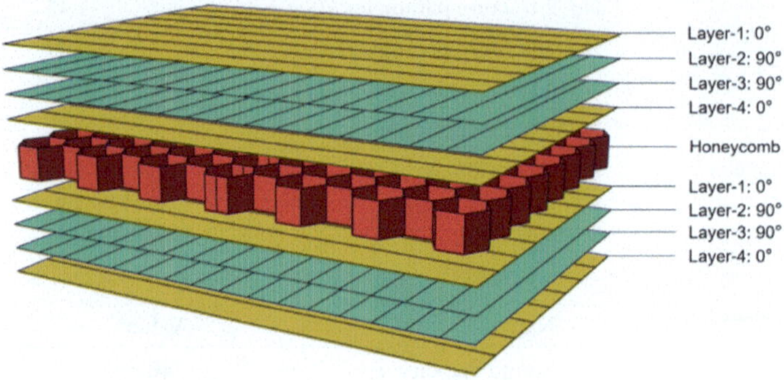


**Fig. 5** Lamination naming rules

of each layer panel when the impact velocity is 20 m/s. The red area indicates that the structure is damaged; the blue color indicates that the structure has not received damage. The ply of panel is named as shown in Fig. 5.

With an impact velocity of 20 m/s, there are only four failure modes in the honeycomb sandwich structure: simultaneous tensile failure of the matrix and fibers, fiber tensile failure, matrix tensile failure, and the matrix compression failure. In the area affected by the impact, the main performance is the simultaneous failure of the fiber and matrix, the cross-shaped area centered on the impact point, the main failure mode is the matrix tensile failure, and the direction of the failure is perpendicular to the fiber layup direction.

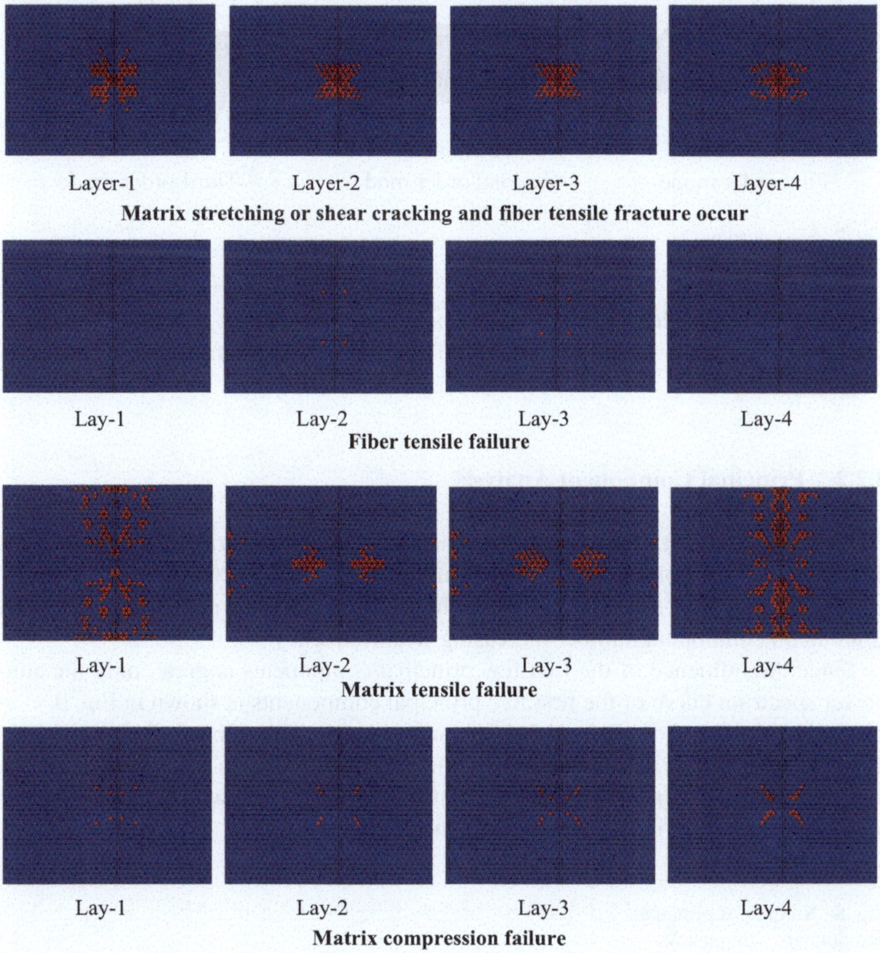


**Fig. 6** Damage cloud map of each layer under different damage modes

## *4.2 Acoustic Simulation Results*

### 4.2.1 Modal Analysis

Boundary conditions used in the modal analysis are four-sided clamped. The modal and natural frequencies are inherent properties of the structure. Each natural frequency corresponds to a modality. While the frequency of the external excitation load is close to the natural frequency of a certain order of the structure, the result will produce a corresponding deformation. Since the structure only has different impact

**Fig. 7** First third mode

energies, the low-order modes of each model have similarities. Only the first three-order modes with an impact velocity of 20 m/s and the natural frequencies are given (Fig. 7).

### 4.2.2 Principal Component Analysis

Principal component analysis is performed on the PSD matrix of the input load by using Schur decomposition with a truncation error of 0.0001. The input power spectral density load can be extracted as the 53-order principal component. Figure 8 is the main component number increasing with frequency.

Since the influence of the first few principal components is great, only the auto power spectrum curve of the first five principal components is shown in Fig. 9.

As can be seen from Fig. 9, the first-order principal component has a very significant contribution at low and intermediate frequency and is slightly low at high frequency. The remaining main components are lower in the low frequency and have a larger contribution in the middle frequency band.

**Fig. 8** Number of principal components—frequency map

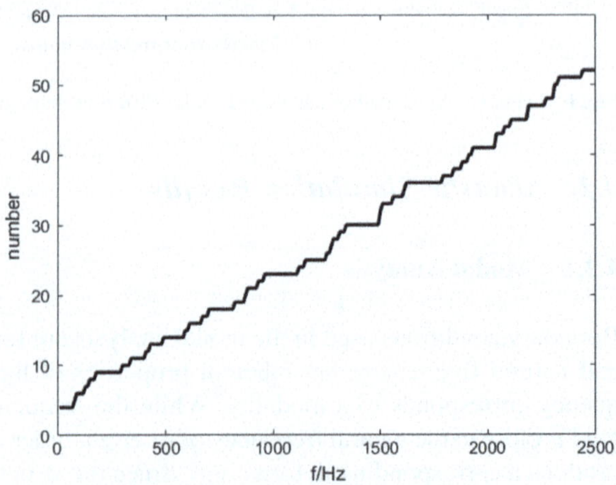

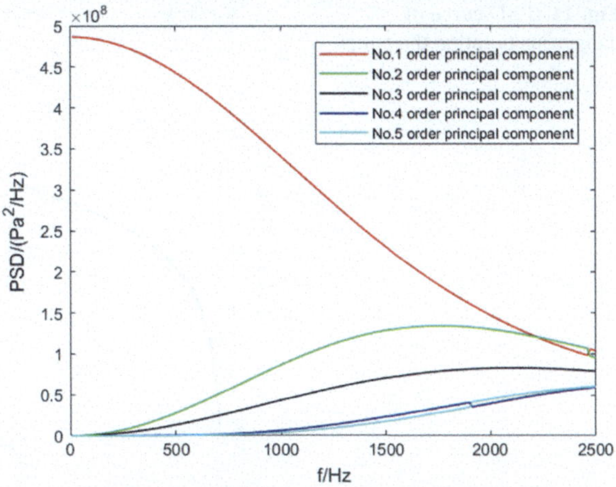


**Fig. 9** Auto-power spectral density curve of the first five orders of principal components

1840.0Hz 2142.4Hz 2299.9Hz

**Fig. 10** Displacement cloud map of the first three natural frequencies under the first-order principal component

### 4.2.3 Acoustic-Vibration Coupling Analysis

According to the modal-based acoustic-vibration coupling analysis, the response of the structure at each frequency can be obtained. Figure 10 shows the impact velocity of 20 m/s and the displacement cloud map of the first three natural frequencies under the first-order principal component. The cloud image shows only 1/4 of the structure.

Since the structure is fixed on four sides and is impacted at the center, the portion with a large displacement is located at the center of the plate. The displacement gradually decreases toward the edge of the plate. The low-order mode contributes a lot to the displacement. It can be easily seen that the displacement of the first-order natural frequency is about 100 times larger than the second order or third order.

### 4.2.4 PSD Curve

The PSD curve of the dangerous point with the impact velocity of 20 m/s is obtained by numerical analysis. Due to the different materials, the PSD curves of the dangerous points of the panel and the honeycomb are, respectively, shown in Figs. 11 and 12.

The PSD curve of the panel and the honeycomb has the same trend. The PSD curve peak value of the panel dangerous point is 721.44 $Pa^2/Hz$, while the honeycomb is

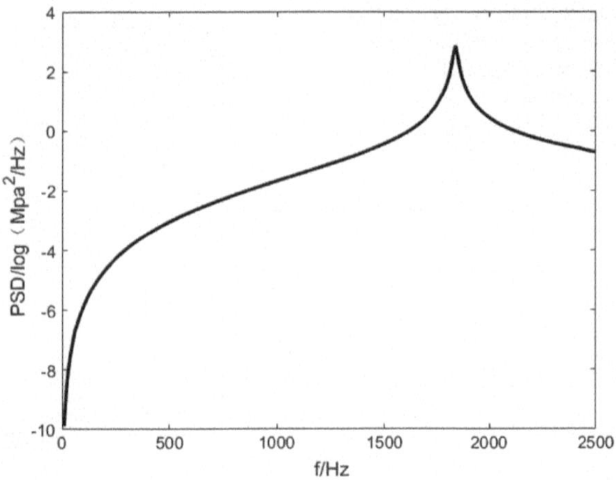


**Fig. 11** PSD curve of dangerous point on the panel

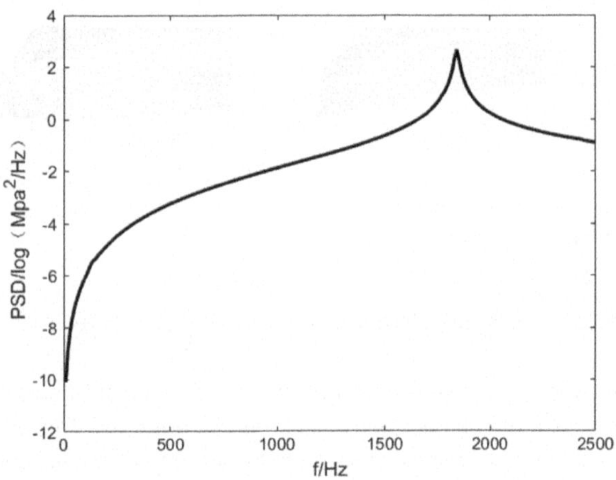


**Fig. 12** PSD curve of dangerous point on the honeycomb

455.59 $Pa^2/Hz$. The peak appears at the first-order natural frequency. The PSD curve shows small jitters in the natural frequencies of the remaining steps. It can be seen that the first-order natural frequency of the structure plays an important role in the acoustic lifetime.

## *4.3 Acoustic Fatigue Life*

### 4.3.1 Acoustic Fatigue Life of Structure Under Different Impact Velocities

Table 5 and Fig. 13 show the structural-acoustic fatigue life at different impact velocities under 163 dB white noise loading.

**Table 5** Lifetime at different velocities

| Impact velocity (m/s) | Impact energy (J) | Panel life (cycle) | Honeycomb life (cycle) |
|---|---|---|---|
| 0 | 0 | 1616.451 | 36.410 |
| 5 | 1.25 | 1605.813 | 35.940 |
| 10 | 5 | 1587.377 | 34.552 |
| 15 | 11.25 | 1096.320 | 11.266 |
| 20 | 20 | 267.552 | 8.087 |

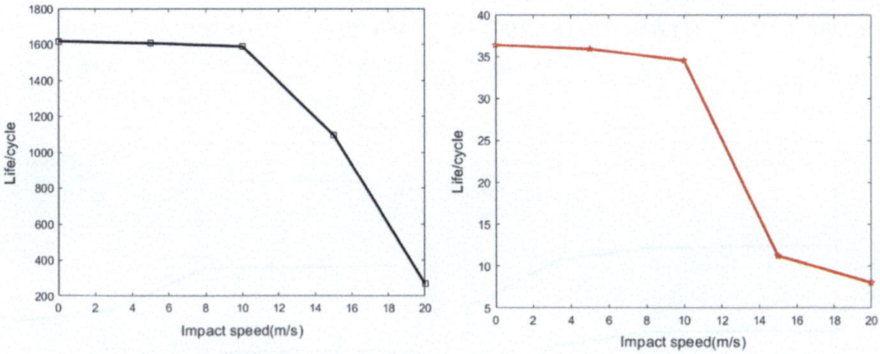


Panel life    Honeycomb life

**Fig. 13** Lifetime

Because the panel is made by composite material and the thickness is greater than the wall thickness of the honeycomb, the life of the dangerous point on the panel is much longer than the honeycomb. But the trend of the panel and the honeycomb is the same. While the impact velocity is low, the influence on the acoustic fatigue life is very small. But when the impact velocity exceeds 10 m/s, the acoustic fatigue life of the structure falls off. It can be seen that impact damage has a very serious effect on the life of the structure.

### 4.3.2 Acoustic Fatigue Life of Structures Under Different Noise Loads

In order to determine if the conclusions above are applicable under different load conditions, the structural fatigue life at different loads of 154–163 dB with increments of 3 dB is given below. Tables 6, 7, and Fig. 14 show the acoustic fatigue life of the structure under different sound loads and different impact velocities. The units not marked in the table are cycles

It can be seen from Fig. 14 that the life curves of structures under different acoustic loads have similarities. The structural life decreases by an order of magnitude with every 3 dB increment of noise load at the same impact velocity.

**Table 6** Lifetime of panel under different velocities and different sound loads

| Panel | 0 m/s | 5 m/s | 10 m/s | 15 m/s | 20 m/s |
|---|---|---|---|---|---|
| 157 dB | 160123.579 | 160087.364 | 158421.148 | 123213.215 | 28915.124 |
| 160 dB | 17220.584 | 16989.294 | 16624.487 | 11842.218 | 3000.235 |
| 163 dB | 1616.451 | 1605.813 | 1587.377 | 1096.320 | 267.552 |

**Table 7** Lifetime of honeycomb under different velocities and different sound loads

| Honeycomb | 0 m/s | 5 m/s | 10 m/s | 15 m/s | 20 m/s |
|---|---|---|---|---|---|
| 157 dB | 3710.125 | 3615.025 | 3582.215 | 1200.541 | 799.254 |
| 160 dB | 371.021 | 355.992 | 350.225 | 108.589 | 82.104 |
| 163 dB | 36.410 | 35.940 | 34.552 | 11.266 | 8.087 |

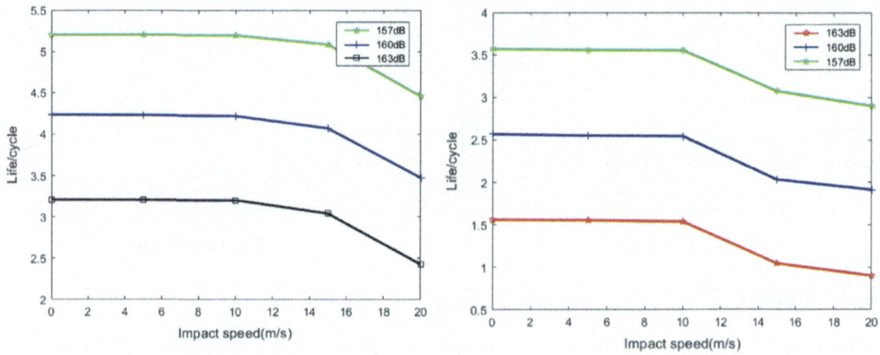


Panel life Honeycomb life

**Fig. 14** Life table in logarithmic coordinate system

# 5 Conclusions and Outlook

(1) The acoustic fatigue life of the honeycomb sandwich structure is insensitive to low energy impacts. While the impact of the energy exceeds 5 J, the life span decreases greatly.
(2) With every 3 dB increment of acoustic load, the life of the honeycomb sandwich structure is reduced by an order of magnitude.
(3) Using the power spectral density method to calculate the structural-acoustic fatigue life, only need to obtain the response power spectral density curve of the structure and give the corresponding probability density function of the broadband and narrowband, then calculate the acoustic fatigue life of various forms of random processes. It is easy to be complied.

In the future research, the acoustic fatigue life assessment method can be verified and improved by combining the acoustic fatigue experiment. The relationship between honeycomb sandwich structure and acoustic fatigue life of different sizes and different layers can be discussed.

## References

1. Liu M, Wu Z (2016) Application of composite honeycomb sandwich structure in aircraft. Sci Technol Rev 34(8):21–25
2. White RG (1990) Developments in the acoustic fatigue design process for composite aircraft structures. Compos Struct 16(1):171–192. https://doi.org/10.1016/0263-8223(90)90071-L
3. Vaicaitis R (1987) Acoustic fatigue—a Monte Carlo approach. In: AIAA/ASME/ASCE/AHS 28th SDM Conference, Paper 87-0916l, Monterey, CA, April 1987. https://doi.org/10.2514/6.1987-916
4. Zhao X, Jiang D, Zhang Q (2009) Application of rain-flow counting method in the analysis of load spectrum. Sci Technol Rev 27(3):67–73. https://doi.org/10.1109/CLEOE-EQEC.2009.5194697
5. Hong N (1991) A modified rainflow counting method. Int J Fatigue 13(6):465–469. https://doi.org/10.1016/0142-1123(91)90481-d
6. Jin Y, Li L (2003) Sonic fatigue life prediction of aeroengine structure. J Aerosp Power 18(3):373–377. https://doi.org/10.3969/j.issn.1000-8055.2003.03.013
7. Xu F, Xiao S (1996) The power spectral density method for the estimation of the sonic fatigue life. J Mech Strength (4):38–42
8. Wang M (2009) Research on life analysis method for structure vibration fatigue. Nanjing University of Aeronautics and Astronautics. https://doi.org/10.7666/d.d076113
9. Tovo R (2002) Cycle distribution and fatigue damage under broad-band random loading. Int J Fatigue 24(11):1137–1147. https://doi.org/10.1016/s0142-1123(02)00032-4
10. Abrate S (1998) Impact on composite structures. Low-Velocity Impact Damage (4):135–160. https://doi.org/10.1017/cbo9780511574504.005
11. Fangyu C, Li Z, Yihao T (2019) Damage and residual compressive strength of multi-layer composite laminates after low velocity impact. Int J Crashworthiness, 1–7. https://doi.org/10.1080/13588265.2018.1478925
12. Bai C, Zhou J, Yan G (2011) Effects of sound field on thin-wall cylindrical structure dynamic characteristics. J Mech Eng. https://doi.org/10.3901/jme.2011.05.078
13. Cheng L (1994) Fluid-structural coupling of a plate-ended cylindrical shell: vibration and internal sound field. J Sound Vib 174(5):641–654. https://doi.org/10.1006/jsvi.1994.1299
14. Craggs A (1979) Coupling of finite element acoustic absorption models. J Sound Vib 66(4):605–613
15. Reifsnider KL (1991) Fatigue of composite material. Compos Mater Series 4(1775):11–77. https://doi.org/10.1007/978-3-7091-2544-1_6
16. Tipping ME, Bishop CM (1999) Probabilistic principal component analysis. J R Stat Soc Ser B (Statistical Methodology) 61(3):611–622. https://doi.org/10.1134/s1063771017030071
17. Li Z, Zhan F (2010) Virtual lab acoustics acoustics simulation calculation advanced application example. National Defense Industry Press

# Combining Autoencoder with Similarity Measurement for Aircraft Engine Remaining Useful Life Estimation

**Mengni Wang, Yuanxiang Li, Honghua Zhao and Yuxuan Zhang**

**Abstract** Remaining useful life (RUL) estimation is very important for the maintenance of aircraft engines. We can evaluate the current condition of an aircraft engine and predict its RUL by constructing its degradation curve. However, the degradation curve is often difficult to obtain due to the unobserved degradation patterns. Currently, many researchers estimated RUL by setting a fixed RUL target function or making assumptions about how an engine degrades. But in the real world, degradation models of aircraft engines are generally individualized. To obtain personalized degradation curves and predict RUL accurately, we propose a method based on autoencoder and similarity measurement. First, an autoencoder trained with normal data is adopted to extract degradation curves of aircraft engines and build a degradation model template library. Then, we measure each test object with all template curves to get similarities and corresponding RULs based on a sliding window and complexity-invariant distance. At last, the estimated RUL can be obtained by calculating the weighted average of highly relevant corresponding RULs. We conduct the proposed method on the aircraft engine dataset provided by NASA. The experimental results demonstrate that our method can utilize the information of multi-sensor data to generate personalized degradation curves effectively and estimate RUL more accurately.

**Keywords** Aircraft engine · Remaining useful life · Autoencoder · Degradation curve · Similarity measurement

---

M. Wang · Y. Li (✉) · Y. Zhang
Shanghai Jiao Tong University, Shanghai, China
e-mail: yuanxli@sjtu.edu.cn

M. Wang
e-mail: moniwong@sjtu.edu.cn

Y. Zhang
e-mail: yuxuanzhang@sjtu.edu.cn

H. Zhao
Eastern Airlines Technic Co., Ltd., Shanghai, China
e-mail: hhzhao@ceair.com



Z. Jing (ed.), *Proceedings of the International Conference on Aerospace System Science and Engineering 2019*, Lecture Notes in Electrical Engineering 622,
https://doi.org/10.1007/978-981-15-1773-0_14

# 1 Introduction

Remaining useful life (RUL) estimation is an important part of prognostics and health management. With the fast development of industry and manufacturing, sensors are widely used to collect running data from complex systems, which are very beneficial to evaluate their states and estimate their RULs. This promotes the popularity of condition-based maintenance rather than time-based maintenance, for it can not only ensure the safety of equipment but also lead to significant financial savings. Such a goal can be reached by predicting the RUL of target object based on establishing its degradation model with collected multi-sensor data.

RUL estimation methods mainly consist of model-based methods, data-driven methods and experience-based methods [1]. Model-based methods evaluate the system state by building a physical model. A physics-based model for bearing prognostics is proposed in [2], by which we can get spall growth trajectory and calculate time to failure according to operating conditions, and then reduce prediction uncertainty based on self-adjusting. Data-driven methods usually learn degradation model by running data to estimate RUL, where hidden Markov model [3], Gaussian process regression [4] and neural networks [5] are frequently used. Experience-based approaches predict the system RUL based on historical failure cases, building a hidden relationship among current system states, current lives and recorded failure models. In [6], RUL is obtained by evaluating the instance similarity, which is related to the usage and maintenance history. Since the difficulty in modeling assumptions and capturing features that directly reflect system change, model-based methods are limited, while due to the available of large amount of running data, data-driven methods and experience-based methods have been greatly developed.

In the past few years, the artificial neural network (ANN) has become noticeable. Many researchers tend to estimate RUL based on ANN such as convolutional neural network (CNN), long short-term memory (LSTM) [7] and recurrent neural network (RNN). Most of them predict RUL by setting a fixed RUL target function or making assumptions about how an engine degrades. The piece-wise model is commonly employed [8–10], in which a set of instances are assumed to begin with the same RUL and then start linear degradation after a certain cycle. In [8], a piece-wise RUL target function is adopted to obtain assumed trajectory of RUL, and then LSTM model is obtained by the training data to find optimal parameters. A RUL estimation method based on CNN is applied in [10]. This method first sets a piece-wise linear RUL target function to get target RUL, then CNN is applied on sensor data split by a sliding window.

However, these fixed RUL target functions or degradation models are always too restricted and unreasonable for complex systems. The reasons are as follows: (1) The lifetime of each engine is different. (2) Each engine begins to work with unknown wear which is basically different from each other. (3) With diverse operating conditions and faults, degradation curves do not follow a fixed shape. (4) The accuracy of RUL estimation is greatly affected by the selection of degradation point.

Thus, this paper uses the autoencoder to obtain the individualized degradation mode of each engine. We first transform reconstruction errors to state variations of the complex system by autoencoder, and then filter them to get degradation curves. Furthermore, with many degradation curves of run-to-failure systems, we can set up a degradation model template library to match degradation models of test engines. To get RUL, we compute the similarities and corresponding RULs between test engine and all instances in the library by a sliding window, based on the main idea of case-based learning [11]. Our experiment is conducted on public C-MAPSS dataset provided by NASA, and the result shows that our method can improve the accuracy of RUL estimation in comparison with other fixed RUL methods.

The contribution of this paper can be summarized as follows:

- Extract the personalized degradation curve by autoencoder with recorded multi-sensor data.
- Analyze the characteristics of degradation curve and propose an effective framework to calculate RUL based on similarity measurement without any degradation assumptions.

The rest of the paper is organized as follows. Section 2 introduces the related work, including time series anomaly detection based on autoencoder and complexity-invariant distance [12]. Section 3 presents our method, describing how to extract degradation curve and measure similarity between test curve and model curves. Section 4 shows the experimental results on C-MAPSS dataset. The last section summaries the paper and discusses the future work.

## 2 Related Work

### *2.1 Time Series Anomaly Detection*

Autoencoder is proposed for the process of high-dimensional data [13], which is widely used in the field of data compression, representative learning and signal denoising. The main idea of autoencoder is to reconstruct original data by some sparse high-level features or in other words to learn an equation to make output same as input. The structure of autoencoder is shown in Fig. 1. It consists of an encoder and a decoder. The encoder maps original high-dimensional data to a hidden low-dimensional vector while the decoder aims to convert this vector to a same series as input. For $N$-dimensional data $x = \{x^i | i = 1, 2, \ldots, N\}$, the reconstruction error can be obtained by (1)

$$e = \sum_{i=1}^{N} \left\| x^i - g\left(f\left(x^i\right)\right) \right\| \tag{1}$$

where $g_\theta(\cdot)$ is the decoder function, and $f_\theta(\cdot)$ is the encoder function.

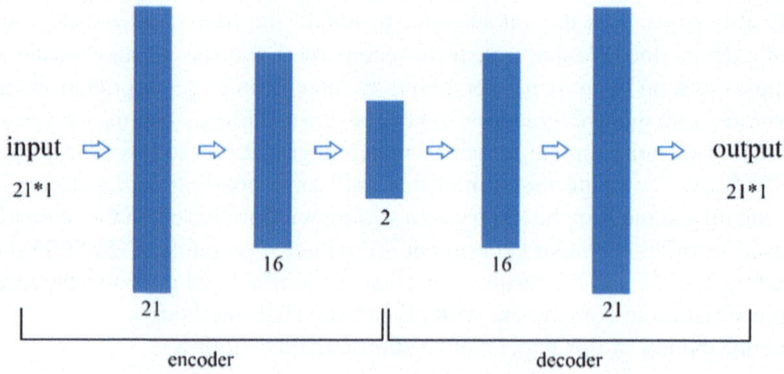


**Fig. 1** Structure of autoencoder

Autoencoder-based approaches for time series anomaly detection have been proposed in [14, 15]. Due to the unrecorded factors or variables, it is difficult to detect anomalies by mathematical models or prediction models. So, researchers adopt autoencoder-based architecture to reconstruct normal data behavior and denote anomaly as large reconstruction error, for the autoencoder only knows the representation of normal data.

## 2.2 *Complexity-Invariant Distance*

For case-based learning, the method of similarity measurement or in other words the way to calculate distance between cases is particularly important. Common distance calculation methods include Euclidean distance and dynamic time warping (DTW), but they are not very applicable to current situation. Euclidean distance is often effective, but it tends to assign a complex object to a simpler class [12]. DTW is usually used to compare two sequences of different lengths, in which one series may be warped by stretching or shrinking its time axis [16]. To evaluate the similarity between target curve and model curve, here complexity-invariant distance (CID) is adopted. CID improves the accuracy of classification and clustering by taking complexity differences between two sequences into consideration [12]. For physical intuition, the complexity of a sequence can be defined by its length, since complex series is always longer than a simpler series after stretched. On the basis of Euclidean distance, CID just adds a correction factor compared with it. Assume that $Q$ and $C$ are two series with $n$ steps, then main calculation formulas of CID can be listed as follows:

$$\mathrm{CID}(Q, C) = \mathrm{ED}(Q, C) \times \mathrm{CF}(Q, C) \tag{2}$$

$$\mathrm{CF}(Q, C) = \frac{\max(\mathrm{CE}(Q), \mathrm{CE}(C))}{\min(\mathrm{CE}(Q), \mathrm{CE}(C))} \tag{3}$$

$$\mathrm{CE}(Q) = \sqrt{\sum_{i=1}^{n-1} (q_i - q_{i+1})^2} \tag{4}$$

where $\mathrm{CID}(Q, C)$ calculates the CID between $Q$ and $C$, complexity correction factor of $Q$ and $C$ is defined as $\mathrm{CF}(Q, C)$, and $\mathrm{CE}(Q)$ represents the complexity estimation of time series $Q$.

## 3 Method

### *3.1 Degradation Curve Extraction*

Since autoencoder can fuse high-dimensional data and reconstruct them effectively by representative learning, we utilize an autoencoder trained with normal data to extract degradation curve, so that it can reconstruct normal cycles with small errors while high error is generated at the occurrence of anomalous subsequence. Because each engine start running normally and some faults occur at an unknown point during operating, we take the sensor data collected from several initial cycles as normal data. Assume that matrix $X^i = \{x_1^i, x_2^i, \ldots, x_n^i\}$ represents all recorded multi-sensor data of engine $i$ with $n$ running cycles. Then, autoencoder is trained with initial data $x_1^i \sim x_m^i$ to minimize the loss and reconstruct $X^i$ to get reconstruction error $\mathrm{e}^i = \{e_1^i, e_2^i, \ldots e_n^i\}$, which can be considered as the deviation from normal state. Since test engines are not run-to-failure, normalization to 0–1 on $e^i$ will change the relative shape between test curve and model curve. So, here we just filter it to get health indicator $\mathrm{HI}^i = \{h_1^i, h_2^i, \ldots, h_n^i\}$ as well as degradation curve, where 0 represents normal and increase in amplitude means degradation in state. After extracting all run-to-failure degradation curves of engines in training subsets, a template library will be established.

### *3.2 Similarity Measurement*

According to the main idea of case-based learning, we need to calculate the CID between two series to get the similarity between them. A sliding window of the same length as test series is used to calculate CID between test series and all segments of each degradation curve, with a step length of 1. Then, the similarity $\mathrm{sim}(Q^i, C^j)$ between test engine $i$ and model engine $j$ can be obtained by (5).

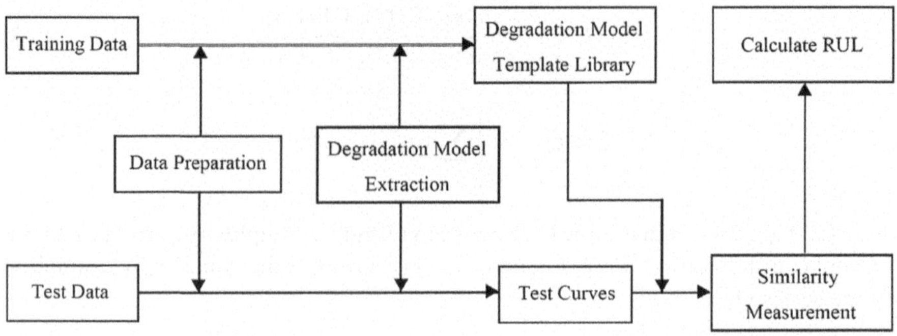


**Fig. 2** Flowchart of our method

$$\text{sim}\left(Q^i, C^j\right) = \max\left\{\frac{1}{\text{CID}\left(Q^i, C_1^j\right)}, \frac{1}{\text{CID}\left(Q^i, C_2^j\right)}, \ldots, \frac{1}{\text{CID}\left(Q^i Q, C_n^j\right)}\right\} \quad (5)$$

where $C_k^j, k = 1, 2, \ldots, n$ are the segments of the same length as $Q^i$, split from model curve $C^j$.

### *3.3 The Proposed Architecture*

The flowchart of our method is shown in Fig. 2. First, we preprocess the original multi-sensor data to satisfy the latter experiment requirements. Second, we use an autoencoder trained with normal data to obtain the state variations of engines as degradation curves which are same as their HIs and build a degradation model template library. Then, we measure each test object with all degradation curves in the template library to get similarities and corresponding RULs based on CID. At last, we can get the estimated RUL by calculating the weighted average of highly relevant corresponding RULs.

## 4 Experiments

### *4.1 C-MAPSS Turbofan Engine Dataset and Performance Evaluation*

Commercial modular aero-propulsion system simulation is a tool for simulating a realistic turbofan engine, and Saxena et al. [17] modeled a series of turbofan engines as well as recorded their run-to-failure data as C-MAPSS dataset. As shown in

**Table 1** C-MAPSS dataset

| Sub dataset | FD001 | FD002 | FD003 | FD004 |
|---|---|---|---|---|
| Number of training engines | 100 | 260 | 100 | 249 |
| Number of test engines | 100 | 259 | 100 | 248 |
| Operating conditions | 1 | 6 | 1 | 6 |
| Fault modes | 1 | 1 | 2 | 2 |

Table 1, the dataset contains four subsets covering different operating conditions and fault modes, and then they are further divided into training and test subsets. Each engine operates normally at the beginning with unknown different initial wear and then develops faults at some different points during running. The faults keep growing until the system fails. Engines in training subset fail at the last cycle while in test subset recording data ends at some cycle which is prior to final failure. Each subset includes 26 columns, containing engine id, running time (in cycles), three-dimensional operating condition settings and 21-dimensional sensor data. The number of remaining cycles to failure of each test engine is given in another text file.

In some cases, predicting failure early is better than late. Late prediction may lead to accidents because condition-based maintenance is too late to perform, while early failure warning will not pose life threatening. So, an unbalanced evaluation indicator score is employed to penalize late prediction. Its definition is denoted as (6).

$$s = \begin{cases} \sum_{i=1}^{n} e^{-\left(\frac{d^i}{a_1}\right)} - 1 \text{ for } d^i < 0 \\ \sum_{i=1}^{n} e^{-\left(\frac{d^i}{a_2}\right)} - 1 \text{ for } d^i \geq 0 \end{cases} \tag{6}$$

where $s$ is the computed score, $n$ is the number of test engines, $d^i = \text{RUL}_{\text{est}}^i - \text{RUL}_{\text{true}}^i$ (Estimated $\text{RUL}^i$—True $\text{RUL}^i$) is the estimation error of engine $i$, $a_1 = 13$ and $a_2 = 10$.

We can know from (6) that if there are some instances whose $d$ is large enough, then the score will grow exponentially. Another evaluation index is root mean square error (RMSE), which is given in (7). Compared with the score, it penalizes estimation errors equally.

$$\text{RMSE} = \sqrt{\frac{1}{n}\sum_{i=1}^{n} d_i^2} \tag{7}$$

where $d^i = \text{RUL}_{\text{est}}^i - \text{RUL}_{\text{true}}^i$ (Estimated $\text{RUL}^i$—True $\text{RUL}^i$) is the estimation error of engine $i$, and $n$ is the number of test engines.

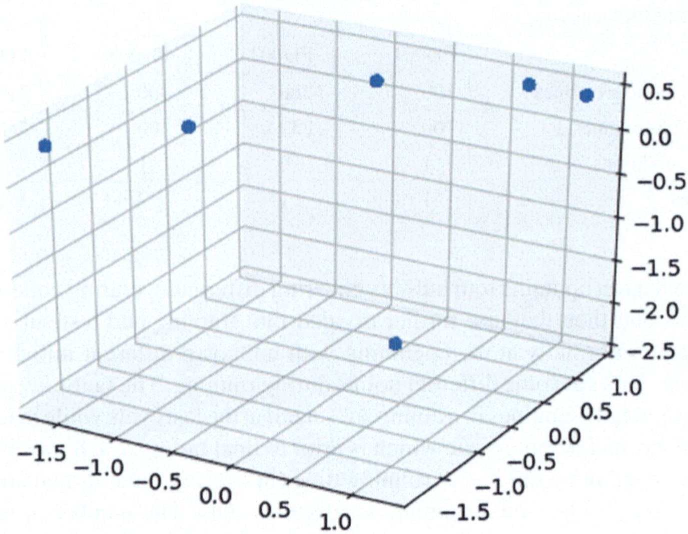


**Fig. 3** Clustering result of operating conditions based on K-means

## 4.2 Data Normalization

The raw data are collected from many different sensors with various value scales, and it is not beneficial to use these data to make estimation directly. To convert the raw data to desirable form, data preparation is necessary. In order to compare our results with previous studies, here we choose the same z-score normalization as follows:

$$z = \frac{x - \mu}{\sigma} \tag{8}$$

where $\mu$ is the mean, and $\sigma$ is the corresponding standard deviation.

However, engines in subsets FD002 and FD004 run in six different operating conditions, and it should be taken into consideration that the sensor parameters can be affected by various operating conditions. This problem can be solved by clustering all running data into six species based on operating conditions and conducting z-score normalization on cycles which belong to same condition separately. The clustering result of operating conditions based on K-means is shown in Fig. 3.

## 4.3 RUL Estimation

After reconstructing all engines in dataset, we can see that engines degrade variously and their lifetimes are individualized, which verify fixed RUL target function or degradation model is too restricted for complex systems. Figure 4 shows reconstruc-

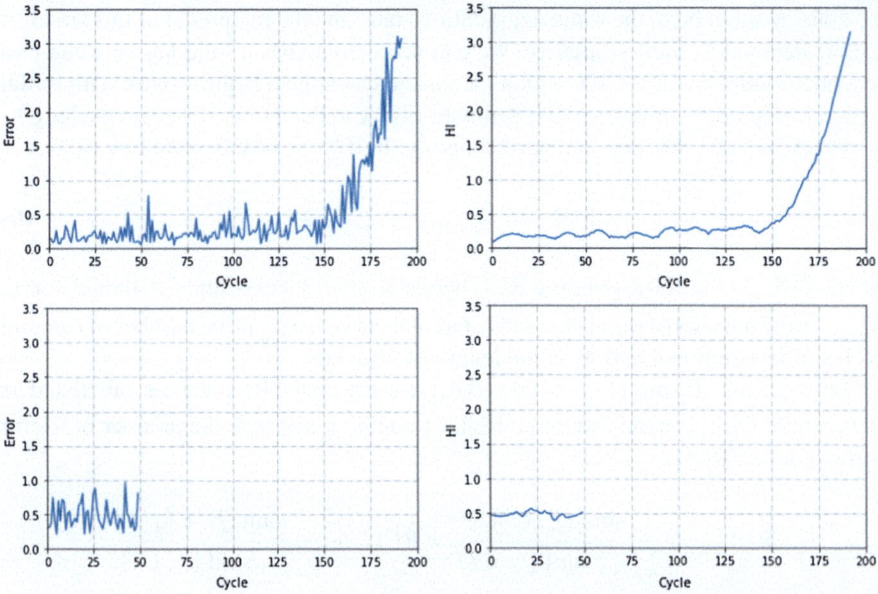


**Fig. 4** Reconstruction error (top left) and degradation curve (top right) of training engine, reconstruction error (bottom left) and degradation curve (bottom right) of test engine

tion errors and degradation curves of engines in training set and test set. Degradation curve is obtained by filtering the corresponding reconstruction error. We can see that fast degradation begins around 150th cycle in this training engine, and degradation curve of test engine starts at about 0.5, which indicates a slight deviation from the healthy state.

Figure 5 shows the process of getting similarity and corresponding RUL between test curve and model curve. A sliding window of the same length as test series is used to calculate CID with a step length of 1. Two sequences with smallest distance

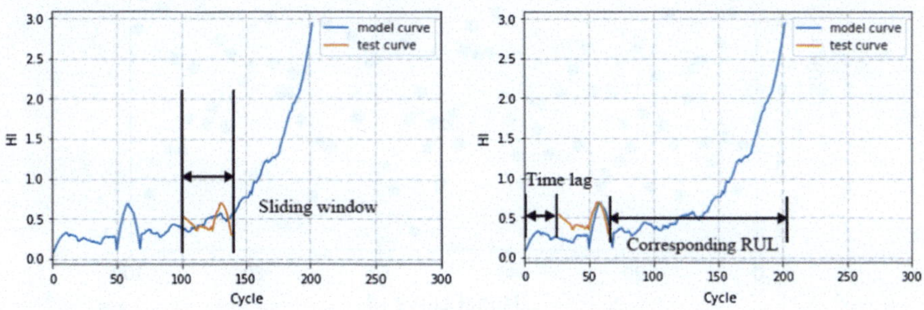


**Fig. 5** Segments in sliding window are combined to calculate CID with a step length of 1, and $\mathrm{RUL}_j^i$ is calculated after finding the most similar segment

are considered to be in the same degradation state, and the reciprocal of this smallest CID is denoted as their similarity. We can see there exists a time lag between two sequences with smallest CID, which means the test engine begins to run with initial wear and degrades similarly to the template engine at that point. Based on the lengths of two curves and the time lag, corresponding $\mathrm{RUL}_j^i$ can be calculated by (9).

$$\mathrm{RUL}_j^i = \mathrm{N}_{\mathrm{model}}^j - N_{\mathrm{test}}^i - N_t \tag{9}$$

where $\mathrm{RUL}_j^i$ is the corresponding RUL between *i*-th test curve and *j*-th model curve, $\mathrm{N}_{\mathrm{model}}^j$ is the number of running cycles of model curve *j*, $N_{\mathrm{test}}^i$ is the number of running cycles of test engine *i*, and $N_{\mathrm{t}}$ is the length of time lag.

After getting all $\mathrm{sim}\left(Q^i, C^j\right)$ and $\mathrm{RUL}_j^i$, the estimated RUL can be calculated as (10), where $\mathrm{RUL}_{\mathrm{est}}^i$ is the estimated RUL of engine *i*, and *m* is the number of useful corresponding RULs.

$$\begin{aligned}\mathrm{RUL}_{\mathrm{est}}^i = & \frac{\mathrm{sim}\left(Q^i, C^1\right)}{\sum_{j=1}^m \mathrm{sim}\left(Q^i, C^j\right)} \times \mathrm{RUL}_1^i + \frac{\mathrm{sim}\left(Q^i, C^2\right)}{\sum_{j=1}^m \mathrm{sim}\left(Q^i, C^j\right)} \\ & \times \mathrm{RUL}_2^i + \cdots + \frac{\mathrm{sim}\left(Q^i, C^m\right)}{\sum_{j=1}^m \mathrm{sim}\left(Q^i, C^j\right)} \times \mathrm{RUL}_m^i\end{aligned} \tag{10}$$

In order to achieve the results with high confidence, only those corresponding RULs whose similarities are not less than 70% of the maximum similarity are used. An example of one test engine's similarity distribution with all model curves is shown in Fig. 6.

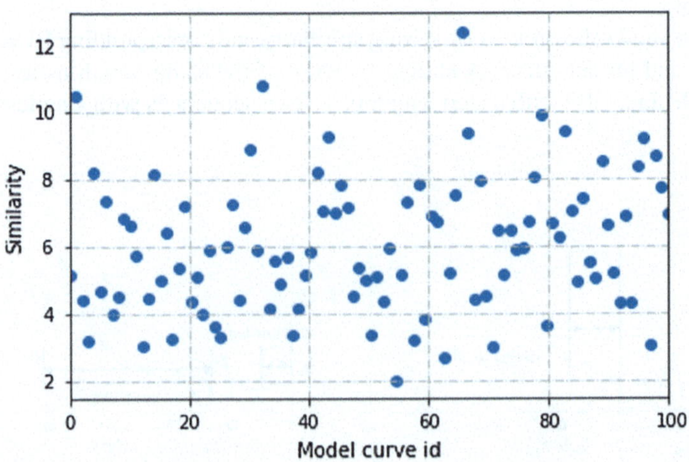


**Fig. 6** Example of one test engine's similarity distribution with all model curves. It is obvious that some model curves are far from the test curve

**Table 2** Score comparison on C-MAPSS dataset

| Sub dataset | FD001 | FD002 | FD003 | FD004 |
|---|---|---|---|---|
| SVR [10] | $1.38 \times 10^3$ | $5.90 \times 10^5$ | $1.60 \times 10^3$ | $3.71 \times 10^5$ |
| RVR [10] | $1.50 \times 10^3$ | $1.74 \times 10^4$ | $1.43 \times 10^3$ | $2.65 \times 10^4$ |
| CNN [10] | $1.29 \times 10^3$ | $1.36 \times 10^4$ | $1.60 \times 10^3$ | $7.89 \times 10^3$ |
| Deep LSTM [8] | $3.38 \times 10^2$ | $4.45 \times 10^3$ | $8.52 \times 10^2$ | $5.55 \times 10^3$ |
| Our method | $3.24 \times 10^2$ | $3.25 \times 10^3$ | $6.73 \times 10^2$ | $4.12 \times 10^3$ |

**Table 3** RMSE comparison on C-MAPSS dataset

| Sub dataset | FD001 | FD002 | FD003 | FD004 |
|---|---|---|---|---|
| SVR [10] | 20.96 | 42.00 | 21.05 | 45.35 |
| RVR [10] | 23.80 | 31.30 | 22.37 | 34.34 |
| CNN [10] | 18.45 | 30.29 | 19.82 | 29.16 |
| Deep LSTM [8] | 16.14 | 24.49 | 16.18 | 28.17 |
| Our method | 15.09 | 21.34 | 15.23 | 24.68 |

### *4.4 Performance Comparison*

As a public dataset, there are many studies on C-MAPSS dataset. Here, we compare our results with some researches which are conducted with the fixed RUL target function and same normalization method as well as evaluation indexes. As shown in Tables 2 and 3, our proposed method makes progress in both RMSE and score. This is because the unsupervised learning autoencoder can capture degradation information adequately and similarity measurement with CID can make full use of historical information.

## 5 Discussion and Conclusion

In this paper, we combine autoencoder and similarity measurement to estimate RUL of aircraft engines. An autoencoder is trained to capture hidden correlation of multi-sensor data and extract degradation curve of engines. During estimation, a sliding window and CID are used to calculate the similarity between test curves and each model curve, making full use of historical information. The experiment result proves that our method can obtain individualized degradation models of engines adequately and improve the accuracy of aircraft engine RUL estimation.

Though comparison with other researches shows our approach is effective, it should be noted that it is still limited. Since we need to build a degradation curve template library, training objects need to be run-to-failure, or in other words remaining cycles at certain time are necessarily known.

Further improvements are still available. In the process of degradation model extraction, we just use a basic autoencoder architecture, and more complex structure such as sparse autoencoder and denoising autoencoder can be tried to improve representative learning ability. In addition, different distance calculation methods can also be explored to make estimation more robust.

## References


1. Wang T (2010) Trajectory similarity based prediction for remaining useful life estimation. University of Cincinnati, Cincinnati
2. Marble S, Morton BP (2006) Predicting the remaining useful life of propulsion system bearings. In: Proceedings of the IEEE aerospace conference, pp 1–8
3. Giantomassi A, Ferracuti F, Benini A, Longhi S, Petrucci A (2011) Hidden Markov model for health estimation and prognosis of turbofan engines. In: ASME 2011 international design engineering technical conferences & computers and information in engineering conference, pp 1–6
4. Liu DT, Pang J, Zhou J, Peng Y (2012) Data-driven prognostics for lithium-ion battery based on gaussian process regression. In: Proceedings of IEEE conference on PHM, May 2012, pp 1–5
5. Ren L, Sun Y, Wang H, Zhang L (2018) Prediction of bearing remaining useful life with deep convolution neural network. IEEE Access 6:13041–13049
6. Xue F, Bonissone P, Varma A, Yan W, Eklund N, Goebel K (2008) An instance-based method for remaining useful life estimation for aircraft engines. J Fail Anal Prev 8(2):199–206
7. Hochreiter S, Jürgen Schmidhuber (1997) Long short-term memory. Neural Comput 9(8):1735–1780
8. Zheng S, Ristovski K, Farahat A, Gupta C (2017) Long short-term memory network for remaining useful life estimation. In: 2017 IEEE international conference on prognostics and health management (ICPHM). IEEE, pp 88–95
9. Heimes FO (2008) Recurrent neural networks for remaining useful life estimation. In: International conference on prognostics and health management, PHM 2008. IEEE, pp 1–6
10. Babu GS, Zhao P, Li XL (2016) Deep convolutional neural network based regression approach for estimation of remaining useful life. In: International conference on database systems for advanced applications. Springer, pp 214–228
11. Aamodt A, Plaza E (1994) Case-based reasoning: foundational issues, methodological variations, and system approaches. AI Commun 7(1):39–59
12. Batista GEAPA, Keogh EJ, Tataw OM, Souza VCMA (2014) CID: an efficient complexity-invariant distance for time series. Data Min Knowl Disc 28(3):634–669
13. Rumelhart DE (1986) Learning representations by back-propagating errors. Nature 323(6088):533–536
14. Park D, Hoshi Y, Kemp CC (2018) A multimodal anomaly detector for robot-assisted feeding using an LSTM-based variational autoencoder. IEEE Robot Autom Lett 3(3):1544–1551
15. Malhotra P, Ramakrishnan A, Anand G, Vig L, Agarwal P, Shroff G (2016) LSTM-based encoder-decoder for multi-sensor anomaly detection. In: Presented at ICML 2016 anomaly detection workshop, July 2016
16. Salvador S, Chan P (2007) Toward accurate dynamic time warping in linear time and space. Intell Data Anal 11(5):561–580
17. Saxena A, Goebel K, Simon D, Eklund N (2008) Damage propagation modeling for aircraft engine run-to-failure simulation. In: 2008 International conference on prognostics and health management, Denver, CO., pp 1–9

# Development of Ryerson's Hyperloop Pod Systems Using a Modular Approach

**Mohammed M. Khan**

**Abstract** Current modes of transportation tend to be slow or expensive or a combination thereof. The Hyperloop is the next mode of transportation that aims to shift this notion by being fast, and inexpensive after being commercialized. In order to drive a revolutionary change within the transportation industry to better interconnect cities, our work focused on the development of various Hyperloop pod systems using an innovative, and a modular approach. Modularity and standardization allowed for the rapid development of Hyperloop systems including MagLev and MagDrive systems for propulsion, mechanical and pneumatic systems for braking, and health monitoring and control systems for guidance, navigation, and control.

**Keywords** Hyperloop · SpaceX · Pod · SRAD · Modular · Systems · Propulsion · Braking · Control

## 1 Introduction

Ever since the release of the Hyperloop Alpha document, great strides have been undertaken within the Hyperloop sphere [1]. This has been made possible by the research and developmental efforts being carried out by universities and the industry. Current high speed travel is greatly impeded by air resistance. With aerodynamic drag increasing with the square of speed, a significantly large power input is required to go faster. This can be seen as power requirements need to be met with the cube of speed. By placing the pod in a low-pressure environment, this aerodynamic drag can be greatly reduced. This coupled with the implementation of contactless propulsion systems will aid in high-speed travel that would otherwise have resulted in contact friction.

---

M. M. Khan (✉)
Department of Aerospace Engineering, Ryerson University, 350 Victoria Street, Toronto, ON M5B 2K3, Canada
e-mail: mohammed.mohiuddin.k@ryerson.ca



Z. Jing (ed.), *Proceedings of the International Conference on Aerospace System Science and Engineering 2019*, Lecture Notes in Electrical Engineering 622,
https://doi.org/10.1007/978-981-15-1773-0_16

## 2 Hyperloop Tube

As the Hyperloop concept utilizes a low-pressure environment, the system is proposed to be closed loop. The drawing of a hard to near hard vacuum has been avoided due to the difficulties that would be faced in order to maintain them. In addition, operating a low-pressure environment would allow for the implementation of commercial off-the-shelf (COTS) pumping systems.

The development of Ryerson's Hyperloop Pod has been based on the track built by SpaceX. SpaceX took it upon itself to build a Hyperloop tube in order to accelerate the development of Hyperloop while providing university design teams the much needed resources to test, and validate their designs.

The Hyperloop tube at SpaceX is approximately a mile long steel tube fitted with two airlocks on either end. An internal guide rail spanning the entire length of the tube is placed within it, and is meant to be utilized by the pods. As the operating environment is a low-pressure environment, vacuum pumps have been fitted at regular intervals to maintain the tube within a predefined pressure range. With a pump down time of approximately 45 minutes to 1 hour, the temperature within the tube is based on the time of day, and the weather as it is located outside with no built-in thermal cooling methods.

## 3 Testing and Validation Challenges

With the SpaceX's tube being located at their HQ in Hawthorne, California, an early issue that needed to be addressed had to do with the testing and validation of the systems. The pod being designed along with all systems needed to operate within a low-pressure environment, using the track specifications as dictated by SpaceX at a varying temperature range.

As a university team, the possibility of traveling to Hawthorne more than once in a year was out of question mainly due to logistical and economic difficulties. To overcome this, a possibility that was looked into was to design and build an exact replica of the guide rail used by SpaceX. In doing so, testing could be conducted without the need to make multiple trips to SpaceX. In addition, it would open up the doors to an iterative design process where even minor modifications in the design can be tested and validated relatively quickly.

Being a downtown university, square footage is a sought after, and an expensive commodity. As a result, the possibility of acquiring the required square footage to lay down a track along with its accompanying safety measures was next to impossible. Therefore, alternative methods for testing and validating the systems were looked into.

# 4 The Modular Test Rig

For development of the Hyperloop pod, the entire vehicle can be categorized into three main systems: Propulsion (PRP), Structures (STR), and Guidance, Navigation and Control (GNC) with a few of their roles being listed in Table 1.

In order to overcome the aforementioned space limitations, an innovative approach was taken where a modular test rig capable of accommodating tests from PRP, STR, and GNC was designed. The development of such a rig as shown in Fig. 1 allowed for the validation of said systems individually, and while being integrated in conjunction with each other.

Furthermore, the need to lay down a guide rail was completely eliminated by making the test rig to rotate. By doing so, an infinite track was created while occupying very little square footage. Through this approach, a total of approximately 15 $ft^2$ was used rather than approximately 450 $ft^2$ that would have been needed to lay down

**Table 1** Hyperloop pod systems breakdown and developmental responsibilities

| System | Development responsibilities |
|---|---|
| Guidance, navigation, and control (GNC) | Health monitoring system, control, telemetry, pod keep alive command |
| Propulsion (PRP) | MagLev, MagDrive, power systems |
| Structures (STR) | Chassis, braking, thermal management, and vehicle dynamics |

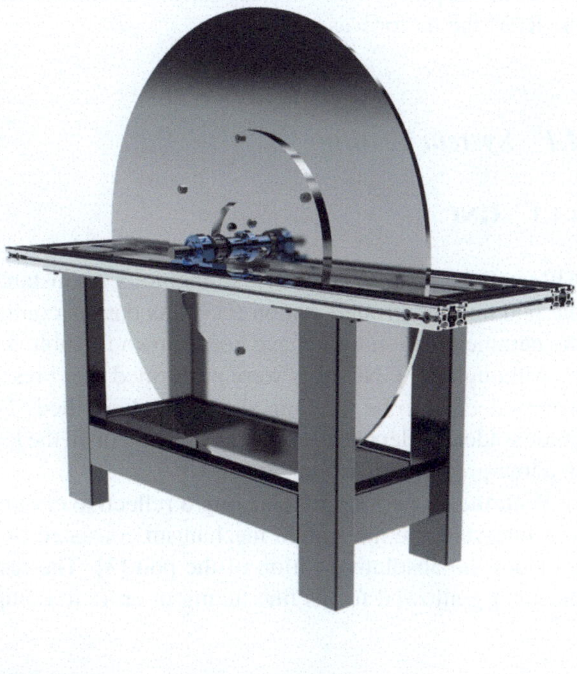

**Fig. 1** Modular test rig

**Table 2** Breakdown of the modular test rig and material selection

| Component | Purpose | Material selection |
|---|---|---|
| Support frame | Allows for the modular test rig to be bolted to the ground | High strength steel |
| System mounting frame | Allows for the integration of the track to the rig while providing a modular method of testing various systems | Aluminum (Al 6105-T5) |
| Track | Allows for characterization of system behaviors, such as PRP, STR, and GNC | Replication of the guide rail at SpaceX |
| Motor | Provides a means to rotate the track at a specified RPM | N/A |

around 150 ft of the test track. The test rig itself can be broken down into: support frame, system mounting frame, track, and the motor as shown in Table 2.

The manufacturing of the track using the same material as that of SpaceX's allowed for the characterization of the developed systems. It ensured similar behavior would be observed had they been conducted at SpaceX. This was critical to generate the much needed data for the MagLev, MagDrive, Eddie current braking, and GNC systems as they were dependent on the material properties of the track being used. In addition, using a large enough diameter for the track on the test rig provided a sufficient testing area. The incorporation of standard aluminum extrusions around track provided a method of support for the various systems that required testing all the while the pillow block bearings allowed for the track to rotate smoothly no matter the RPM the motor was set at.

## *4.1 System Testing*

### 4.1.1 GNC

Although the pod is completely autonomous, a constant stream of data flows between the pod and the ground station (GS). As data is continuously transmitted, some of the parameters monitored have been shown in Table 3.

Although the GNC tests were performed in a pressurized environment, all electronics needed to be vacuum, and thermal certified. This included the use of space grade solder, solder paste, and x-ray imaging of all the joints to guarantee performance in a low-pressure environment [2].

With the Hyperloop tube having a reflective circumferential tape placed at regular intervals, the navigation mechanism consisted of a stripe counting sensor that provides the absolute position of the pod [3]. The testing of the stripe counter on the test rig allowed for its fine tuning in order to compensate for reflectivity of the

**Table 3** Sample parameters monitored by GNC

| Parameter | Data transmission |
| --- | --- |
| Pod position, velocity, and acceleration | To GS |
| Pod health | To GS |
| Pod keep alive command | To pod |
| Pod manual control | To GS/To pod |
| Ambient tube temperatures and pressures | To GS |
| Pod component specific temperatures | To GS |
| Pod ground clearance height | To GS |
| Pneumatic air tank and piston pressures | To GS |
| Brake positions | To GS/To pod |

**Table 4** GNC pod test and run states

| State | Description |
| --- | --- |
| Fault | Automatically stops the pod when any non-nominal, unknown, or unsafe circumstances are observed |
| Safe to approach | Indicated that the pod is stationary, in standby condition, and is safe to be approached by humans |
| Ready to launch | Test systems and prepare systems for launch |
| Launch | Accelerate pod to a predefined target speed |
| Levitation | Deploy MagLev system upon reaching predefined target speed |
| Braking | Bring the pod to a safe halt |
| Crawling | Move the pod at a very low speed using manual control from GS |

material as the laser needed to be reflected off it. Furthermore, the lighting set up present within the tube could be accounted for in order to generate accurate results. In addition to validating the method used to determine the location of the pod, some of the pod states shown in Table 4 could also be tested.

### 4.1.2 Structures Testing

When designing the braking system, key considerations included the distance it takes to stop, braking effectiveness, reliability of the system, and the resulting temperature change of the track. As the pod undergoes multiple state changes as outlined Table 4, the ability to deploy the brakes is present in all of them. As a result, multiple tests were needed to characterize them.

The pod contains two independent pneumatic braking systems while each system has the ability to deploy two brake pistons simultaneously. To ensure the stability of the pod during all braking phases, the braking systems were placed at the front and rear of the pod. To validate this design, a simplification was made where only one of the systems was mounted on the test rig. This was done to obtain braking data for

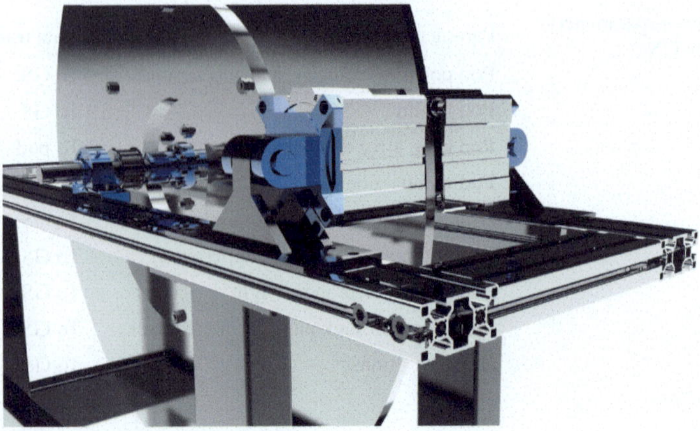

**Fig. 2** Test rig set up with brakes mounted on either side

the worst-case scenario, i.e. one system is inoperable. This set up can be observed in Fig. 2 where the two pistons are mounted to the mounting frame on either side of the test track.

For these tests, Table 5 outlined some of the parameters that were used. Data gather showed that this braking set up would be sufficient in bringing the pod to a safe stop within the predefined braking distance. This was found to be sufficient in combination with one set being inoperable while still resulting in a factor of safety (FS) of two.

After the determination that the braking system was capable of bringing the pod to a stop, its thermal characteristics were determined. Results showed that a temperature change of 11.895 °C would occur as a result of the heat dissipation between the brake pads and the track itself. Conduction was assumed to be the only source of heat dissipation as the operating environment would be a low pressure one. Furthermore, as convection and radiation account for a fraction of the value compared to conduction, they were ignored.

**Table 5** Worst-case braking test parameters

| Parameter | Value |
|---|---|
| Minimum coefficient of kinetic friction | 0.30 |
| Maximum velocity | 60 m/s |
| Operational systems | 1 of 2 |
| Operational pistons | 2 of 4 |
| Scenario | Worst case |

### 4.1.3 Propulsion Testing

In order to develop a propulsion system capable of propelling and levitating through a true contactless system, they had to be researched and developed from the ground up. MagLev and MagDrive were the two main student researched And developed (SRAD) systems of the propulsion system [4].

MagLev is derived from the magnetic levitation system that is used to support the pod along the normal direction at high speeds [5]. It uses a Halbach array to induce a repulsive magnetic field to suspend the pod at a particular height above the track. Some of the main advantages of incorporating such a system were found to be:

- Contact or frictional drag was reduced when compared to wheels at high speeds
- A resulting improvement in the pod's acceleration profile
- The design of a completely static system with no moving parts led to a much simpler design overall.

With the MagLev system being SRAD, it needed to be validated thoroughly to show that a sufficient force or lift could be generated in order to levitate the pod. To do so, a dual axis force transducer (DAFT) was developed to be tested on the modular test rig as shown in Fig. 3. The DAFT was capable of measuring the lift and drag forces as a function of the test rig's velocity.

Due to unforeseen issues, the test apparatus was only able to run up to a peak velocity of 8 m/s while holding the rotations at a constant speed reliably. Although this issue was encountered, tests were performed using the DAFT as shown in Fig. 4 with a singular North–South period configuration of N42 magnets measuring 1″ by 1″ by 0.5″. In Fig. 3, the readings taken between 0 and 4 clearly illustrate the ramp up in the test rig's speed. Nearing peak velocity, the lift to drag ratio was observed to be 1, and can be seen in Fig. 5 as well.

Using the results obtained from testing, a theoretical profile was generated to show the behavior had the speed been increased further based on the tests conducted

**Fig. 3** Dual axis force transducer (DAFT)

**Fig. 4** Test rig showcasing one of the two SRAD DAFTs mounted on it

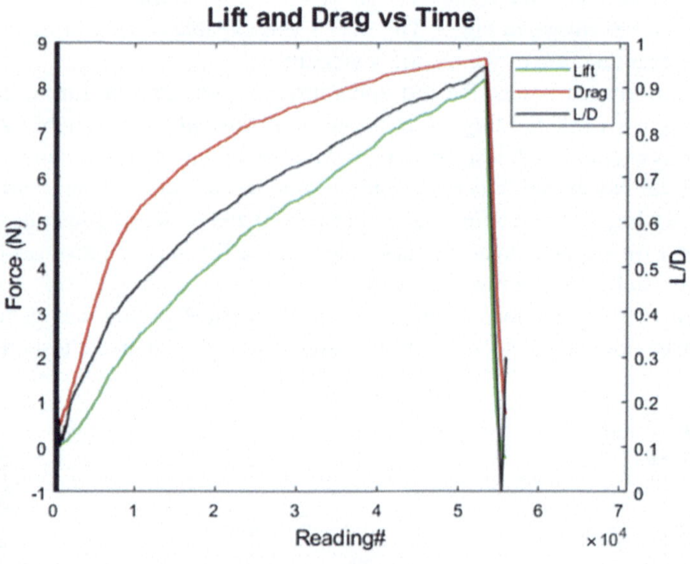


**Fig. 5** Testing results of lift and drag versus readings

by MIT [6]. Figure 4 also showcases the crossover point where drag begins to depreciate considerably as lift continues to increase. This occurs up a certain point after which both the parameters being to plateau. Using the observed trends, a relationship was developed along with it being used to calculate the lift at levitation speed. The relationship has been tabulated by MIT as shown in Table 6. This estimate will be refined further based on tests resulting from the test rig going even faster Fig. 6 [6].

**Table 6** Design parameter performance effect

| Design parameter | Effect on lift | Effect of L/D | Effect on weight |
|---|---|---|---|
| Array wavelength | Slight decrease | Square root increase | Increase |
| Number of periods | Linear increase | No effect | Increase |
| Array width | Linear increase | No effect | Increase |
| Array thickness | Squared increase | Negligible | Increase |
| Magnet grade | Increase | No effect | No effect |
| Back iron thickness | Slight increase | Negligible | Increase |
| Nominal gap height | Inverse square decrease | No effect | No effect |

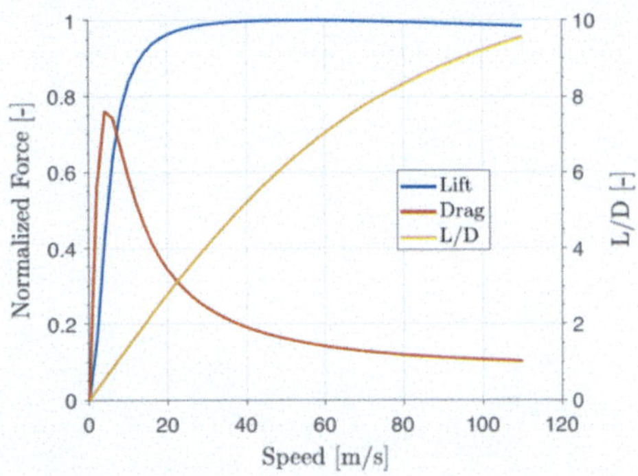


**Fig. 6** Simulation of lift and drag versus speed

## 5 Conclusion

Having the modular test rig ensured that the SRAD systems could be validated with relative ease. Their characteristics and behaviors could be replicated without the need to utilize SpaceX's track during the various developmental phases. With the test rig not being in a low-pressure environment, research still needs to be conducted on how the braking system would behave when operated in one. Further testing needs to be conducted using more magnet combinations after resolving the issues encountered during MagLev tests.

## References

1. SpaceX (2013) Hyperloop Alpha, Hawthorne
2. Reliability Prediction of Electronic Equipment 1990
3. SpaceX (2019) 2019 SpaceX Hyperloop Pod Competition, Hawthorne
4. Bhamidi SP (2005) Design of a single sided linear induction motor (SLIM) using a user interactive computer program, Missouri
5. Boldea I (2017) Linear electric machines, drives, and MAGLEVs handbook. CRC Press, Boca Raton
6. MIT Hyperloop Final Report, 2017

# Revisited: Machine Intelligence in Heterogeneous Multi-Agent Systems

**Kaustav Jyoti Borah and Rajashree Talukdar**

**Abstract** Machine-learning techniques have been widely applied for solving decision-making problems. Machine-learning algorithms perform better as compared to other algorithms while dealing with complex environments. The recent development in the area of neural network has enabled reinforcement learning techniques to provide the optimal policies for sophisticated and capable agents. In this paper, we would like to explore some algorithms people have applied recently based on interaction of multiple agents and their components. We would like to provide a survey of reinforcement-learning techniques to solve complex and real-world scenarios.

**Keywords** Machine learning · Heterogeneous systems · Multi Agents · Q learning

## 1 Introduction

An entity that recognizes its ambience with the help of sensors and uses its effectors to act upon that environment is called an agent [1]. Multi-agent systems, a subfield of machine learning, are used in many intelligent autonomous systems to make it smarter. To coordinate with other independent agents' behaviour in multiple agent systems environment, ML techniques aims to provide principles for construction of complex systems. When the agents are not dependent of one another in such a way that they can have approach to the environment independently. Hence, they need to embrace new circumstances. Therefore, learning and exploring about the environment demand the inclusion of a learning algorithm for each agent. Some amount of interaction becomes mandatory amongst the different agents in a multi-agent system for them to act like a group. However, Internet technology becomes useful for the application of software agents which gets high importance in academia and commercial institutes.

K. J. Borah (✉)
Department of Aerospace Engineering, Ryerson University, Toronto, Canada
e-mail: kborah@ryerson.ca

R. Talukdar
Department of Computer Science and Engineering, SMIT, Majitar, India

 
Z. Jing (ed.), *Proceedings of the International Conference on Aerospace System Science and Engineering 2019*, Lecture Notes in Electrical Engineering 622,
https://doi.org/10.1007/978-981-15-1773-0_17

Based on the heterogeneity of agent systems, multi-agent systems can be further subdivided into two categories (1) Homogeneous and (2) Heterogeneous. Homogeneous MAS includes agents that all have the same characteristics and functionalities, while heterogeneous MAS includes agents with variety of features, that means the heterogeneous systems are composed of many subagents or subcomponents which are not uniform throughout. It is a distributed system and contains many other hardware and software that works together in cooperative fashion to solve a critical problem. Examples of heterogeneous systems are smart grids, computer networks, stock market, power systems distributions in aircraft, etc.

A good advantage about multi-agent learning is that, the performance of the agent enhances every moment. Multi-agent learning systems are known for the exchange of informations between their agents and interacting with the environment. All the algorithms in machine learning which were developed can be transferred to settings where there are multiple, interdependent and interacting learning agents [2]. However, they may need alteration to take into account about other agents in the environment [3, 4]. In our paper, we focussed on different machine-learning techniques. The paper organizes as follows: Section. 3 describes the environments in multi-agent system, Sect. 4 describes general heterogeneous system architecture, Sect. 5 describes machine-learning techniques followed by Sect. 6 applications and future work and Sect. 7 is conclusion.

## 2 Ongoing Research

This research is carried out during my Ph.D. studies so far has been focused primarily on multi-agent systems and machine learning.

## 3 Multi-Agent System Environments

Agents can operate in many different types of environments. The main categories are summarized below, in mutually excluding pairs, based on the definitions provided by [5, 6]

1. *Static environments*: Based on agents action, the environment reacts.
2. *Dynamic environments*: If there is no input from an agent, still the environment can change, to potentially unknown states.
3. *Fully observable environments*: At each time step, the full state of environment is available.
4. *Partially observable environments*: All the state of environment is not available, only some part will be available.
5. *Deterministic environments*: To get the next state of environment, the agent has to take actions in the current state.

6. *Stochastic environments*: Current state action can lead to another state if needed.
7. *Stationary environment*: a stationary environment does not evolve over time and has a predefined set of states.
8. *Non-stationary environment*: a non-stationary environment evolves over time and can lead an agent to previously unencountered states.

## 4 General Heterogeneous Multi-Agent Architecture

The figure here shows general multi-agent scenario for heterogeneous communicating agents. These agents are communicating to each other to fulfil a global goal. In the figure above, we have "m" group of heterogeneous agents where $A = \{A^1, A^2, \ldots, A^m\}$ are the group of agents. Inside group 1, we assume that all agents are homogeneous. All the arrow marks with blue colours show a communication link between the agents meaning all agents can communicate (Fig. 1).

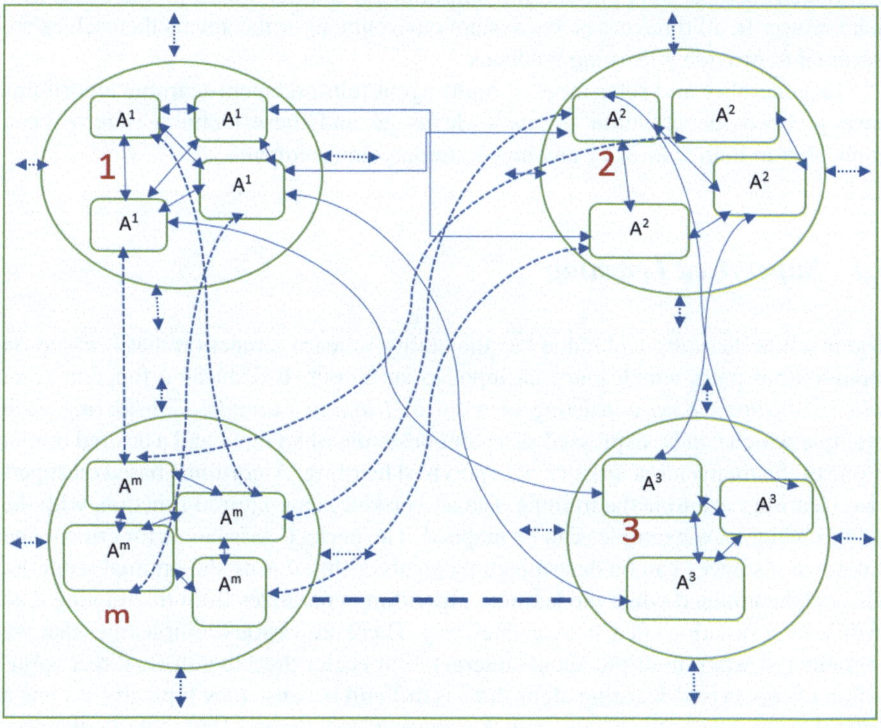


**Fig. 1** General heterogeneous architecture

# 5 Machine-Learning Techniques

Artificial Intelligence has a well-known subfield under its domain which is identified widely as multi-agent systems. It makes a prominent influence on devising and resolving extremely difficult framework problems. In multi-agent systems; an agent has no clear definition and, therefore, as defined previously, the agents are considered as entity with goals, action and domain knowledge in the environment. The agents will function as a "behaviour" and although the capability to coordinate the behaviours of autonomous agents is a recent concept yet development in the field is quick by building upon the existing work conducted earlier in the field of distributed artificial intelligence (DAI) [7].

The perspective of machine learning is mostly based on the type of the reward or return that critic can furnish as a response to learner. The three main techniques [2] that were widely discussed previously were (1) Supervised learning (2) Unsupervised learning and (3) Reinforcement learning. In supervised learning, the correct result of the system is provided by the critic where as in unsupervised learning, no response is guaranteed at all. While in reinforcement learning; actor-critic techniques, a newly introduced method, is applied to the output of the learner which produced an ultimate return. In all three cases, the system environment or the agents themselves are assumed to provide a learning feedback.

The generality and robustness of multi-agent reinforcement-learning algorithms have grabbed the attention for its wide usage and these techniques have been applicable in both stationary and non-stationary environments.

## *5.1 Supervised Learning*

The machine-learning technique has the ability to learn a function that is based on input–output pairs which maps an input to an output. It deduces a function from *labelled-training data* consisting of a set of *training example.* Considering each example which can be expressed as a pair consisting of an input and a desired output is one of the main characteristics of supervised learning. Algorithms based on supervised learning examine the training data and produces an inferred function, with the help of which new examples can be mapped. The perfect class labels for critical and unknown instances can be determined by an algorithm during the optimal scenario. This can be attained when the learning algorithm generalizes from the training data to unknown situations in a "reasonable" way. There are various complexities that are encountered when multiple agents interact with each other; therefore, direct application of supervised-learning algorithms is difficult because they typically assume a critic that can confer the agents with the "correct" behaviour [2] for a particular specific situation. Supervised-learning method was used by Sniezynski [8] for the fish bank game. The interaction in heterogeneous agents' environment is improved by Garland and Alterman [9] with the use of learning-coordinated procedures. Williams

**Table 1** Supervised learners compared to learning aspects [2]

| | Cen/Decentralized | Coordination | Learning | Final goal |
|---|---|---|---|---|
| Sniezynski [8] | Centralized | No | 2-agents | Selfish |
| Garland and Alterman [9] | Decentralized | Yes | All-agents | Cooperative |
| Williams [10] | Decentralized | Yes | All-agents | Cooperative |
| Gehrke and Wojtusiak [11] | Decentralized | Yes | All-agents | Cooperative |
| Airiau et al. [12] | Decentralized | Yes | All-agents | Cooperative |

[10] and his team delineated inductive learning methods to understand individual's attitude's in the area of semantic Webs. A rule induction technique was applied by Gehrke and Wojtusiak [11] in their research for route planning. Learning abilities are incorporated into BDI model by Airiau et al. [12], in which decision-tree learning is utilized to enhance plan applicability testing. Table 1 shows few important characteristics of above-supervised learners compared to the learning aspects [2].

## *5.2 Unsupervised Learning*

In this case, there is no explicit concepts are given. Many researchers have already reviewed the unsupervised-learning algorithms applied to multi-agent-based systems and conclude that, this is not a suitable technique for multi-agent applications. Main approach in multiagent is to increase the overall performance, but unsupervised learning is supposed to be an aimless approach for this application. Hence, minimal research has been witnessed in this area. However, unsupervised-learning algorithms were used by the researchers to find a solution to the auxiliary issues that help agent through its learning [2, 13, 14].

## *5.3 Reinforcement Learning*

Reinforcement learning stands different from the other two methods mentioned above, in a way that it is a method in which the paradigm of the agent is matched exactly, whereas for the supervised and unsupervised learning, the learner must be provided with healthy data [15]. Autonomous agent that has no prior knowledge behaviour of the system or of the environment is assigned by reinforcement learning by gradually enhancing its performance based on given returns as the learning task is carried out [14]. Hence, a huge percentage of people working in this area used reinforcement-learning algorithms. As stated in [2], the literature available for reinforcement-learning algorithms of multi-agent systems can be divided into two different subsets: (1) Estimating value function-based methods and (2) Stochastic search-based methods in which behaviours are learnt directly by the agents without involving value functions and it focusses on evolutionary computation [2]. Single and multiple agents' algorithms [2] are available for learning methods based on estimate value functions. A Markov decision process also known as MDP is known to the environment in reinforcement learning as most of reinforcement-learning algorithms uses dynamic programming approach. There are many algorithms based on reinforcements learning. Comparison of various reinforcement-learning algorithms are [2] (Tables 2 and 3).

**Table 2** Comparison of various reinforcement-learning algorithms [2]

| Researchers | Cen/Decentralized | Coordination b/w agents | Learning | Final goal |
|---|---|---|---|---|
| Bowling and Veloso [16] | D | Yes | All-agents | Cooperative and competitive |
| Barto et al. [17] | C | No | All-agents | Selfish |
| Sutton [18] | C | No | All-agents | Selfish |
| Moore and Atkeson [19] | C | No | All-agents | Selfish |
| Greenwald and Hall [20] | D | Yes | All-agents | Competitive |
| Kononen [21] | D | Yes | One-agent | Cooperative |
| Lagoudakis and Parr [22] | C | No | One-agent | Selfish |
| McGlohon and Sen [23] | D | Yes | All-agents | Cooperative |
| Qi and Sun [24] | D | Yes | All-agents | Cooperative |
| Puterman [25] | C | No | All-agents | Selfish |
| Watkins and Dayan [26] | C | No | All-agents | Selfish |
| Bertsekas [27] | C | No | All-agents | Selfish |

**Table 3** Standard algorithms for reinforcement-learning [2]

| Algorithm | Description | Model | Policy | Action space | State space | Operator |
|---|---|---|---|---|---|---|
| Monte Carlo | Every visit to monte Carlo | Model free | Off | Discrete | Discrete | Sample-means |
| Q-Learning | State–action reward state | Model free | Off | Discrete | Discrete | Q-value |
| SARSA | State–action reward state action | Model free | On | Discrete | Discrete | Q-value |

### 5.3.1 Monte–Carlo Method

Monte–Carlo Method [28] obtains value function producing the episode repeatedly and it keeps a note about the average return at each state or each state–action pair. Thus, the calculation of the state-value function is carried out as follows:

$$V_{\pi}^{\mathrm{MC}}(s) = \lim_{i \to +\infty} E\left[r^{i}(s_t)|s_t = s, \pi\right] \tag{1}$$

where $r^{i}$ $(s^{t})$ symbolizes observed returns at state $s^{t}$ in episode $i$th. Similarly, the value function of state–action pair is given as follows:

$$Q_{\pi}^{\mathrm{MC}}(s, a) = \lim_{i \to +\infty} E\left[r^{i}(s_t, a_t)|s_t = s, a_t = a, \pi\right] \tag{2}$$

MC method is known as model free as prior knowledge about transitional probabilities is not mandatory in this method.

However, for convergence to take place, this method is based on two vital assumptions [28]

(1) it has a large number of episodes and
(2) visiting every state and every action should be carried out for a large number of times. Making this "exploration" a feasible one, usage of e-greedy strategy in policy improvement should be made:

$$R_S : \pi \to \pi' = \psi'(s) = \left\{ \begin{array}{l} 1 - \epsilon + \frac{\epsilon}{|\Delta_\Pi(s)|}, a_i = a_j \Lambda j = \arg\max Q_\Pi(s \cdot a_k) \\ \frac{\epsilon}{|\Delta_\Pi(s)|}, \forall a_i \in \Delta_\pi \wedge a_i \neq a_j \end{array} \right\} \tag{3}$$

where $|\Delta_{\pi}(s)|$ depicts number of candidate action taken in state $s$ and $0 < \epsilon < 1$. Usually, the division of MC algorithm are made into two groups: on-policy and off-policy. In on-policy method, application of policy $\pi$ is carried out for both evaluation and exploration purpose. Therefore, the policy $\pi$ should be stochastic or soft. But off-policy uses different policy $\pi' \neq \pi$ in order to produce the episodes and hence, $\pi$ can be deterministic. Off-policy method is more attractive due to its simplicity, but the stability of on-policy method is more when dealing with continuous state space problems and when applying it together with function approximator, such as neural networks.

### 5.3.2 Temporal Difference Method

Temporal difference (TD) is a model-free approach that learns from its experiences [28]. This method in order to make an update does not wait for the episode to get over. Every step is updated within the episode by leveraging 1-step Bellman equation; therefore, a faster convergence is possibly feasible for this method:

$$U_1 : V^i(s_t) \leftarrow aV^{i-1}(s_t) + (1-\alpha)(r_{i+1} + \gamma V^{i-1}(s_{t+1})) \tag{4}$$

where $\alpha$ is a step size parameter and $0 < \alpha < 1$. It makes use of previously calculated values $V^{i-1}$ for updating of the current one's $V^i$, which is known as bootstrapping method, which has the advantage of learning fast over the non-bootstrapping methods in most of the cases. TD learning is also divided into two major categories: On-policy TD learning (Sarsa) and Off-policy TD learning (Q-learning). In Sarsa algorithm, estimating value function of state–action pair based on:

$$U_2 : Q^i(s_i, a_t) \leftarrow \alpha Q^{i-1}(s_{t,}a_t) + (1-\alpha)(r_{t+1} + \gamma Q^{i-1}(s_{t+1}, a_{t+1})) \tag{5}$$

On the other hand, single-step optimality Bellman equation is used by Q-learning to carry out the update, i.e., direct approximation of value function of optimal policy is done by Q-learning:

$$U_3 : Q^i(s_t, a_t) \leftarrow \alpha Q^{i-1}(s_{t,}a_t) + (1-\alpha)\left(r_{t+1} + \gamma \max_{a^j_{t+1}} Q^{i-1}(s_{i+1}, a^j_{t+1})\right) \tag{6}$$

It can be noticed that the operator max in update rule substitutes for a deterministic policy and this strongly explains why Q-learning is off-policy.

Tabular structure is used by these two methods in order to store the value function of each state or each function pair. However, for solving complicated problems which consist of a number of stages, it becomes insufficient due to lack of memory. Therefore, actor-critic (AC) method is introduced to overcome these limitations. AC includes two memory structures for an agent [28]: actor structure is utilized for selection of appropriate action in accordance to the observed state and transfers the same to critic structure for evaluation. Critic structure uses the following equation as a TD error to decide future tendency of a selected action.

$$\delta(a_t) = \beta\big(r_{t+1} + \gamma V_{(s_{t+1})}\big) - (1-\beta)V_{(s_t)} \tag{7}$$

### 5.3.3 Q-Learning

In order to learn agent systems or sub-systems, Q-learning is the widely used reinforcement-learning based approach. The final objective of a Q-learning is to evaluate a state of action that is so-called policy that results in maximum utility for the agent. After one action is executed, there will be reward for each policy that is generated. In simple word, the environment is explored in which a reward is observed by experimenting the different actions such that the agent learns in the process. In a finite Markov decision process, an optimal action selection can be identified by Q-learning provided indefinite time for exploration is given and a party random policy. The result of Q-learning, i.e., the policy, may be observed as a table in which a numeric value is assigned to each state–action pair ($s$, $a$), that gives an estimate of the (possibly long-term) reward to be received when implementing $a$ in $s$. After receiving a reward, an agent performs the required numerical value updating of the state–action pair, based on the reward and on the evaluated best reward to be obtained in the new state. Now with time, the agent is capable of enhancing its estimates of the rewards to be obtained for all state–action pairs. The reader can find the Q-learning algorithm in more details [29, 30].

There are some advantages of Q-learning which includes (1) Absolute convergence towards the optimum and (2) Natural applicability to agent systems because of the coupling of learning and exploration. However, it has drawbacks that include (1) Designing numerical suitable reward function can be a nontrivial task, (2) Convergence to optimal and (3) No explanation is given for any action preferences in learning the result.

## 6 Applications and Future Work

One major application in this area is development of an algorithm which can perform very well in airport ground handling management system. In this communication, the ground handling fleet management problem [31] at airports is taken into consideration with the objective of enhancing aircraft service at arrival and departure terminals, while we consider the ground cost issues. After collaborative discussion between ground management, airport authorities and airline, a machine-learning technique will address to solve this multi-agent-based problem.

Another example involves prediction in the stock market. Portfolio management in the stock trading [32] results in a successful handling and presents a theory-based foundation for a stock trading system. The overall portfolio management tasks include extracting the user profiles, collecting information on initial portfolio position of user, on behalf of the user, it observes the environment and suggesting better decisions, so the investment goals of the user meet the requirements, and making decision suggestions to meet the investment goals of the user. Based on the requirement analysis, Davis [32] presented a framework for a multi-agent system for stock trading (MASST). The primary issues addressed, it provides with different information sources and by interacting the agents and providing decision making for investors in the stock market. The candidate agents are identified and also the tasks that the agents perform. Agent interaction and exchange of information and knowledge between agents also have been described.

## 7 Conclusion

This research addressed some machine-learning techniques for agents and multi-agent systems [2]. Furthermore, we also have provided two examples for reader as an application of multi-agent-based systems that are being developed at Ryerson University, Toronto, Canada. While machine-learning methods applied to multi-agent systems still require more attention to prove their practicality. ML for multiagents is still a relatively new research area, and there are lot of open issues that require further development, some of them have already been mentioned.

**Acknowledgements** I would like to thank my wife Priyanka Talukdar, research scholar, department of Civil Engineering of IIT-Guwahati (India) for her valuable suggestions in shaping this paper. This survey was funded by Natural Sciences and Engineering Research Council (NSERC) Canada and my supervisor in Ryerson University, Canada.

## References

1. Shoham Y, Leyton-Brown K (2009) Multiagent systems algorithmic, game-theoretic, and logical foundations. Cambridge University Press
2. Khalil KM, Abdelaziz M, Nazmy TT, Salem ABM (2015) Machine learning algorithms for multi agent systems. In: Proceedings of the international conference on intelligent information processing, security and advanced communication—IPAC'15
3. Yang Z, Shi X (2014) An agent-based immune evolutionary learning algorithm and its application. In: Proceedings of the intelligent control and automation (WCICA), pp 5008–5013
4. Qu S, Jian R, Chu T, Wang J, Tan T (2014) Computational reasoning and learning for smart manufacturing under realistic conditions. In: Proceedings of the Behavior, Economic and Social Computing (BESC) Conferences, pp 1–8
5. Marinescu A (2016) Prediction-based multi-agent reinforcement learning for inherently non-stationary environments. PhD thesis, Computer Science, University of Dublin, Trinity College
6. Russell S, Norvig P (2003) Artificial intelligence: a modern approach. Prentice Hall
7. Stone P, Veloso M (2008) Multiagent systems: a survey from a machine learning perspective. Auton Robot 8(3):345–383
8. Sniezynski B (2009) Supervised rule learning and reinforcement learning in a multi-agent system for the fish banks game. In: Theory and novel applications of machine learning
9. Garland A, Alterman A (2004) Autonomous agents that learn to better coordinate. Auton Agent Multi-Agent Syst 8:267–301
10. Williams A (2004) Learning to share meaning in a multi-agent system. Auton Agent Multi-Agent Syst 8:165–193
11. Gehrke JD, Wojtusiak J (2008) Traffic prediction for agent route planning. In: Proceedings of the international conference on computational science, pp 692–701
12. Airiau S, Padham L, Sardina S, Sen S (2008) Incorporating learning in BDI agents. In: Adaptive Learning Agents and Multi-Agent Systems Workshop (ALAMAS + ALAg-08)
13. Kiselev A (2008) A self-organizing multi-agent system for online unsupervised learning in complex dynamic environments. In: Proceedings of the Twenty-Third AAAI Conference on Artificial Intelligence, pp 1808–1809
14. Sadeghlou M, Akbarzadeh TMR, Naghibi SMB (2014) Dynamic agent-based reward shaping for multi-agent systems. In: Proceedings of the Iranian Conference on Intelligent Systems (ICIS), pp 1–6

15. Lewenberg Y (2017) Machine learning techniques for multiagent systems. In: Proceedings of the Twenty-Sixth International Joint Conference on Artificial Intelligence (IJCAI-17), pp 5185–5186
16. Bowling M, Veloso M (2002) Multiagent learning using a variable learning rate. Artif Intell 136:215–250
17. Barto AG, Sutton RS, Anderson CW (1983) Neuronlike adaptive elements that can solve difficult learning control problems. IEEE Trans Syst Man Cybern 5:843–846
18. Sutton RS (1990) Integrated architectures for learning, planning, and reacting based on approximating dynamic programming. In: Proceedings of the Seventh International Conference on Machine Learning (ICML-90), Austin, US, pp 216–224
19. Moore AW, Atkeson CG (1993) Prioritized sweeping: reinforcement learning with less data and less time. Mach Learn 13:103–130
20. Greenwald A, Hall K (2003) Correlated-Q learning. In: Proceedings of the Twentieth International Conference on Machine Learning (ICML-03), Washington, US, pp 242–249
21. Kononen V (2005) Gradient descent for symmetric and asymmetric multiagent reinforcement learning. Web Intell Agent Syst 3:17–30
22. Lagoudakis MG, Parr R (2003) Least-squares policy iteration. Mach Learn Res 4:1107–1149
23. McGlohon M, Sen S (2004) Learning to cooperate in multi-agent systems by combining Q-learning and evolutionary strategy. In: Proceedings of the world conference on lateral computing
24. Qi D, Sun R (2003) A multi-agent system integrating reinforcement learning, bidding and genetic algorithms. Web Intell Agent Syst 1:187–202
25. Puterman ML (2008) Markov decision processes: discrete stochastic dynamic programming, 1st edn. Wiley
26. Watkins CJCH, Dayan P (1992) Q-learning. Mach Learn 8:279–292
27. Bertsekas DP (2001) Dynamic programming and optimal control, 2nd edn. Athena Scientific
28. Nguyen TT, Nguyen ND, Nahavandi S (2019) Deep reinforcement learning for multi-agent systems: a review of challenges, solutions and applications. retrieved from arXiv:1812.11794v2 [cs.LG] 6 Feb 2019
29. Mitchell T (1997) Machine learning. McGraw-Hill, New York
30. Kaebling LP, Littman ML, Moore AW (1996) Reinforcement learning: a survey. J Artif Intell Res 4
31. Fitouri Trabelsi S, Alberto NCC, Gustavo ZCL, Mora-Camino F (2013) AN operational approach for ground handling management at airports with imperfect information. In: 19th International conference on industrial engineering and operations management, Valladolid, Spain, July 2013
32. Luo Y, Davis D, Liu K (2002) A multi-agent framework for stock trading. School of Computing. Staffordshire University, Stafford ST18 0DG, UK, Department of Computer Science, University of Hull, HU6 7RX, UK

# Dynamics of Partial Space Elevator with Parallel Tethers and Multiple Climbers

**Gangqiang Li and Zheng H. Zhu**

**Abstract** This paper proposes a novel concept of partial space elevator with parallel tethers and multiple climbers. The parallel tethers impacting on the dynamic response of partial space elevator is investigated based on a high-fidelity and accurate model of PSE. The model is developed based on the nodal position finite element method in the arbitrary Lagrangian–Eulerian description. The results show that the tethers collide when the transient motion of climbers is not the same, such as the movement direction of climbers is opposite and the time delay between climbers. It also found that the multiple climbers of each other may aggravate the libration motion of PSE without predesigned time shift. The results show the trajectories of climber are very important and should be well designed to avoid the collision of tethers and assure the safety operation of load transfer.

**Keywords** Partial space elevator · Nodal position finite element method · Arbitrary Lagrangian–Eulerian · Parallel tethers · Multiple climbers · Variable-length element

## 1 Introduction

Partial space elevator (PSE) is an attractive alternative to the classical space elevator due to its ability for long-range transfer of payloads between two satellites at low cost [1, 2]. PSE has been a good demonstration for space elevator technology, and the construction of a PSE is relatively easy to achieve in the near term [3].

Past decades have witnessed many efforts to the dynamics of PSE, and they can be categorized in terms of number of climbers, one climber case and multiple climbers. For the case of one climber, Lorenzini [4] and Lorenzini et al. [5] initially proposed the

G. Li · Z. H. Zhu (✉)
Department of Mechanical Engineering, York University, 4700 Keele Street, Toronto, ON M3J 1P3, Canada
e-mail: gzhu@yorku.ca

G. Li
e-mail: lgq1984@yorku.ca



Z. Jing (ed.), *Proceedings of the International Conference on Aerospace System Science and Engineering 2019*, Lecture Notes in Electrical Engineering 622,
https://doi.org/10.1007/978-981-15-1773-0_18

concept of space elevator system attached to International Space Station and studied its dynamics based on the two-piece dumbbell model. Later, the transient motion of climber impacting on the dynamics of PSE is investigated [6–8], and the study of system parameters of PSE, such as the mass ratio of climber to satellite, tether length, and climber's speed, influence on the dynamic characteristics of PSE is conducted. After that, the trajectory of climber is optimized to maximize the suppression of the libration angle of PSE based on the two dumbbell models [9, 10], and the feasibility study of the trajectory of climber applied to the PSE with flexible tether is conducted [10]. Recently, a stabilization method is proposed to suppress the libration motion of PSE by canceling the generated Coriolis force of moving climber through the regulation of deployment/retrieval speed at end satellites [11]. Here, just to name a few, the detailed review of the dynamics of PSE with one climber can be found in our previous work [12, 13]. For the case of multiple climbers, the concept of multiple climbers with proper phase shift is proposed to alleviate the residual libration of PSE by one climber [2]. The background mechanism is that the induced Coriolis force of one moving climber is canceled by another climber with a pre-planned trajectory. From this viewpoint, the simultaneous operation of dual climbers impacting on the dynamic response of PSE is investigated, where one climber moves from bottom to upper and the other one moves from upper to bottom [14]. The results show the simultaneous operation of two climbers which provide a positive effect in suppressing the libration of PSE. Furthermore, different moving patterns of multiple climbers are detailed investigated [15]. To suppress the libration of PSE with multiple climbers, the concept of optimization of middle climber relative to the other climbers is firstly proposed, and the results show it is feasible.

In the current paper, to take advantage of multiple climbers and increase the efficiency of payload transfer, a novel concept of PSE with parallel tethers and multiple climbers is proposed. Among them, the parallel tethers are connected by a transverse support structure with large rigidity at end satellites, and multiple climbers are supposed to operate along each tether. The transient motion of climber impacting on the libration motion of PSE is investigated by a high-fidelity model of PSE based on the nodal position finite element method.

## 2 Mathematical Formulation of Partial Space Elevator

### 2.1 Coordinate Systems

Consider a PSE with parallel tethers moving in an arbitrary orbit initially in a central gravity field as shown in Fig. 1. The PSE consists of main satellites, subsatellites, and multiple parallel elastic and flexible tethers connecting these satellites, and multiple climbers are moving up and down along the tether. For simplicity, the numbering of tether is counted from left to right, and the movement directions of climbers (in the case of multiple climbers) belong to the same tether are supposed to be the same. The

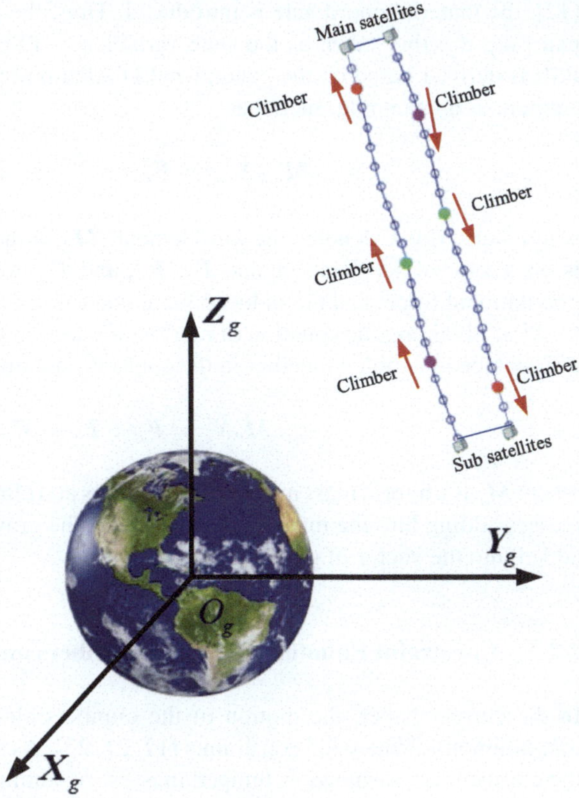


**Fig. 1** Schematic of PSE with parallel tethers (two tethers) and multiple climbers

satellites and climbers are modeled as lumped masses with their attitude dynamics ignored, given the extremely large ratio of tether length over the dimensions of satellites and climbers. The motion of the PSE is described by three sets of coordinate systems: the global inertial frame (*OXYZ*), the local frame of element (*otbn*), and the orbital frame ($O'X_oY_oZ_o$). The detailed definition of these coordinate systems can be found in our previous works [10, 12, 13].

## 2.2 *Nodal Position Finite Element Formulation of Partial Space Elevator*

### 2.2.1 Nodal Position Finite Element Formulation in Arbitrary Lagrangian Eulerian (ALE) Description

In the current paper, the nodal position finite element method is used to describe tether dynamics with large rotation and displacement [16–18]. Here, to easily describe and incorporate the motion of multiple climbers [13] and tether deployment or retrieval

[12], the material coordinate is introduced. Thus, the position and material coordinates are together taken as the state variables of PSE. The equation of motion of PSE is derived based on the generalized D'Alembert principle [17–20]. Taking $k$th element as an example, such that

$$\boldsymbol{M}_{e,k}\ddot{\boldsymbol{X}}_{e,k} = \boldsymbol{F}_{e,k} + \boldsymbol{F}_{g,k} - \boldsymbol{F}_{p,k} \tag{1}$$

where subscript $k$ denotes the $k$th element. $\boldsymbol{M}_e$ is the element's mass matrix, $\ddot{\boldsymbol{X}}_e$ is the vector of acceleration, and $\boldsymbol{F}_e$, $\boldsymbol{F}_g$, and $\boldsymbol{F}_p$ are the vectors of elastic force, gravitational force, and the induced force due to variation of material coordinate.

After obtaining the equation of motion of one element, the equation of motion of PSE can be assembled together in the conventional finite element.

$$\boldsymbol{M}_e\ddot{\boldsymbol{X}}_e = \boldsymbol{F}_e + \boldsymbol{F}_g - \boldsymbol{F}_p \tag{2}$$

where $\boldsymbol{M}_e$ is a hybrid mass matrix with satellites and climbers are supposed as lumped masses adding into the mass matrix of tether. The gravity of lumped masses is also added into the vector of gravity of tether.

### 2.2.2 Constraint Equations Between Climbers and Tether

In the current paper, the motion of the climbers along the tether is enforced by non-holonomic kinematic constraints [12, 21, 22]. As stated in the previous section, the climbers are assumed as lumped masses. Assuming there are $m$ parallel tethers ($m \geq 2$), each tether has $w$ climbers ($w \geq 1$). The movement of climber is implemented by the moving nodes and variable-length elements. There are two steps to implement the movement of climber: (i) moving nodes with the time-varying material coordinate are assigned to the initial position of the climbers and (ii) adjacent elements connecting to the moving nodes are defined as the variable-length elements. The material coordinates of the moving nodes follow pre-defined trajectories of the climbers. Accordingly, constraint equations describing the motion of the climber are defined as follows:

$$\begin{aligned} C_{1,j}^{kl}(p,t) &= \dot{p}_j^{kl} - \dot{p}_{j,\text{desired}}^{kl} = 0 (k = 1, 2, \ldots, w \text{ and } l = 1, 2, \ldots, m) \\ \text{or } \boldsymbol{C}_1(p,t) &= \boldsymbol{0} \end{aligned} \tag{3}$$

where subscript $j$ denotes node numbering of the moving node, and superscripts $k$ and $l$ denote the index of climber and tether, respectively. $p_{j,\text{desired}}^{kl}$ denotes the pre-defined trajectory of the $k$th climber which belongs to $l$th tether, and the overhead dot represents the first-order derivative with respect to time.

### 2.2.3 Constraint Equations of Normal Nodes

Except for the moving nodes, the rest nodes are the normal nodes, and their material coordinates are kept constant. Accordingly, the following constraint equations should be satisfied.

$$C_{2,k}(p,t) = \dot{p}_k = 0(k = 1, j-1, j+1\ldots(n+1-mw)) \text{ or } \boldsymbol{C}_2(p,t) = \boldsymbol{0} \quad (4)$$

### 2.2.4 Constraint Equations of Transverse Support Structure

As shown in Fig. 1, multiple parallel tethers are connected at end satellites by the transverse support structure with super high rigidity, for simplicity, the rigidity is assumed as infinite, and the length is kept constant. The constraint equations of constant length are as follows:

$$\begin{cases} C_{3,1} = D_1^2 - \left[(X_{1,j} - X_{1,k})^2 + (Y_{1,j} - Y_{1,k})^2 + (Z_{1,j} - Z_{1,k})^2\right] \\ C_{3,2} = D_2^2 - \left[(X_{n+1,j} - X_{n+1,k})^2 + (Y_{n+1,j} - Y_{n+1,k})^2 + (Z_{n+1,j} - Z_{n+1,k})^2\right] \\ \begin{pmatrix} j = 1,2,\ldots,m \\ k = 1,2,\ldots,m \\ j \neq k \end{pmatrix} \end{cases} \text{ or } \boldsymbol{C}_3(p,t) = \boldsymbol{0} \quad (5)$$

where $D_1^2$ and $D_2^2$ are the square of constant distance connecting two tethers.

### 2.2.5 Equations of Motion for the PSE with Parallel Multiple Tether and Multiple Climbers

The equations of motion for the whole PSE can be obtained by combining Eqs. (2–5) as follows:

$$\begin{cases} \boldsymbol{M}_e\ddot{\boldsymbol{X}}_e + \sum_{j=1}^{3} \boldsymbol{C}_{j,\boldsymbol{X}_e}^{\mathrm{T}}\boldsymbol{\lambda}_j = \boldsymbol{F}_e + \boldsymbol{F}_g - \boldsymbol{F}_p \\ \boldsymbol{C}_1(p,t) = \boldsymbol{0} \\ \boldsymbol{C}_2(p,t) = \boldsymbol{0} \\ \boldsymbol{C}_3(p,t) = \boldsymbol{0} \end{cases} \quad (6)$$

where $\boldsymbol{C}_{j,\boldsymbol{X}_e}^{\mathrm{T}}(j = 1,2,3)$ denotes the Jacobian matrix of constraint equations with superscript T representing the transpose of matrix, and $\boldsymbol{\lambda}_j(j = 1,2,3)$ denotes the vector of Lagrange multipliers corresponding to these constraint equations.

### 2.2.6 Merging and Dividing of Elements

The detailed information of the process of merging and dividing of element can be found in Refs. [10–13]. Here, only a brief introduction will be given. There are two types of elements are defined, the constant-length and variable-length elements. The length of variable-length element varies as the moving node (climber) moves along the tether. To avoid its length excessively long or excessively short, special attention should be paid. Four parameters are defined to implement the process of dividing and merging of elements meanwhile which suppresses the oscillation caused by the node removal. They are the standard element length $L_s$, the upper ($L_{\max}$) and lower ($L_{\min}$) bounds of an element length, and the acceptance criteria $\delta_e$. These detailed parameters will be given in the simulation part.

## 2.3 Libration Motion of PSE with Parallel Tethers

As shown in Fig. 2, the libration motion of the PSE can be easily expressed in the orbital coordinate system [1, 2, 6, 7, 23]. The detailed definition of orbital coordinate system ($O'X_oY_oZ_o$) and transformation matrix from global coordinate system to the orbital coordinate system can be found in our previous work [13, 24]. The results of PSE are shown in the orbital coordinate system relative to the origin point. The libration motion of PSE with single tether cannot be directly used [6, 7, 13]. For the PSE with parallel tethers, the libration motion of PSE is defined as follows: a series of virtual libration angels are defined by straight lines connecting the main

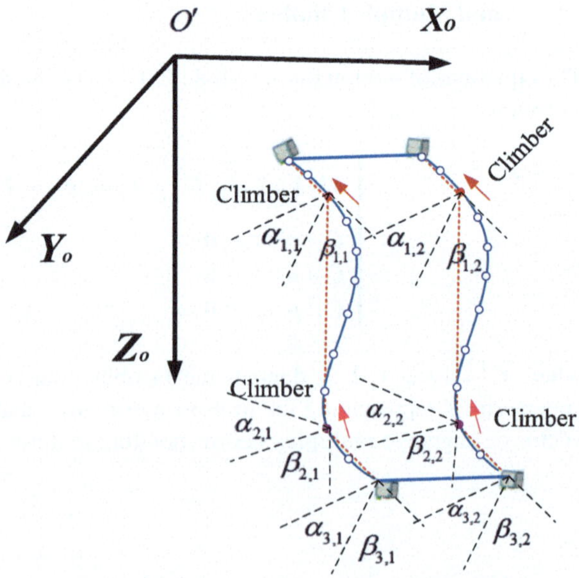


**Fig. 2** Definition of libration angles of PSE with parallel tethers and multiple climbers

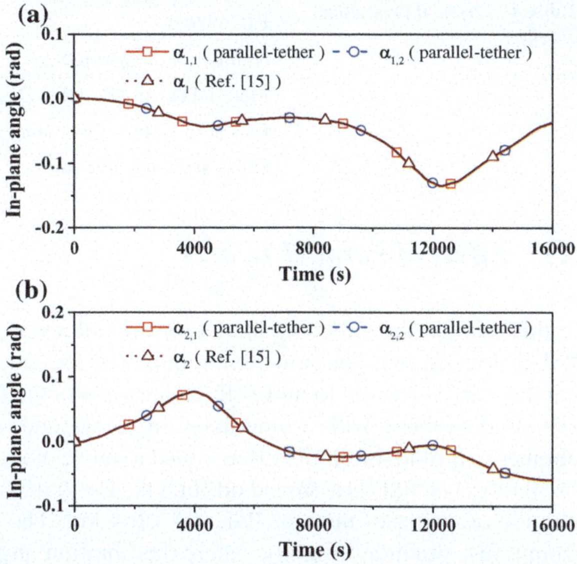


**Fig. 3** Comparison results of libration angles of PSE in case A

satellite, the climbers, and the subsatellite; see the red-dotted lines in Fig. 2. Taking the case of two climbers and two parallel tethers as example, there are six-dotted lines. The in-plane angles $\alpha_{i,j}(i = 1, \ldots, 3 \text{ and } j = 1, 2)$ and out-of-plane angles $\beta_{i,j}(i = 1, \ldots, 3 \text{ and } j = 1, 2)$ are calculated as follows:

$$\begin{aligned} \alpha_{i,j} &= \tan^{-1}\left(\boldsymbol{R}_{X_o,i,j}/\boldsymbol{R}_{Z_o,i,j}\right), \\ \beta_{i,j} &= \tan^{-1}\left[-\boldsymbol{R}_{Y_o,i,j}/\left(\boldsymbol{R}_{Z_o,i,j}\cos\alpha_{i,j} + \boldsymbol{R}_{X_o,i,j}\sin\alpha_{i,j}\right)\right] \end{aligned} \tag{7}$$

where $\boldsymbol{R}_{i,j} = \left(\boldsymbol{R}_{X_o,i,j}, \boldsymbol{R}_{Y_o,i,j}, \boldsymbol{R}_{Z_o,i,j}\right)^{\mathrm{T}}$ is the vector of a dotted line expressed in the orbital coordinate system with the subscript $i$ $(1 \sim 3)$ representing the sequence of these three lines which belongs to the $j$th $(1 \sim 2)$ tether (Figs. 2 and 3).

## 3 Results and Discussion

In the current paper, the backward Euler formulation is employed together with the Newton–Raphson iteration method to solve this nonlinear differential algebraic Eq. (6) [12, 13, 25]. Among them, the maximum iteration number and error tolerance of each iteration step are set as 100 and $10^{-11}$, respectively. For simplicity, the PSE is assumed to have two tethers, the physical properties of the tether material of PSE are listed in Table 1, and other parameters will be given in the simulation part.

**Table 1** Physical properties of tether

| Parameters | Values |
|---|---|
| Number of parallel tethers | 2 |
| Material density of tether (kg/m$^3$) | 1440 |
| Elastic modulus of the tether ($10^9$ N/m$^2$) | 72 |
| Cross-sectional area (m$^2$) | $2.0 \times 10^{-6}$ |

## *3.1 Effect of Parallel Tethers*

In this section, the effect of parallel tethers influencing on the dynamic response of PSE is investigated. The numerical simulations are carried out at two scenarios, where climbers are supposed to move simultaneously in the first scenario and climbers are supposed to move with a time delay in the second scenario. Each tether has one climber only, and the climber is assumed to move along the tether at a constant speed of 50 m/s. The PSE is assumed orbiting the Earth in a circular orbit initially, and the altitude of center of mass of PSE is 43,164 km. The climber is supposed to move from a rest situation instantly, where the libration angle and velocity are zero. The physical parameters of PSE are used, $L = 1000$ km, $m_m = 10^7$ kg, $m_s = 10000$ kg, $m_c = 1500$ kg. As listed in Table 2, both upward and downward transfer motions of climbers are considered. The initial position of climber is locating $0.10\ L$ away from the end spacecraft. In this section, each tether is discretized into two variable-length elements without consideration of adding and subtracting of nodes, which leads the degrees of freedom of the PSE model is constant. The numbering of element starts from left tether to right tether, and it starts from upper to bottom in each tether. The element is counting consecutively. The time-step size is 0.001 s.

In the first scenario, four cases are carried out as listed in Table 2, where all the climbers move simultaneously. To investigate the effect of parallel tethers influencing the dynamic response of PSE, the results of PSE with parallel tethers are compared with results of PSE with a single tether. First, the comparison results for cases A and B are shown in Figs. 4, 5, 6, and 7, respectively. It can be easily found that the libration angles of the comparison results match very well as expected. It shows the difference of libration angles between the parallel tether and single tether which can be ignored when the climber's trajectories are the same. Furthermore, the same tendency of the variation of geometrical configuration of PSE is observed, as seen in Figs. 5 and 7. Second, the comparison results for cases C and D are shown in

**Table 2** Cases with simultaneous movement of climbers

| Name | Motion of climber (Tether 1) | Motion of climber (Tether 2) |
|---|---|---|
| Case A | Upward | Upward |
| Case B | Downward | Downward |
| Case C | Upward | Downward |
| Case D | Downward | Upward |

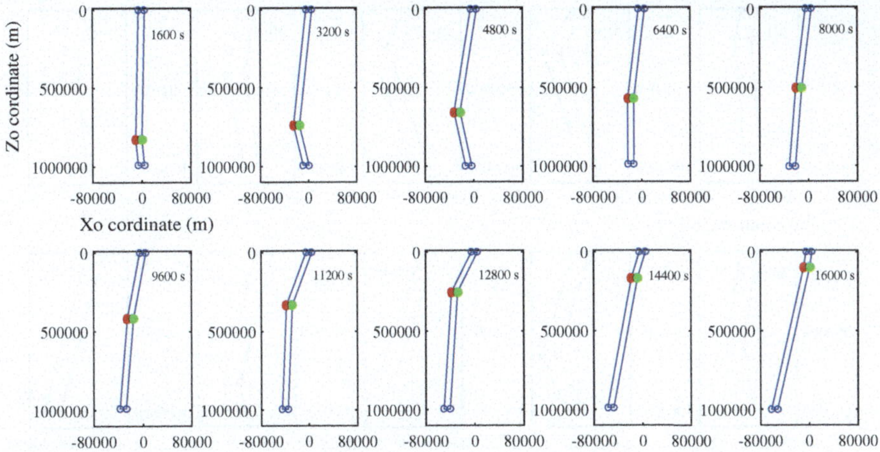


**Fig. 4** Geometric configuration of PSE in case A

**Fig. 5** Comparison results of libration angles of PSE in case B

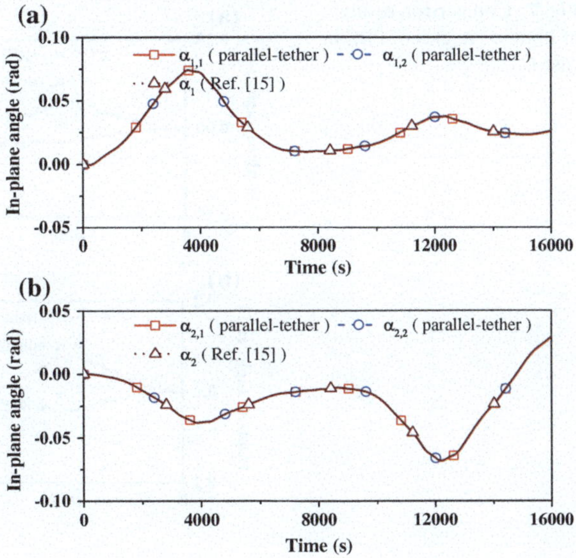


Figs. 8, 9, 10, and 11. A significant difference is observed from the results of PSE with parallel tether comparing with the results of single tether when the movement directions of two climbers of each tether are opposite. For example, as shown in Figs. 8 and 10, the difference in the libration angles increases dramatically as times going on. Two reasons can be attributed: (i) the Coriolis forces of moving climbers are opposite and (ii) the reaction forces caused by the rigid structure connecting two tethers. For example, as shown in Figs. 8 and 11, the variation of tension of PSE with transverse connecting structure varies different from the results of single tether.

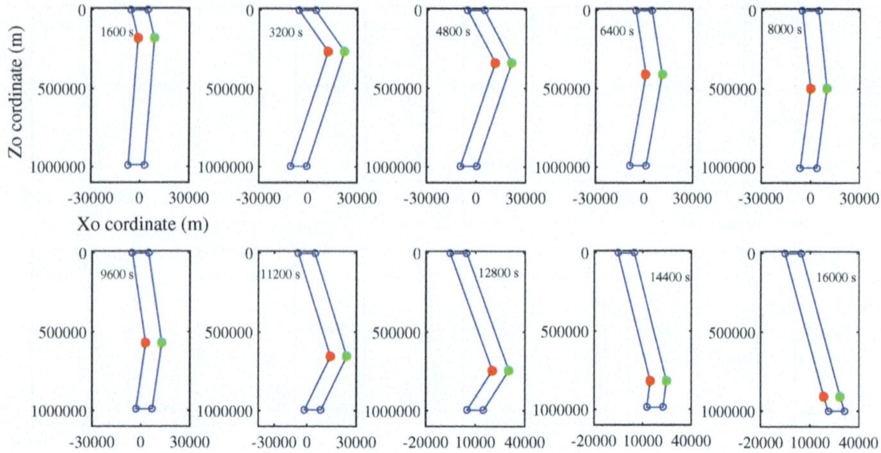


**Fig. 6** Geometric configuration of PSE in case B

**Fig. 7** Comparison results of libration angles of PSE in case C

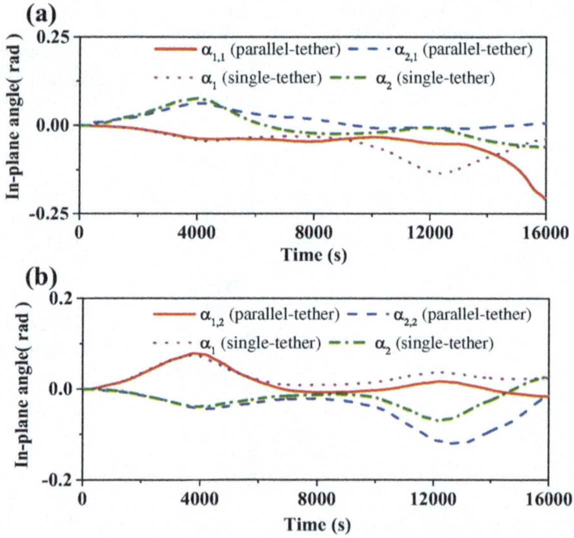


Because there is a constraint caused by the connecting structure that result the two tethers interact each other. The same phenomenon is found by observing geometrical configuration of PSE; see Figs. 9 and 11. For example, the distance between two parallel tethers increases for case C. However, for case D, the distance decreases until two tethers tangle together. The tangling of two tethers should be stopped to happen from the design viewpoint. The reason is that the direction of Coriolis forces of two moving climbers is the same and points outward direction. For case D, they are opposite and point inward direction. It reveals a fact, from the viewpoint of avoiding

**Fig. 8** Comparison results of tension in case C

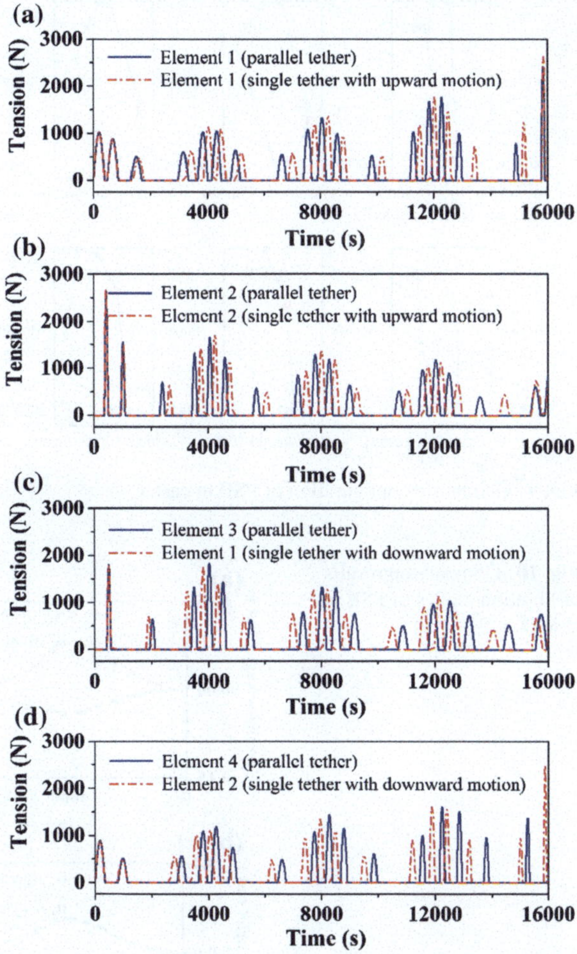


of tether collision, the movement direction of climber for PSE with parallel tethers is already determined once the rotational direction of PSE is determined. For example, the rotation axis of PSE points to the positive $Z$-axis direction, the climber of left tether should be designed to move upward, and the climber of right tether should be designed to move downward (Fig. 12).

In the second scenario, the climbers are supposed to move non-simultaneously. From simplicity, consider these two climbers starting in a different time. As listed in Table 3, another three numerical simulations are carried out. For simplicity, the climber of left tether moves firstly, after 200 s, the climber of the second tether starts to move. All the other parameters and initial conditions are the same as the cases A, B, and C. The results are compared with the results of the cases A, B, and C.

As shown in Fig. 13, the difference in the libration angles of PSE is not obvious at the starting stage. However, the difference becomes noticeable as times going on.

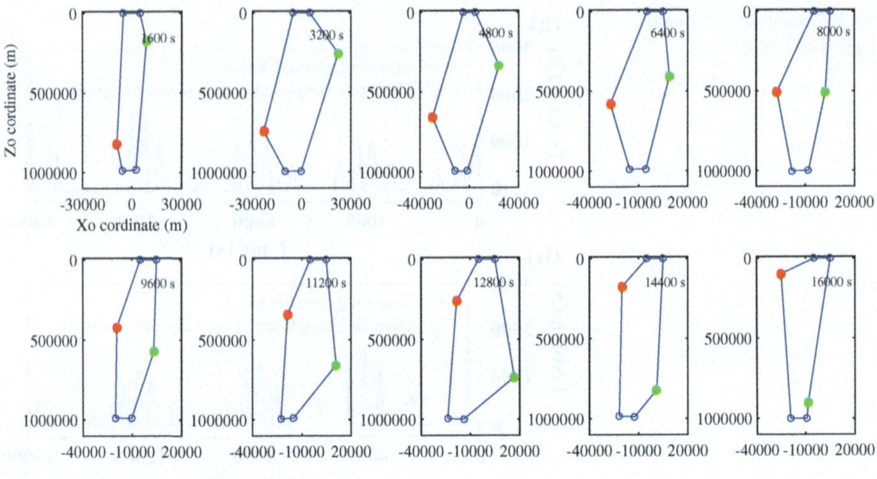


**Fig. 9** Geometric configuration of PSE in case C

**Fig. 10** Comparison results of libration angles of PSE in case D

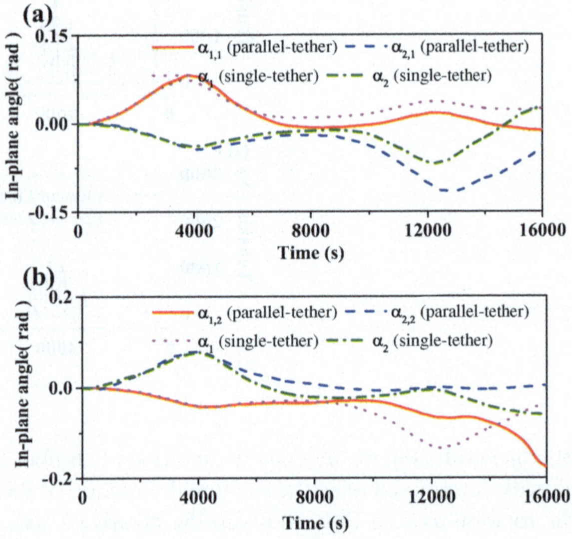


The same phenomenon can be observed from the geometric configuration of PSE as shown in Fig. 14. The reason for this phenomenon is that two tethers interact with each other caused by the non-simultaneous motions of climber.

In this section, we can conclude that dynamic behavior of PSE with parallel tethers is the same as the results of single tether when the motion of climber of each tether is the same. On the contrary, the noticeable difference in the dynamic behavior of PSE is found due to the interaction between two tethers when the motions of climber of each tether, including the movement direction and the time shift, are different.

**Fig. 11** Comparison results of tension in case D

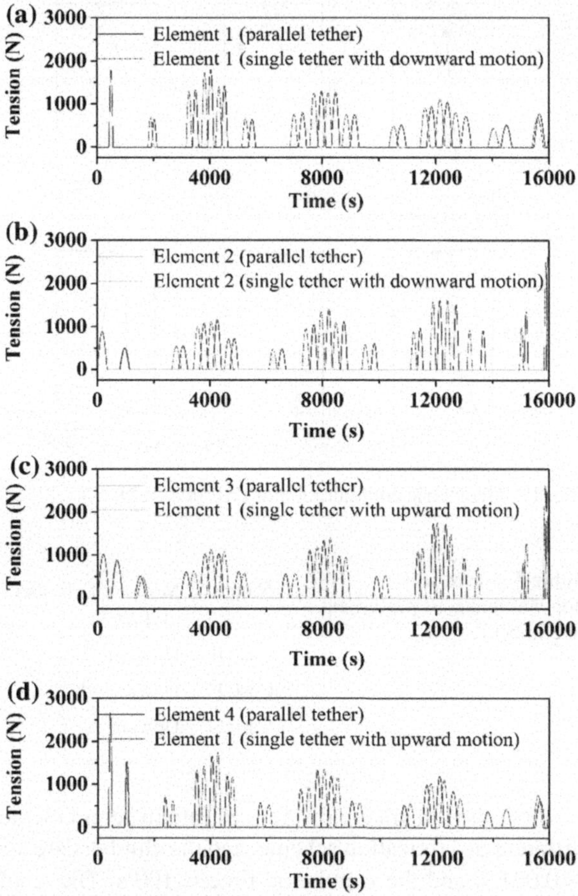


## 3.2 *Effect of Tether Discretized Scheme*

In this section, the effect of tether discretization scheme on the dynamic response of PSE of parallel tethers is investigated. Here, each tether has one climber only, and the physical parameters, initial conditions, and motion of climber are set the same as case C. As listed in Table 4, one tether was discretized into two, four, six, eight, and twelve elements, and the other tether was discretized into the same element. Therefore, the total number of elements of PSE is double the time of one tether. To observe the trend of convergence due to the discretization scheme, a new parameter is defined to show the difference in the position of climbers $R_{j,k} = \frac{\sqrt{(X_{c,k}-X_{c,j})^2+(Y_{c,k}-Y_{c,j})^2+(Z_{c,k}-Z_{c,j})^2}}{\sqrt{X_{c,j}^2+Y_{c,j}^2+Z_{c,j}^2}} \times 100\%$ with subscripts $j, k =$ 4, 8, 12, 18, and 24 ($j \neq k$) representing the different discretization schemes of PSE.

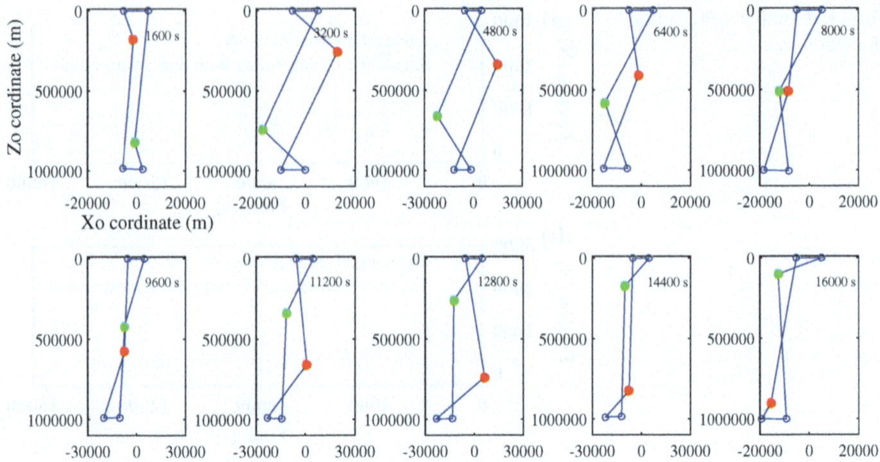


**Fig. 12** Geometric configuration of PSE in case D

**Table 3** Cases with non-simultaneous movement of climbers

| Name | Motion of climber (Tether 1) | Motion of climber (Tether 2) |
|---|---|---|
| Case E | Upward | Upward (200 s later) |
| Case F | Downward | Downward (200 s later) |
| Case G | Upward | Downward (200 s later) |

First, the tethers are discretized into different element to investigate the effect of different discretization scheme, and the climbers are keeping still. The time-step size is 0.001 s, and the simulation time is 100 s. The results of the ratio $R_{j,k}$ are listed in Table 5. It can be easily seen that the difference in the ratio $R$ decreases as the number of elements increases. For example, as listed in Table 5, $R_{16,24}$ is several times smaller than $R_{4,8}$. However, the computational cost increases significantly for case J; however, the accuracy does not improve a lot. Thus, after the trade-off analysis between the computational cost and accuracy, the PSE is discretized into sixteen elements with eight elements of each tether in the following simulations.

Second, as presented in Sect. 2.2.6, the process of merging and dividing of elements occurs if the tether is discretized into more than two elements. Accordingly, the degree of freedom of the PSE model varies. For the PSE with parallel tethers, the setting of control parameters is different from the setting parameters of PSE with single tether [12, 13]. To avoid the processes of dividing and merging of elements of two tethers which happen simultaneously, the standard length $L_{s,j} (j = 1, 2)$ is chosen slightly different for two tethers. For case J, $L_{s,1} = 12850\,m$ and $L_{s,1} = 15350\,m$,. Other control parameters of the process of merging and dividing of element are set the same for both two tethers, such as $L_{\max,j} = 1.65\ L_{s,j}$, $L_{\min,j} = 0.49\ L_{s,j}$, and $\delta_{e,j} = 10^{-2}\,m\ (j = 1, 2)$. The comparison results of cases J and C are shown in

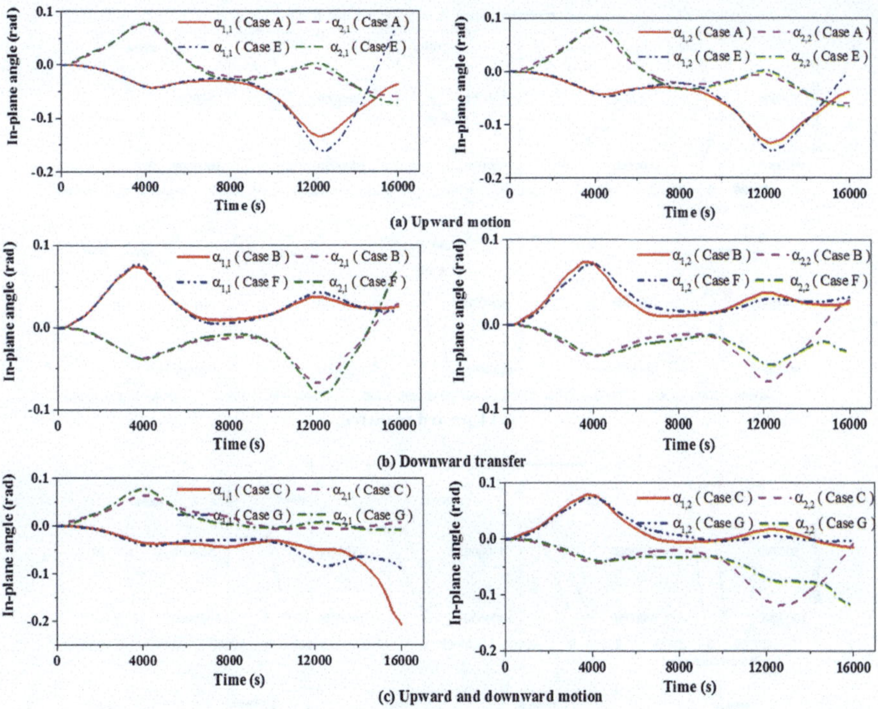


**Fig. 13** Comparison results of libration angles of PSE considering time delay between two climbers

Figs. 15 and 16. As shown in Fig. 15, it can be easily found that the difference in the libration angles between case C and case J is obvious. Because the high-order transverse motion of tether is considered in the multiple-elements tether model. The same phenomenon can be observed from the geometric configuration of PSE; see Fig. 16. It indicates the multiple elements tether model which is more accurate to display the variation of PSE geometric configuration. Moreover, it is noticed that the two tethers of PSE collides together which should be avoided. It indicates that the simultaneous optimization algorithm should be employed for the climbers to avoid the collision of tether.

In this section, it can be concluded that the tether model with multiple elements should be used to describe the geometrical configuration of PSE accurately, and the simultaneous optimization algorithm should be adapted to collision of tether.

## 3.3 *Effect of Multiple Climbers of Each Tether*

In this section, the effect of multiple climbers influence on the dynamic response of PSE is investigated. The physical parameters, initial conditions, and the tether

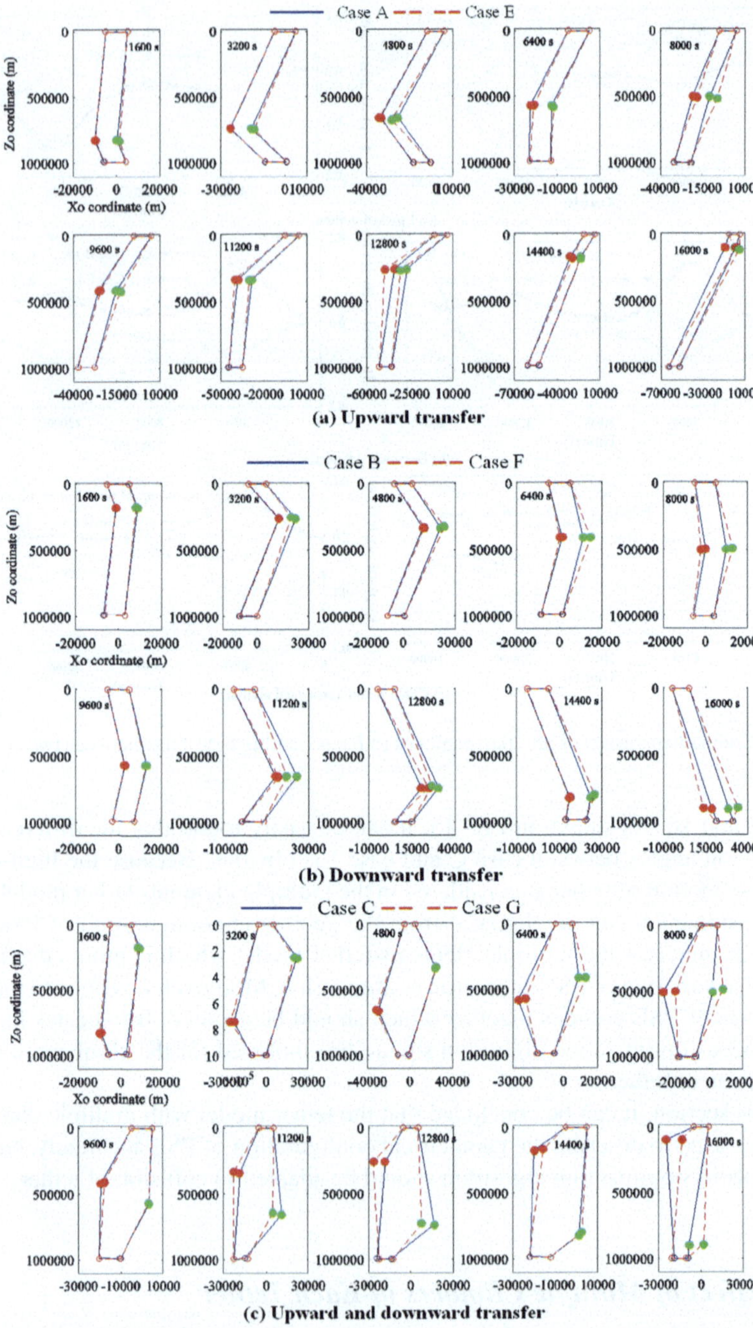


**Fig. 14** Comparison results of geometric configuration of PSE considering time delay between two climbers

**Table 4** Number of elements of each tether

| Name | Number of element (Left tether) | Number of element (Right tether) |
|---|---|---|
| Case H | 4 | 4 |
| Case I | 6 | 6 |
| Case J | 8 | 8 |
| Case K | 10 | 10 |

**Table 5** Ratio of the position of climbers between different discretization schemes

| Time (s) | Ratio $R_{4,8}$ (%) (Climber 1) | Ratio $R_{8,12}$ (%) (Climber 1) | Ratio $R_{12,16}$ (%) (Climber 1) | Ratio $R_{16,24}$ (%) (Climber 1) |
|---|---|---|---|---|
| 10 | 4.38E−06 | 4.58E−07 | 1.21E−06 | 5.71E−08 |
| 20 | 1.69E−05 | 1.78E−06 | 8.22E−07 | 2.18E−07 |
| 30 | 3.69E−05 | 3.83E−06 | 2.43E−07 | 4.57E−07 |
| 40 | 6.33E−05 | 6.37E−06 | 4.57E−07 | 7.43E−07 |
| 50 | 9.44E−05 | 9.14E−06 | 1.20E−06 | 1.05E−06 |
| 60 | 1.28E−04 | 1.19E−05 | 1.92E−06 | 1.34E−06 |
| 70 | 1.63E−04 | 1.43E−05 | 2.57E−06 | 1.61E−06 |
| 80 | 1.96E−04 | 1.63E−05 | 3.09E−06 | 1.82E−06 |
| 90 | 2.26E−04 | 1.77E−05 | 3.48E−06 | 1.98E−06 |
| 100 | 2.51E−04 | 1.86E−05 | 3.73E−06 | 2.08E−06 |
| Time (s) | Ratio $R_{4,8}$ (%) (Climber 2) | Ratio $R_{8,12}$ (%) (Climber 2) | Ratio $R_{12,16}$ (%) (Climber 2) | Ratio $R_{16,24}$ (%) (Climber 2) |
| 10 | 3.71E−05 | 3.89E−06 | 1.32E−05 | 4.85E−07 |
| 20 | 1.50E−04 | 1.52E−05 | 1.65E−05 | 1.85E−06 |
| 30 | 3.31E−04 | 3.26E−05 | 2.14E−05 | 3.89E−06 |
| 40 | 5.70E−04 | 5.44E−05 | 2.74E−05 | 6.35E−06 |
| 50 | 8.55E−04 | 7.80E−05 | 3.38E−05 | 8.97E−06 |
| 60 | 1.17E−03 | 1.01E−04 | 4.01E−05 | 1.15E−05 |
| 70 | 1.50E−03 | 1.22E−04 | 4.56E−05 | 1.38E−05 |
| 80 | 1.81E−03 | 1.39E−04 | 5.01E−05 | 1.56E−05 |
| 90 | 2.10E−03 | 1.51E−04 | 5.32E−05 | 1.69E−05 |
| 100 | 2.34E−03 | 1.58E−04 | 5.48E−05 | 1.76E−05 |

discretization scheme are set the same as in case J, except the parameters of climbers. One simulation case is conducted with each tether having two climbers with the same mass (half of the mass of climber in Case J). As shown in Fig. 17, the distance between two climbers of each tether is 10 km. The comparison results are shown in Figs. 18 and 19.

As shown in Fig. 18, it is easily found that the difference in the libration angles between the comparison cases increases as the times going on. This is caused by

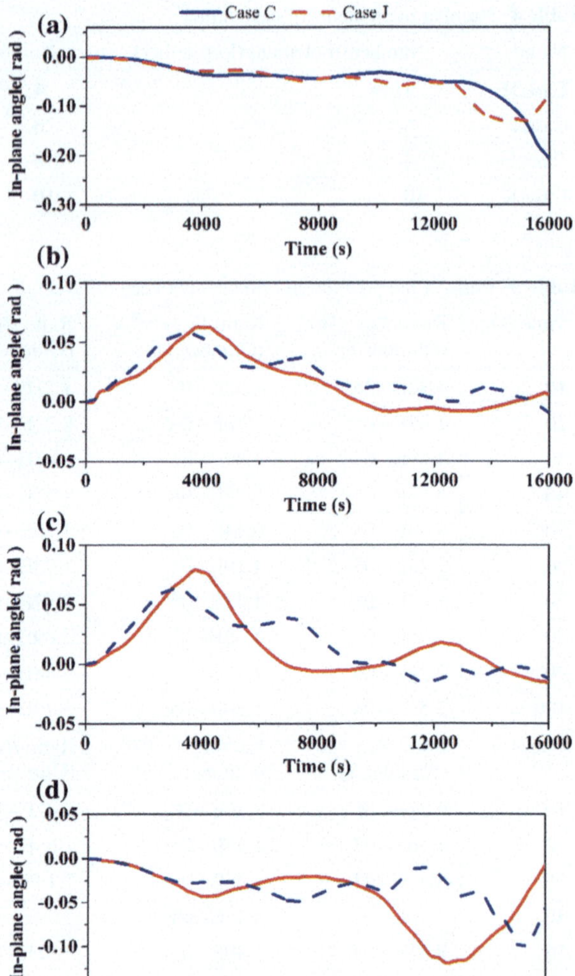


**Fig. 15** Comparison results of libration angles of case C and J **a** $\alpha_{1,1}$ **b** $\alpha_{2,1}$ **c** $\alpha_{1,2}$ **d** $\alpha_{2,2}$

the difference in the Coriolis forces of multiple climbers. As shown in Fig. 18a, d, it is seen that the amplitude of libration angle increases for case J. It indicates the multiple climbers belong to each other without proper time shift may aggravate the libration motion of PSE. Therefore, the phase shift between each climber needs to be designed and optimized. For case L, there is a high frequency of oscillation motion caused by the short distance between climbers; see Fig. 18e. The same phenomenon can be observed from the geometrical configuration of PSE. Moreover, as shown in Fig. 19, it is found that the two tethers collide with each other. It shows the control strategy needs to be explored to avoid the collision of tethers.

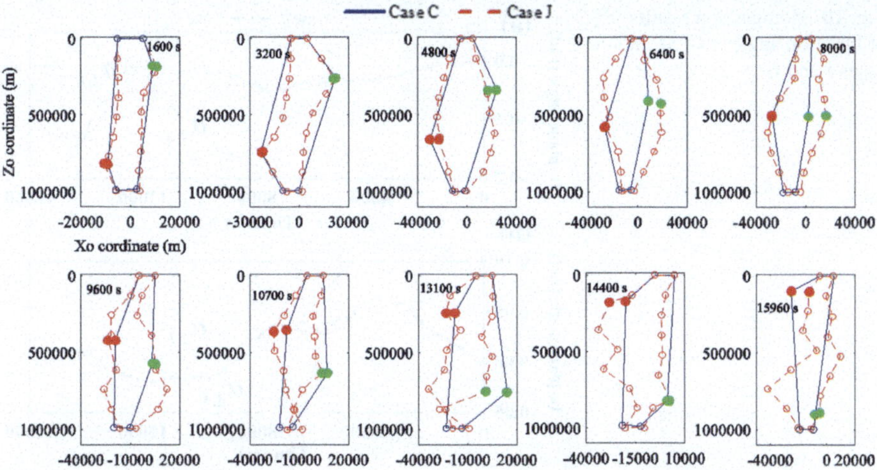


**Fig. 16** Geometric configuration of PSE of case C and J

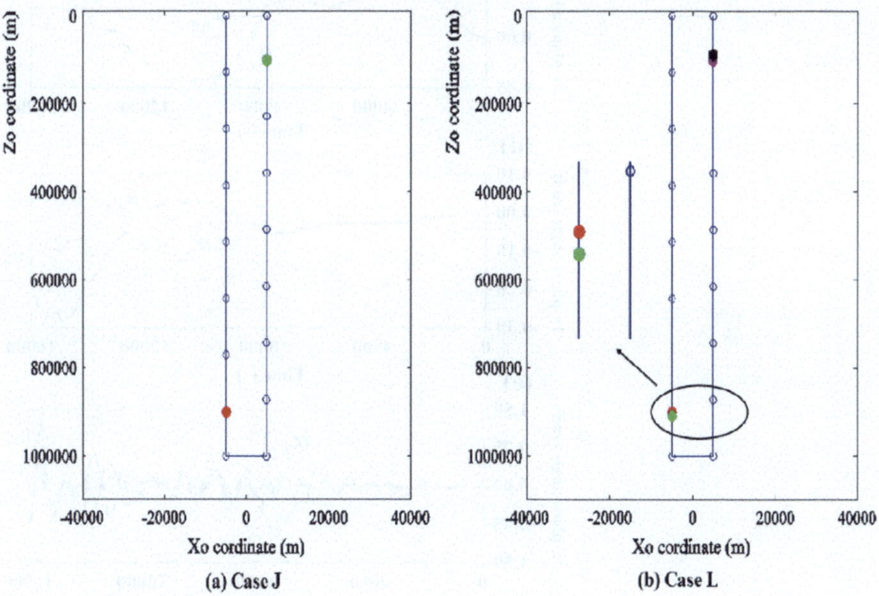


**Fig. 17** Initial mesh of PSE between case J and case L

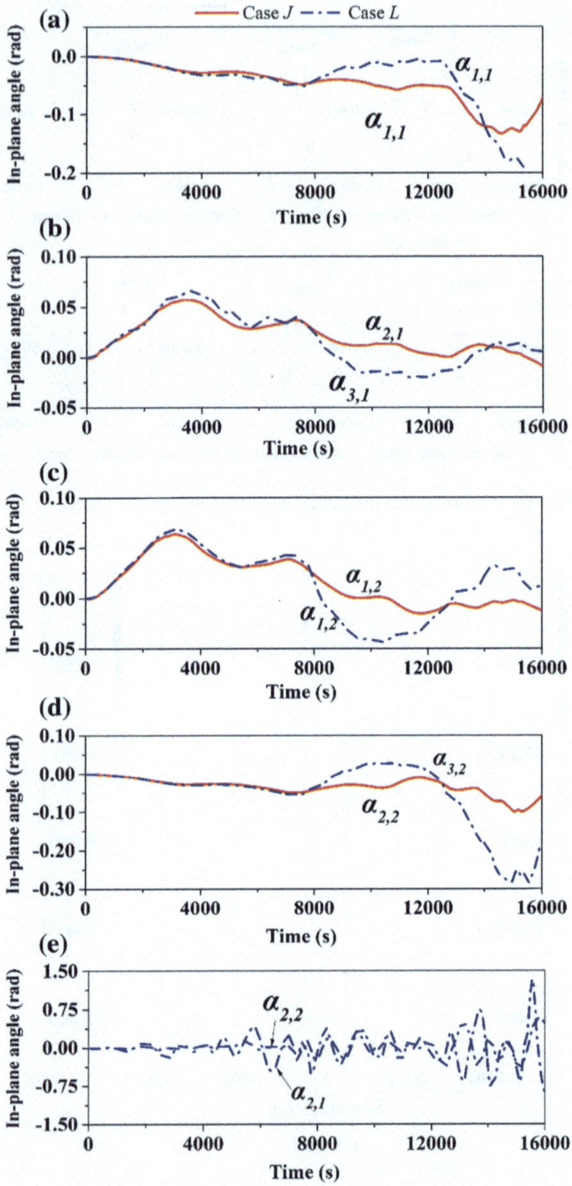


**Fig. 18** Comparison results of libration angles of PSE in cases J and L

## 4 Conclusion

A novel concept of partial space elevator with parallel tethers and multiple climbers is proposed to increase the payload transfer efficiency. A high-fidelity and high-accuracy model of partial space elevator with parallel tethers is developed by the

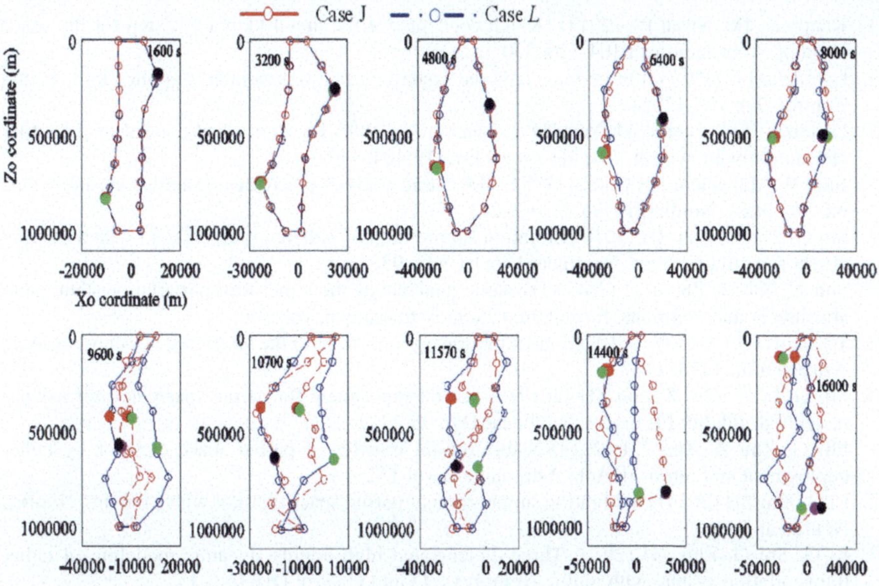


**Fig. 19** Comparison results of geometric configuration of PSE in cases J and L

nodal position finite element method in framework of arbitrary Lagrangian–Eulerian. First, the concept of parallel tethers impacting on the dynamic response of PSE is investigated through the transient motion of climbers, the movement direction, and time delay between climbers. It is found that the tethers collide with each other when the movement direction of climber is opposite. Therefore, for the safety operation of PSE with parallel tethers, the trajectories of climbers should be optimized to avoid the collision of the tether. Second, the transient motion of multiple climbers of each other impacting on the dynamic response of PSE is studied, and it is also found the tethers collide with each other. Moreover, it is found that the multiple climbers may aggravate the libration motion when the time shift between each climber is not well designed.

**Acknowledgements** This work is supported by Discovery Grant (RGPIN-2018-05991) and Discovery Accelerate Supplement Grant (RGPAS-2018-522709) of Natural Sciences and Engineering Research Council of Canada.

## References

1. Woo P, Misra AK (2014) Energy considerations in the partial space elevator. Acta Astronaut 99:78–84
2. Woo P, Misra AK (2010) Dynamics of a partial space elevator with multiple climbers. Acta Astronaut 67:753–763

3. Knapman JM, Swan PA (2014) Design concepts for the first 40 km a key step for the space elevator. Acta Astronaut 104:526–530
4. Lorenzini EC (1987) A three-mass tethered system for micro-g/variable-g applications. J Guid Control Dyn 10:242–249
5. Lorenzini EC, Cosmo M, Vetrella S, Moccia A (1989) Dynamics and control of the tether elevator/crawler system. J Guid Control Dyn 12:404–411
6. Jung W, Mazzoleni AP, Chung J (2014) Dynamic analysis of a tethered satellite system with a moving mass. Nonlinear Dyn 75:267–281
7. Shi G, Zhu Z, Zhu ZH (2018) Libration suppression of tethered space system with a moving climber in circular orbit. Nonlinear Dyn 91:923–937
8. Sun X, Xu M, Zhong R (2017) Dynamic analysis of the tether transportation system using absolute nodal coordinate formulation. Acta Astronautica, vol 139
9. Williams P, Ockels W (2010) Climber motion optimization for the tethered space elevator. Acta Astronaut 66:1458–1467
10. Shi G, Li G, Zhu Z, Zhu ZH (2019) A virtual experiment for partial space elevator using a novel high-fidelity FE model. Nonlinear Dyn 95:2717–2727
11. Shi G, Zhu Z, Zhu ZH (2018) Stable orbital transfer of partial space elevator by tether deployment and retrieval. Acta Astronautica, vol 152
12. Li G, Zhu ZH (2019) On libration suppression of partial space elevator with a moving climber. Nonlinear Dyn
13. Li G, Shi G, Zhu ZH (2019) Three-dimensional high-fidelity dynamic modeling of tether transportation system with multiple climbers. J Guid Control Dyn 0: 1–15
14. Ishikawa Y, Otsuka K, Yamagiwa Y, Doi H (2018) Effects of ascending and descending climbers on space elevator cable dynamics. Acta Astronaut 145:165–173
15. Shi G, Li G, Zhu Z, Zhu ZH (2019) Dynamics and operation optimization of partial space elevator with multiple climbers. Adv Space Res 63:3213–3222
16. Li G, Zhu ZH (2017) Precise analysis of deorbiting by electrodynamic tethers using coupled multiphysics finite elements. J Guid Control Dyn 40:3348–3357
17. Li GQ, Zhu ZH (2015) Long-term dynamic modeling of tethered spacecraft using nodal position finite element method and symplectic integration. Celest Mech Dyn Astron 123:363–386
18. Sun FJ, Zhu ZH, LaRosa M (2011) Dynamic modeling of cable towed body using nodal position finite element method. Ocean Eng 38:529–540
19. Yang S, Deng Z, Sun J, Zhao Y, Jiang S (2017) A variable-length beam element incorporating the effect of spinning. Latin Am J Sol Struct 14:1506–1528
20. Liu J-P, Cheng Z-B, Ren G-X (2018) An Arbitrary Lagrangian-Eulerian formulation of a geometrically exact Timoshenko beam running through a tube. Acta Mech 229:3161–3188
21. Hong D, Tang J, Ren G (2011) Dynamic modeling of mass-flowing linear medium with large amplitude displacement and rotation. J Fluids Struct 27:1137–1148
22. Williams P (2009) Dynamic multibody modeling for tethered space elevators. Acta Astronaut 65:399–422
23. Cohen SS, Misra AK (2009) The effect of climber transit on the space elevator dynamics. Acta Astronaut 64:538–553
24. Li G, Zhu ZH, Cain J, Newland F, Czekanski A (2015) Libration control of bare electrodynamic tethers considering elastic–thermal–electrical coupling. J Guid Control Dyn 39:642–654
25. Hong D, Ren G (2011) A modeling of sliding joint on one-dimensional flexible medium. Multibody Sys Dyn 26:91–106

# Design and Stress Analysis for Aircraft Structure Repair Beyond Specification

**Chen Chen and Kang Li**

**Abstract** The airline structure design engineers often need to design repairs to restore structural integrity for conditions not covered in the Structural Repair Manuals (SRM) provided by the OEM. This is normally called repair beyond specification. Reasons for having to modify SRM repair include: variations in local structural configuration, limitations on space and variations in fastener usage or availability, and so on. As the local design loads are always unavailable for airline design engineers, the principle for repair design is to restore the ultimate static strength and the equivalent fatigue life. Through an example of a chord structure repair design, this paper provides an integrated methodology to design aircraft structure repairs beyond specification. The repair design is based on static analysis. In order to determine the minimum number of fasteners to fulfill the ultimate static strength requirement, the general failure modes for the joint structure and the approach to get its allowable is described. Since the end fasteners transfer more load than the center fasteners when parts are below yield, a joint modeling method is used to calculate the joint load distribution for the operating fatigue load. Based on the load distribution of the critical end fasteners and their geometry properties, the fatigue life could be calculated with a professional Damage Tolerance Analysis (DTA) software AFGROW. The results would be compared to the recommended SRM repair to check whether the fatigue life could be restored. Following the method provided in this paper, engineers could design repairs which can restore the integrity of the damaged aircraft structures.

**Keywords** Repairs beyond specification · Fatigue life · Damage Tolerance Analysis

---

ICASSE Paper.

---

C. Chen (✉) · K. Li
Ameco, No.2 Capital Airport Road, 563, 100621 Chaoyang, Beijing, People's Republic of China
e-mail: chenchen3@ameco.com.cn

K. Li
e-mail: likang@ameco.com.cn

 
Z. Jing (ed.), *Proceedings of the International Conference on Aerospace System Science and Engineering 2019*, Lecture Notes in Electrical Engineering 622,
https://doi.org/10.1007/978-981-15-1773-0_19

## Abbreviations

| | |
|---|---|
| DTA | Damage Tolerance Analysis |
| SRM | Structural Repair Manuals |
| MMPDS | Metallic Materials Properties Development and Standardization |

## 1 Overview

The airline structure design engineers often need to design repairs to restore structural integrity for conditions not covered in the Structural Repair Manuals (SRM). This is normally called repair beyond specification. Reasons for having to modify an SRM repair include: variations in local structural configuration, limitations on space and variations in fastener usage or availability.

As the local design loads are normally unavailable for airline design engineers, the principle for repair design is to restore both the ultimate static strength and the equivalent fatigue life. In other words, the design loads for repair are usually determined either by the ultimate capability of an adjacent joint or by the ultimate strength of the damaged material.

Chord is one of the typical type of aircraft structure which popularly exists in stringers, frames, floor beams, etc., therefore, damage on chord is one of the most common types of damage on aircraft structure. Figure 1 shows an example of chord corrosion damage. Through an example of a chord structure repair design, this paper provides an integrated methodology to design aircraft structure repairs beyond specification.

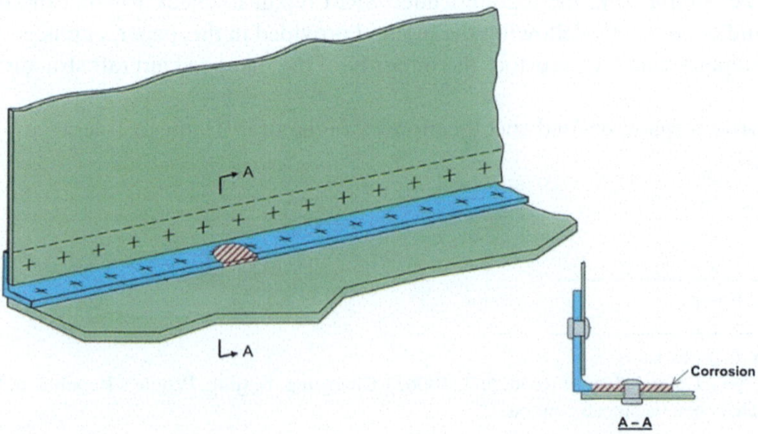


**Fig. 1** Chord corrosion

## 2 Issues to Be Considered

Generally, when the damage is extensive along the width of the structure, the damaged section needs to be totally removed and replaced by a splice.

Following the principle of restoring the ultimate static strength as well as the equivalent fatigue life, two issues need to be taken into account for repair design.

(1) Static issue:

Whether both the repair joint allowable and the splice load capability are equal or higher than the repair design loads?

(2) Fatigue issue:

Whether the fatigue life of the repair joint is equal or higher than the approved repair in SRM?

## 3 Repair Design and Analysis

Structure repair should be designed based on the approved data in SRM as well as the two issues mentioned above.

The basic procedure for aircraft structure repair design is summarized as shown in Fig. 2.

In addition, there are also some other general steps including checking the fastener pitch, row spacing, edge margin, calculating the fastener head clearance, processing corrosion protection with sealing and surface finish, etc. All these steps do not need stress analysis and, therefore, will not be discussed here.

### *3.1 Joint Allowable*

Joint allowable is to be used to calculate the number of fasteners required to restore the ultimate strength.

The basic joint failure modes include (1) Net area tension; (2) Tear out; (3) Bearing; and (4) Fastener shear. The overall joint capability is equal to the lowest allowable load of all the failure modes which is called the critical failure mode.

Normally, there are two ways to determine the joint allowable.

#### 3.1.1 Joint Allowable Equations

According to the MMPDS [1], the equations for the four failure modes can be used alone to accurately estimate the capability of joints if all the following criteria are met:

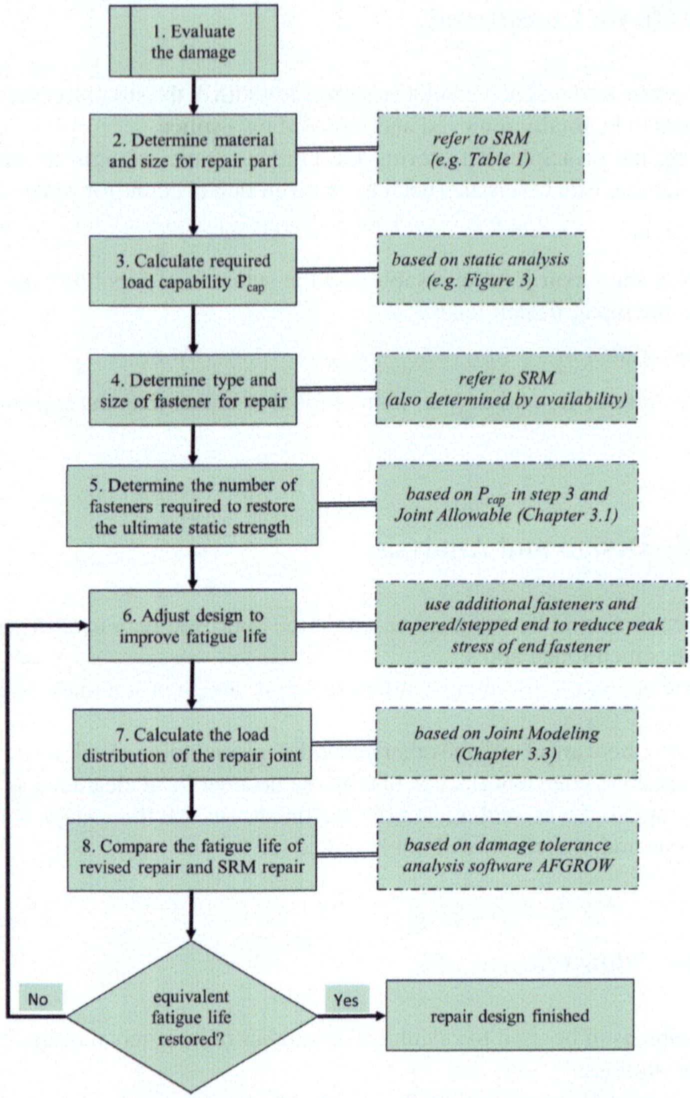


**Fig. 2** General procedure for aircraft structure repair design

(1) The fastener is solid (no hollow shanks or heads);
(2) For single shear, the fastener must be a tension protruding head bolt;
(3) For double shear, the fastener must be a bolt;
(4) The sheet thickness to fastener diameter ratio (t/D) must be equal to or greater than 0.18;

Figure 3 shows the failure modes for joint and is followed by the corresponding joint allowable equations [2].

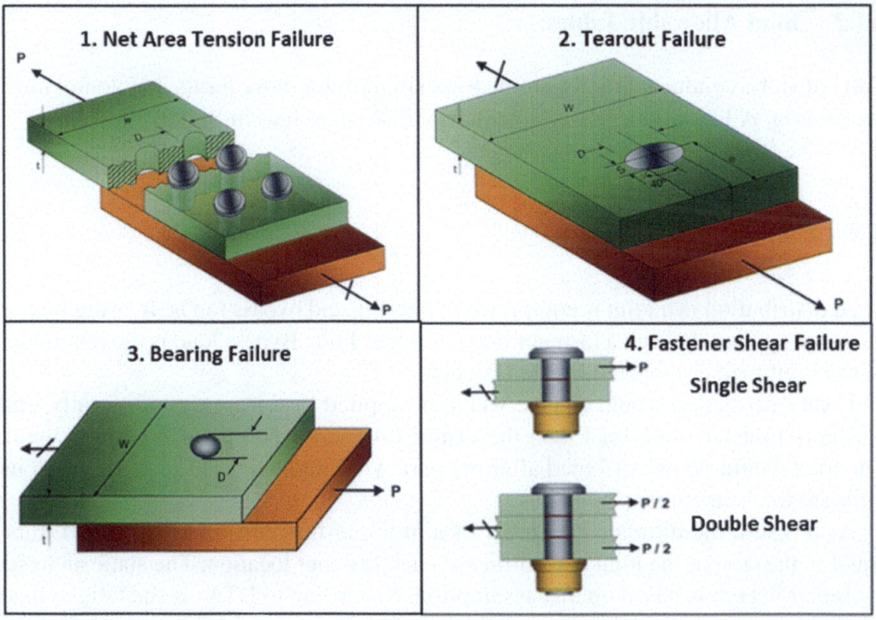


**Fig. 3** Joint failure modes

1. $P_{net} = F_{du} \times A_{net} = F_{du} \times (w - n_{fast} \times D) \times t$
2. $P_{tearout} = F_{su} \times 2 \times (e - S) \times t = F_{su} \times (2 \times e - 0.766 \times D) \times t$
3. $P_{brg} = F_{brg} \times D \times t$
4. $P_{SS} = F_{su} \times A_{shank} = F_{su} \times \pi \times D_{shank}^2/4$
5. $P_{DS} = 2 \times F_{su} \times A_{shank} = 2 \times F_{su} \times \pi \times D_{shank}^2/4$

where

| | |
|---|---|
| $w$ | width of cross section, perpendicular to applied load (in) |
| $n_{fast}$ | number of fasteners in cross section |
| $D$ | hole diameter (in) |
| $t$ | plate thickness (in) |
| $F_{du}$ | lesser of $F_{tu}$ or 1.5 × $F_{ty}$ (psi) |
| $S$ | D/2 × cos 40° = 0.383 × D (in), 40° angle is determined from test |
| $F_{su}$ | shear ultimate allowable (psi) |
| $e$ | edge margin (in) |
| $F_{brg}$ | lesser of $F_{bru}$ or 1.5 × $F_{bry}$ (psi) |
| $D_{shank}$ | fastener shank diameter (in). |

### 3.1.2 Joint Allowable Tables

Tests must be conducted to determine joint strength for those joints that do not meet the criteria. A lot of data is available in MMPDS as well as in SRM.

## *3.2 Load Distribution*

Load distribution in a joint is comprised of bearing and bypass loads. Bearing load is the force applied through a fastener into a fastener hole. Bypass load is the remainder of the joint load not reacted by the fastener.

Load distribution would change when the applied load increases. Generally, end fasteners transfer more load than the center fasteners when parts are below yield. The load would be redistributed after the parts yielding until each fastener reach its ultimate load capability.

As a result, the ultimate capability of a multiple-fastener joint is assumed to be equal to the sum of the joint capabilities at each fastener location. The static analysis for repair design is based on this assumption. Regarding to DTA, as the fatigue load belongs to operating load which is on the level below parts yield, the load distribution needs to be calculated through joint modeling.

## *3.3 Joint Modeling*

Testing has shown that the fatigue life of a joint is a function of both the bearing and the bypass load at a given fastener location. The joint modeling is to be used to get the bearing and bypass load.

Joints may be modeled mathematically using a series of springs as shown in Figure 4. Springs simulate the stiffness of fasteners and plates in load direction.

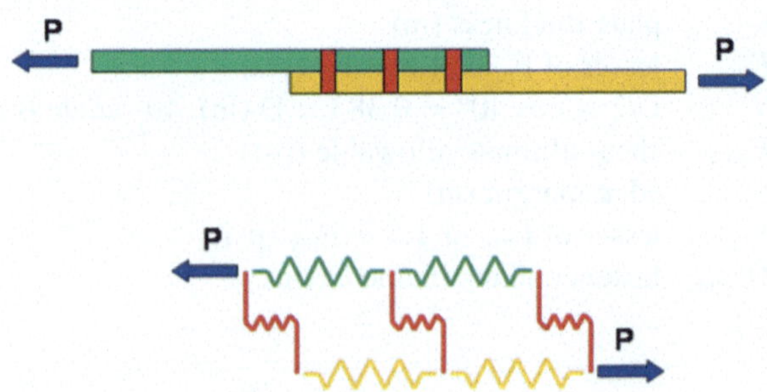


**Fig. 4** Example of a joint model

### 3.3.1 Spring Compliance for Plates

The spring compliance for plate segments between fasteners could be easily calculated with the following equation:

$$C_{\text{plate}} = L/(A \times E)$$

where

$C_{\text{plate}}$ plate spring compliance (in/lb);
$L$ fastener spacing in direction of load (in);
$A$ cross-sectional area of plate between fasteners (in$^2$), $A = w \times t$;
E modulus of elasticity of the plate material (psi);
$w$ fastener spacing perpendicular to load (in);
$t$ plate thickness (in).

### 3.3.2 Spring Compliance for Fasteners

Numerous methods have been proposed for fastener spring compliance calculation. One commonly referenced paper that includes equation for fastener compliance is FAA-AIR-90-01 written by Tom Swift. Other popularly used methods include Huth equation and a non-proprietary Boeing research paper written in 1969 on the stress severity factor concept.

Tom Swift equation [3]: $C_F = [A + B \times (D/t_1 + D/t_2)]/(D \times E)$
where

$C_F$ fastener spring compliance (in/lb)
$A$ 5.0 for aluminum fasteners; 1.666 for steel fasteners
$B$ 0.8 for aluminum fasteners; 0.86 for steel fasteners
$D$ fastener diameter (in)
$E$ modulus of elasticity of the plate material (psi)
$t1$ plate 1 thickness (in)
$t2$ plate 2 thickness (in).

The Tom Swift equation has some limitations such as the fasteners may only be steel or aluminum and the plates must be of the same material.

For material types not covered by the Swift equations, the following equation is preferred [4].

$$C_F = \frac{4(t_i + t_j)}{9G_b A_b} + \frac{t_i^3 + 5t_i^2 t_j + 5t_i t_j^3 + t_j^3}{40E_{bb} I_b} + \frac{1}{t_i}\left(\frac{1}{E_{bb}} + \frac{1}{E_{ibr}}\right) + \frac{1}{t_j}\left(\frac{1}{E_{bb}} + \frac{1}{E_{jbr}}\right)$$

where

$C_F$ fastener spring compliance (in/lb)

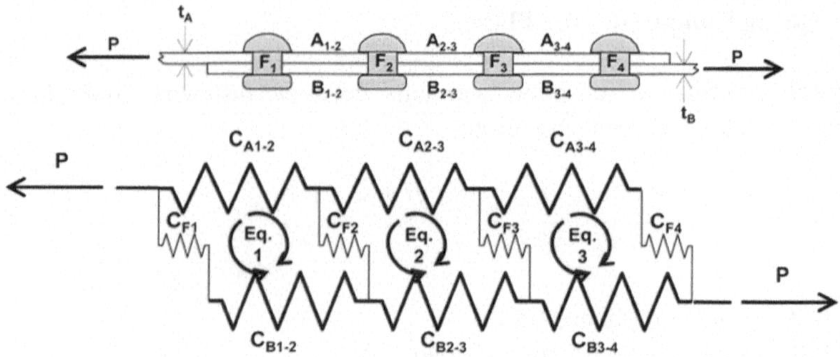


**Fig. 5** Joint modeling deflection equations

$t_i$ thickness of plate i (in)
$t_j$ thickness of plate j (in)
$G_b$ shear modulus of the bolt material (psi)
$A_b$ cross-sectional area of the bolt (in$^2$)
$E_{bb}$ modulus of elasticity of the bolt material (psi)
$I_b$ moment of inertia of the bolt cross section (in$^4$)
$E_{ibr}$ modulus of elasticity of plate i (psi)
$E_{jbr}$ modulus of elasticity of plate j (psi).

### 3.3.3 Load Distribution Calculation

Joint modeling deflection equations are shown in Fig. 5 and the equations are below.

$$C_{A1-2} \times (P - P_{F1}) + C_{F2} \times P_{F2} = C_{F1} \times P_{F1} + C_{B1-2} \times P_{F1} \tag{1}$$

$$C_{A2-3} \times (P - P_{F1} - P_{F2}) + C_{F3} \times P_{F3} = C_{F2} \times P_{F2} + C_{B2-3} \times (P_{F1} + P_{F2}) \tag{2}$$

$$C_{A3-4} \times (P - P_{F1} - P_{F2} - P_{F3}) + C_{F4} \times P_{F4} = C_{F3} \times P_{F3} + C_{B3-4} \times (P_{F1} + P_{F2} + P_{F3}) \tag{3}$$

$$P_{F1} + P_{F2} + P_{F3} + P_{F4} = P \tag{4}$$

Solving these equations can be done manually or with an existing FE software.

# 4 Instruction Case

A fictitious chord repair design is presented in this chapter to describe the whole process. MS Excel is a good tool for the related calculating work.

An upper chord of floor beam has been damaged due to corrosion (Fig. 6). SRM repair is shown in Fig. 7. Since the steel fastener BACB30FN8 is not available and the end fastener F7 would interfere with the seat track attach fitting, a revised repair need to be designed to restore the ultimate static strength and fatigue life.

The dimensions and material of the floor beam upper chord are shown in Fig. 8.

Table 1 is a SRM table which shows the typical factor to calculate the minimum cross-sectional area of repair parts by multiplying the cross-sectional area of the initial part.

Figure 9 shows an Excel spreadsheet including the basic repair design steps based on static analysis. A kind of titanium fastener BACB30NW8K is used to replace the unavailable steel fastener BACB30FN8.

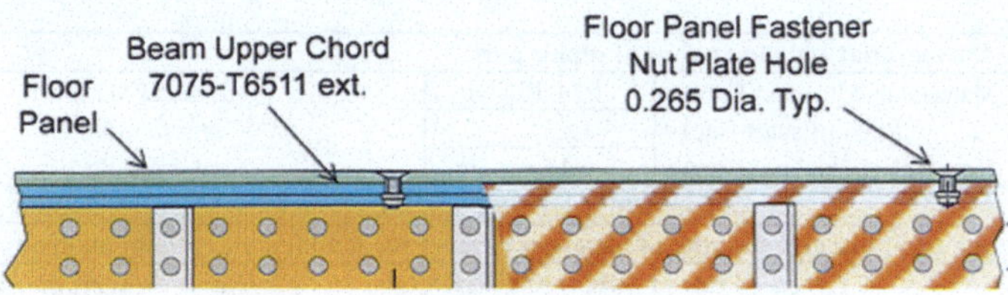


**Fig. 6** Floor beam upper chord corrosion damage

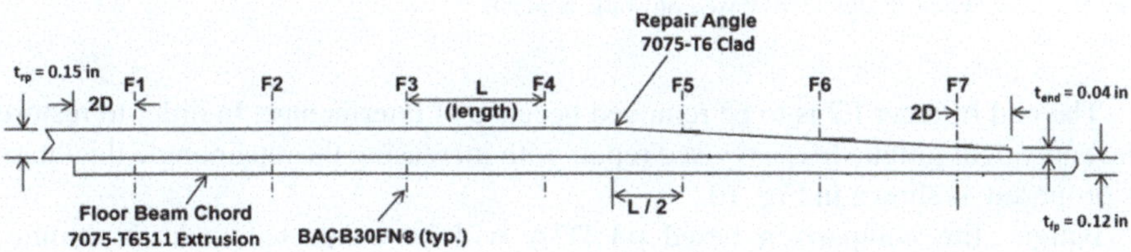


**Fig. 7** SRM repair

**Fig. 8** Material and dimensions

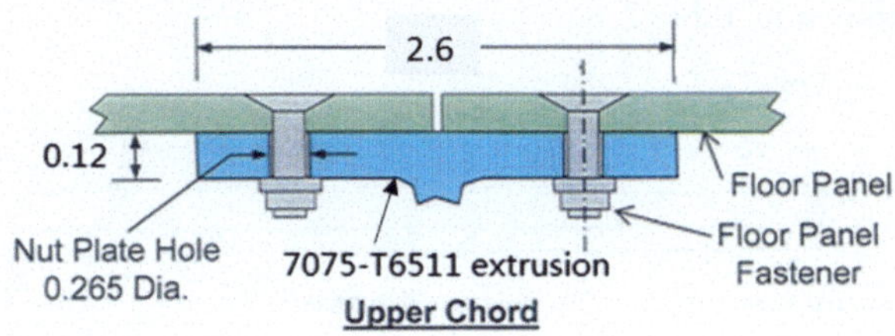

**Table 1** SRM repair material [5]

| Initial section material | Repair part material | Repair material factor |
|---|---|---|
| Extruded 2024-T3 or T3511 | Sheet—clad 2024-T3 | 1.25 |
| Extruded 7075-T6 or T6511 | Sheet—Bare or clad 7075-T6 | 1.35 |
| Extruded 7150-T77511 | Sheet—Bare or clad 7075-T6 | 1.5 |
| Extruded 7150-T77511 | Extruded—7075-T6 | 1.35 |

| **1. Calculate the load-carrying capability of factory part** | | | |
|---|---|---|---|
| Factory Part Thickness (tfp) = | 0.12 | in | |
| Factory Part Width (wfp) = | 1.3 | in | |
| Factory Part Hole Diameter (dfp) = | 0.265 | in | *dfp=0 if no fastener hole exist in factory part* |
| Net Area (Anet) = | 0.1242 | in2 | *Anet = (wfp - dfp) × tfp* |
| Material and Product Form = | 2024 T3 ext. | | |
| Ftu = | 82 | ksi | *refer to MMPDS* |
| Fty = | 74 | ksi | *refer to MMPDS* |
| Design Ultimate Allowable (Fdu) = | 82 | ksi | *Fdu = min (Ftu, 1.5 × Fty)* |
| Load Capability (Pcap) = | 10184.4 | lb | *Pcap = Fdu × Anet* |
| **2. Determine the material and dimensions of repair part** | | | |
| Material and Product Form = | 2024 T3 clad | | *refer to SRM repair* |
| Repair Factor (fac) = | 1.25 | | *refer to table 1* |
| Repair Part Thickness (trp) = | 0.15 | in | *trp = fac × tfp* |
| **3. Determine the type, size and quantity of fasteners to meet the load transfer requirements** | | | |
| Fastener Code = | BACB30NW8K | | *refer to SRM repair* |
| Joint Allowable (Pallow) = | 3235 | lb | *refer to MMPDS/SRM* |
| Number of Fasteners Required (nfast) = | 4 | | *nfast = int(Pcap/Pallow) + 1* |

**Fig. 9** Basic repair design steps based on static analysis

The end fastener F7 is to be removed because of interference. In order to restore the equivalent fatigue life, a revised repair with increasing the repair angle thickness is proposed as shown in Fig. 10.

Fatigue life comparison based on DTA will be conducted under a dummy applied load of 2000 lb. Following the methodology of joint modeling described in Chapter "On Ground Test of an IMU/GNSS Receiver for Atmospheric Re-Entry Vehicle Applications", the bearing and bypass force for each critical location is shown in Table 2.

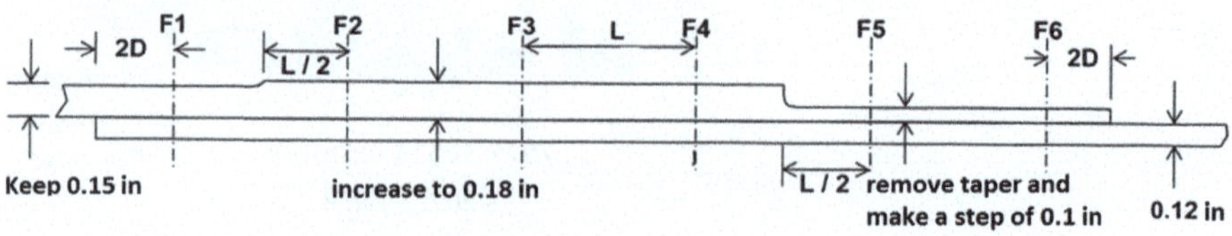


**Fig. 10** Revised repair to eliminate interference

**Table 2** Joint loads for critical locations

| Unit: lb | Factory Part | | | | Repair Part | | | |
|---|---|---|---|---|---|---|---|---|
| | Left Fastener | | Right Fastener | | Left Fastener | | Right Fastener | |
| | $F_{br}$ | $F_{bp}$ | $F_{br}$ | $F_{bp}$ | $F_{br}$ | $F_{bp}$ | $F_{br}$ | $F_{bp}$ |
| SRM Repair | 460 | 0 | 395 | 1605 | 460 | 1540 | 395 | 0 |
| Revised Repair | 440 | 0 | 367 | 1633 | 440 | 1560 | 367 | 0 |

**Table 3** Fatigue life comparison for revised repair and SRM repair

| Fatigue life | Factory part | Repair part |
|---|---|---|
| SRM repair | 21,600 | 46,200 |
| Revised repair | 21,700 | 46,500 |

As shown in Table 2, for the left end fastener location on factory part and right end fastener location on repair part, the loads of revised repair are obviously minor in comparison to SRM repair. Since the material and geometry dimensions are all the same, the equivalent fatigue life of these two locations could definitely be restored. So the DTA would only be implemented for the right end fastener location on factory part and the left end fastener location on repair part (with red borders in Table 2).

The fatigue life comparison is shown in Table 3. Since the fatigue life for the revised repair is higher than SRM repair for both locations, it can be concluded that the equivalent fatigue life is restored.

## 5 Conclusion

Through the instruction case in Chapter "Stall Margin Enhancement of Aeroengine Compressor with a Novel Type of Alternately Swept Blades", a whole process for aircraft structure repair design and analysis are presented. In this procedure, the repair design is based on static analysis to guarantee the ultimate static strength could be restored. Four kinds of joint failure models as well as the joint allowable table are introduced to determine the required fastener number for repair. The character of joint load distribution is presented and the joint modeling method is used to calculate the bearing and bypass loads of a joint under operating fatigue loadcase. DTA conducted in AFGROW provides a way for fatigue life comparison between revised repair and SRM repair under a same invented load. Airline structure engineers can adjust their design according to the results until the equivalent fatigue life could be restored.

In conclusion, this paper provides a feasible methodology to design aircraft structure repair beyond specification.

## References


1. MMPDS, Metallic materials properties development and standardization handbook, FAA
2. Michael C Y NIU (1997) Airframe stress analysis and sizing
3. Swift T (1990) Repairs to damage tolerant aircraft, FAA-AIR-90-01
4. Stress severity factors for axially loaded mechanically fastened joints, Boeing
5. 737–800 SRM, structural repair manual, Boeing

# Concepts for Morphing Airfoil Using Novel Auxetic Lattices

**Zeyao Chen, Xian Wu, Zhe Wang and Jianwang Shao**

**Abstract** In nature, the wings shape of a bird can be adjusted to be suitable for all flight situations providing optimal aerodynamic performance. Unfortunately, wings of traditional aircraft are optimized for only a few conditions, not for the entire flight envelope. Therefore, it is necessary to develop the morphing airfoil with smart structures for the next-generation excellent aircraft. Combined with the actuators, sensors and controller techniques, the smart airfoil will bring a revolution for aircraft. Hence, the design of smart structure which is applicable for the morphing airfoil is the first step, especially the flexible airfoil which exhibits many more changeable degrees than rigid structures. In this paper, the composite structure based on re-entrant quadrangular is designed to be applied in the deformable aircraft. The re-entrance structure can show negative Poisson's ratio performance, also called auxetic, which can offer a great advantage in morphing mechanism. Based on fundamental work about re-entrant quadrangular lattices, the scheme of morphing airfoil is firstly given. Firstly, as shown in the FFT-based homogenization analysis, the enhanced re-entrant lattice outperforms remarkably the original one in stiffness and has similar flexibility. The mechanical characteristics of morphing airfoil with auxetic lattice core are the focus of our paper, which are investigated by using the finite element model. The design loads are extracted from the aerodynamic loads, which are converted to effective nodal loads distributed in the airfoils. The estimated natural model frequencies are given using the model analysis method, which accords with the limits. Furthermore, the compliance performances of the airfoil are investigated under passive and active

Z. Chen · X. Wu (✉) · Z. Wang · J. Shao
School of Automotive Studies, Tongji University, No. 4800 Caoan Road, 201804 Shanghai, China
e-mail: wuxain@tongji.edu.cn

Z. Chen
e-mail: chenzeyao@tongji.edu.cn

Z. Wang
e-mail: wangzhe@tongji.edu.cn

J. Shao
e-mail: shaojianwang@tongji.edu.cn



Z. Jing (ed.), *Proceedings of the International Conference on Aerospace System Science and Engineering 2019*, Lecture Notes in Electrical Engineering 622,
https://doi.org/10.1007/978-981-15-1773-0_20

morphing, respectively. The morphing airfoil with auxetic lattices has the advantages of high deformable, ease of control, variable stiffness, and the ability to bear large amounts of stress. These works offer researchers and designers novel ideas for designing morphing aircraft.

**Keywords** Morphing airfoil · Re-entrance lattices · Auxetic · Stiffness

## 1 Introduction

In the aerospace field, many researches on investigating the morphing aircraft to accommodate the complex flight envelopes have emerged in recent years [1]. The idea is inspired by the nature in which bird, fly or bee can adjust the shape of wings for different flying conditions. Morphing structures can expand the flight envelopes with eliminating flap-type mechanisms, reducing aerodynamic drag and controlling vibration [2]. A variety of methods by using special materials and structures have been developed to provide the capability of morphing airfoil. The morphing materials or structures should easily deform when subjected to driving load with low amplitude and carry the aerostatic or aerodynamic load simultaneously [3], and therefore cannot be made by simply using conventional and readily available materials. "Smart" materials and structures with high flexibility including the cellular material have played a significant role in morphing airfoil [4, 5].

Material with negative Poisson's ratio (NPR), also be named as auxetic material, contracts in compression but expands in tension, which is counterintuitive as most of the natural materials show positive Poisson's ratio [6]. NPR property has been used to enhance mechanical properties like resilience [7], crack resistance [8], fracture toughness [9, 10], sound absorption capacity [11], energy absorption capability [12, 13], etc. When Poisson's ratio approaches $-1.0$, its bulk modulus is much less than its shear modulus which means the material becomes highly compressible [6]. And, skins with large in-plane Poisson's ratio would show anticlastic shapes; however, skins with negative in-plane Poisson's ratio would induce synclastic behavior when bent out of plane [14]. Such characteristics and high strain-energy capability would provide a potential application for morphing structures.

Because of the stiffness and mechanism of cellular materials especially bending-dominated, it is readily applied in morphing structures [15]. A lot of studies about morphing airfoil with cellular material core have been reported [16–19]. The re-entrant hexagonal core shows the highest shear flexibility compare to other cellular materials with positive Poisson's ratio which have been proposed in [20, 21]. However, there is a dilemma between the high stiffness and flexibility which are necessary for morphing airfoil. Therefore, how to obtain flexibility while maintaining the stiffness to carry load is necessary to be further investigated. There is a lot of research on passive morphing, yet there are few reports about the concept of active morphing using cellular material. In this paper, we introduced the active concept for the morphing airfoil with auxetic lattice core.

The paper is categorized into three sections including this introduction. Section 2 shows the homogenized mechanical properties of re-entrant quadrangular lattice and its enhanced form. Section 3 reports how to construct a morphing airfoil using the auxetic lattice. The morphing properties of an airfoil with auxetic lattice core are investigated in Sect. 3. Besides the passive morphing properties, the active morphing mechanism also preliminarily discussed in this section. Concluding remarks are finally exhibited in Sect. 4.

## 2 The Characteristics of Re-Entrant Quadrangle

While designing the morphing structures for engineering, both stiffness and flexibility should be considered; however, they are always the conflicting requirements. In this section, the re-entrant quadrangular lattices are investigated whose structures and corresponding unit cell have been depicted in Fig. 1. New developed FFT-based homogenization method was applied to study both cells. From the homogenized computing, the constitutive stiffness matrix $C^H$ can be obtained. Because of the cubic symmetry of these structures, there are only three independent elastic constants in the homogenized elastic tensor presented in the following.

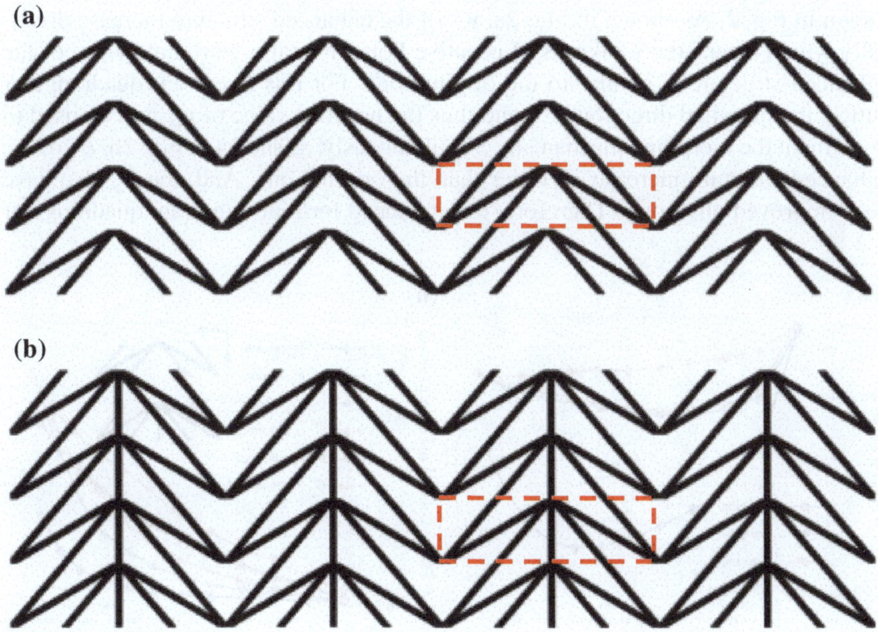


**Fig. 1** Re-entrant quadrangle 4 × 4 lattice and its unit cell: **a** Original; **b** enhanced

$$C^H = \begin{bmatrix} C^H_{11} & C^H_{12} & 0 \\ C^H_{21} & C^H_{22} & 0 \\ 0 & 0 & C^H_{33} \end{bmatrix} \tag{1}$$

From the stiffness tensor (Voigt matrix notation) $C$, the shear modulus and bulk modulus can be derived as following considering the symmetry of the matrix:

$$G^H = C^H_{33} \tag{2}$$

$$S^H = \begin{bmatrix} S^H_{11} & S^H_{12} & 0 \\ S^H_{21} & S^H_{22} & 0 \\ 0 & 0 & S^H_{33} \end{bmatrix} \tag{3}$$

And the effective Poisson's ratio and Young's modulus can be obtained by:

$$v^H_{yx} = -S^H_{12}/S^H_{11}, \quad v^H_{xy} = -S^H_{21}/S^H_{22} \tag{4}$$

$$E^H_x = 1/S^H_{11}, \quad E^H_y = 1/S^H_{22} \tag{5}$$

By varying the width of a strut, we can extend the relative density. At relative density range from 0.05 to 0.65, the mechanical properties of both cells have been shown in Fig. 2. As shown in Fig. 2a, $v_{yx}$ of the enhanced structure increase drastically which means the weakness of negative Poisson's ratio. Fortunately, $v_{xy}$ of the enhanced structure is similar to the original one. For this re-entrant quadrangular lattice, the principal direction is $x$, and thus the negative value of $v_{xy}$ will be used to implement the morphing mechanism. Simultaneously, as shown in Fig. 2b, $E_x$ of the enhanced structure increase ~8 times than the original one. And, the $E_y$ also have been improved obviously. Therefore, the enhanced form of re-entrant quadrangular

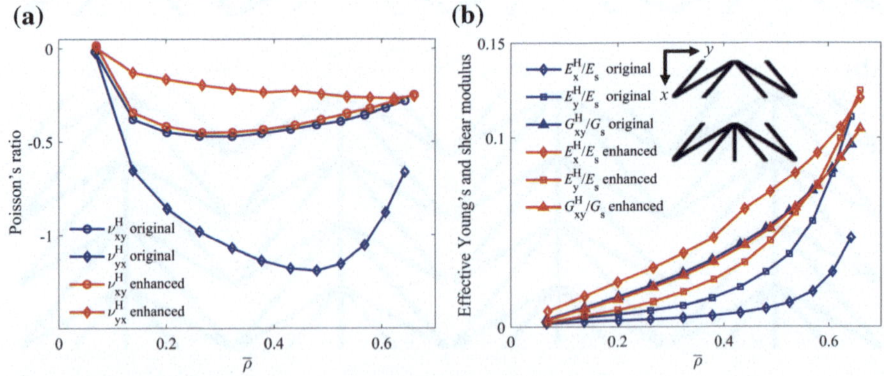


**Fig. 2** Homogenized elastic constants of re-entrant quadrangle lattices: **a** Poisson's ratio; **b** Young's modulus and shear modulus

lattice, maintaining a balance between stiffness and flexibility, is adopted to design the morphing core of the airfoil. The structure is capable of providing good flexibility while maintaining a carrying capability for aerostatic loading.

# 3 Analysis of the Morphing Airfoil

Based on the unit cell of re-entrant quadrangle in Sect. 2, the lattice is mapped into an airfoil profile (NACA-0009 9.0% smoothed). The chord length of airfoil is 1000 mm shown in Fig. 3. The airfoil comprises of the core and frame. The material of core is Carbon-fiber (CC90/ET443 SEAL) material whose Young's modulus is 56.6 GPa and Poisson's ratio is 0.0514, Poisson's ratio is 0.06. The material of frame is aluminum alloy (Al 6061-T051) whose Young's modulus is 70 GPa and Poisson's ratio is 0.33. The finite element model (FEM) in this paper is developed by utilizing the commercially available software Abaqus. The frame of the airfoil is discretized by the 2D solid element (CPS4R). The cellular structure is modeling by the beam element (B21).

## *3.1 Modal Analysis*

We conducted modal analysis for this model. The results can be used to evaluate the dynamic properties of structure. And, by comparing to the modal experiment, the finite element model can be validated. Unfortunately, we have not performed the experiment so far. Furthermore, the modal analysis can be utilized to determine the connectivity of FEM model. Figure 4 shows five normal modes of this morphing airfoil and Fig. 5 exhibits the constrained modes. As shown in both figures, the FEM model is in a good connection. As known, the resonant deformation of 2D airfoil will not be motivated before that of the whole wing as the first modal frequency of 2D airfoil is remarkably higher. It should be noted that the trailing part shows large deformation in every mode. In practical design, the trailing part should be strengthened, yet it is not considered in this paper.

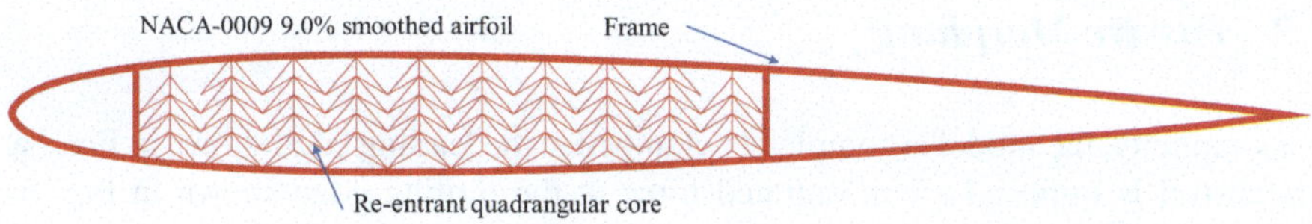


**Fig. 3** Scheme of airfoil with re-entrant quadrangular core

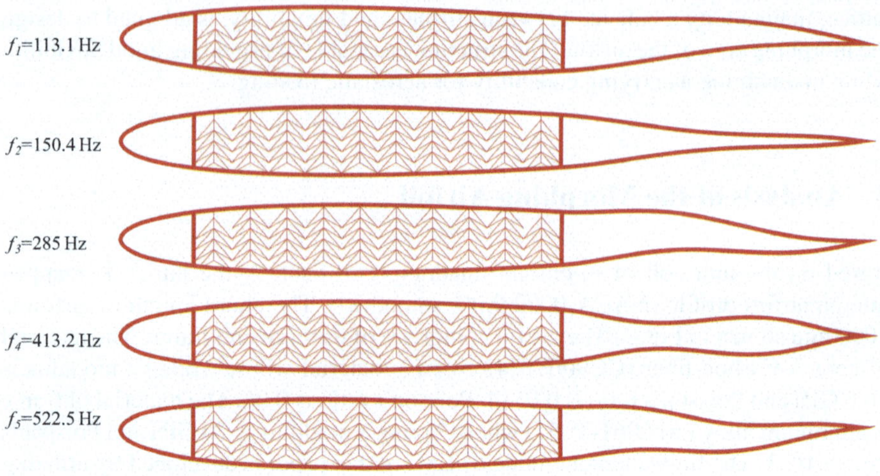


**Fig. 4** Normal modes of morphing airfoil (10 times deformation)

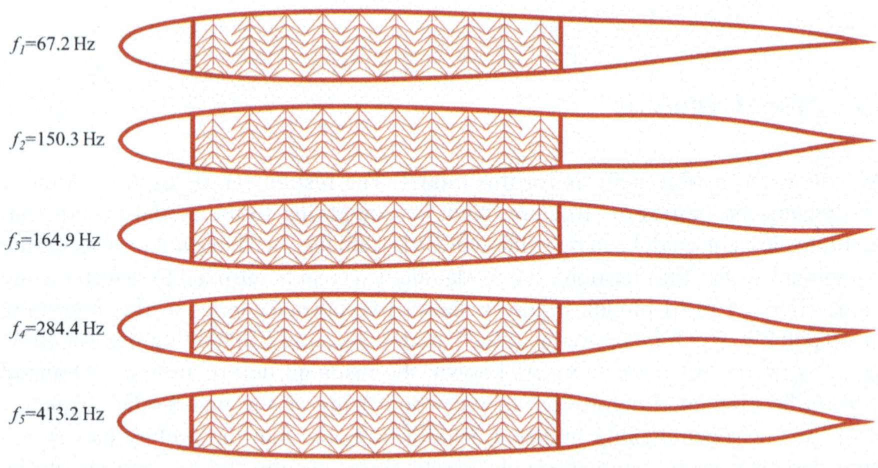


**Fig. 5** Constrained modes of morphing airfoil (10 times deformation)

## 3.2 *Passive Morphing*

For simplicity, the model is completely clamped at the leading edge shown in Fig. 6a. The airfoil is imposed a concentrated force at the trailing edge shown in Fig. 6a while such conditions do not represent the typical aerodynamic load but accord with the experiments carried out in [20, 21]. The linear load respect to the trailing-edge displacement is reported in Fig. 6. It is worth noting that the applied load- trailing-edge displacement curves were generated within the elastic range of the constituent material. The high flexibility in shear of the auxetic lattice while maintaining shear

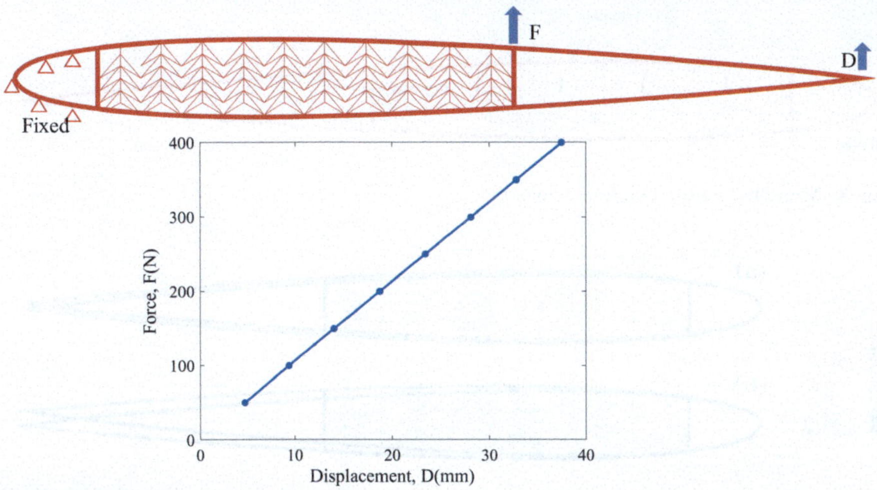


**Fig. 6** Condition and results. **a** The loading and constraints of the airfoil; **b** applied load versus trailing-edge displacement

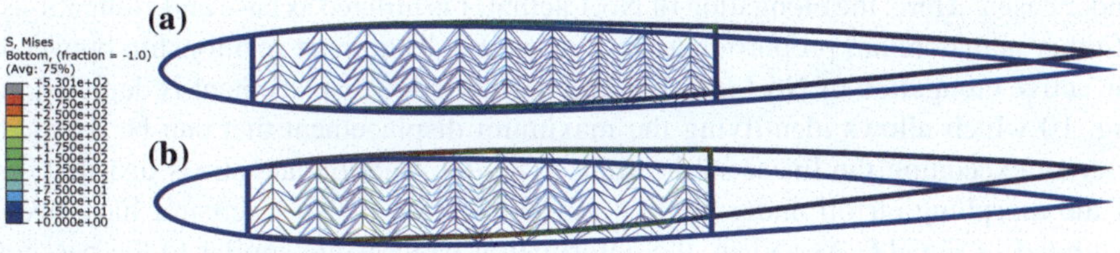


**Fig. 7** Stress distribution **a** 200 N **b** 400 N

stiffness is good for a passive morphing airfoil application. As shown in Fig. 7, the morphing airfoil with auxetic lattice core exhibits high elastic flexibility when it is imposed by the concentrated force. When the passive load is 400 N, the airfoil has a significant deformation in shape.

### *3.3 Active Morphing*

The core of the morphing airfoil is fabricated by the lattice. Therefore, it is easy to replace struts with actuators in this scheme to implement the active morphing. Four actuators are arranged in this model to drive morphing shown in Fig. 8. Here, two cases are conducted including case 1: all actuators have 6 mm elongation; case 2: 4, 6, 8, 10 mm elongation respectively for No.1, 2, 3, 4 actuators. As shown in Fig. 9, case 2 achieves the better property of morphing than case 1 within the same strain limitation. Hence, we use case 2 as the working condition. Simulations under different

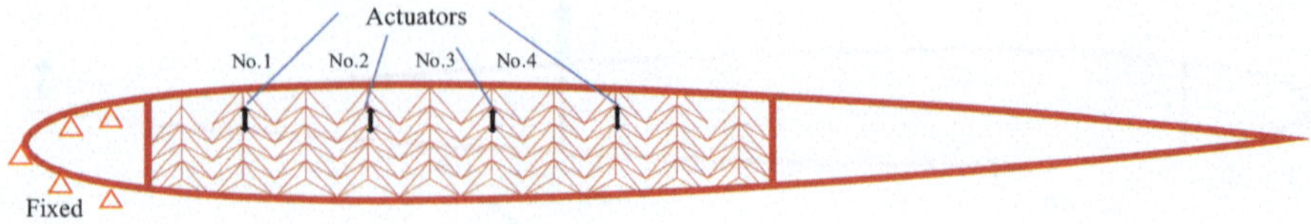


**Fig. 8** Morphing airfoil with four actuators

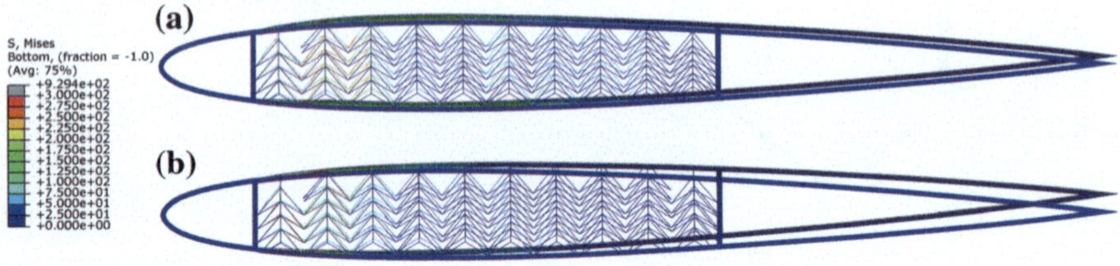


**Fig. 9** Deformation and stress distribution under: **a** Case 1; **b** Case 2

driven scopes are conducted to analysis the morphing properties of the active airfoil under case 2. Here, the elongation of No.1 actuator is utilized as base and elongations of other actuators are proportionate to it as case 2. The linear relationship between the active elongation of No.1 actuator and trailing-edge displacement is depicted in Fig. 10 which allows identifying the maximum displacement that can be reached without exceeding the linear strain limits. The deformation and stress distribution of the morphing airfoil under 3 and 5 mm elongation of No.1 actuator have been exhibited in Fig. 11. As shown, the deformation patterns are similar to the passive condition in the above section.

**Fig. 10** Applied active elongation versus trailing-edge displacement

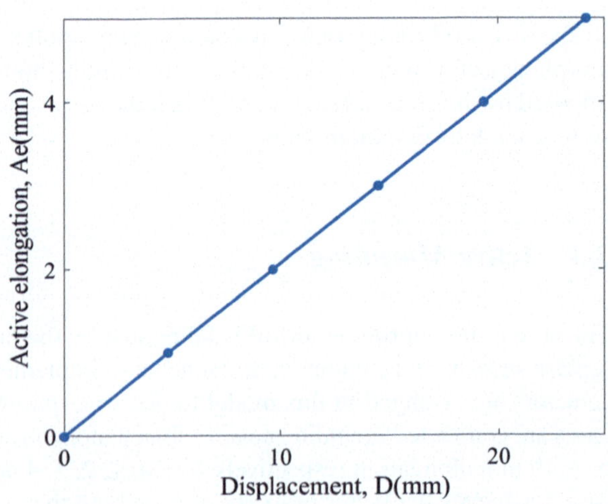

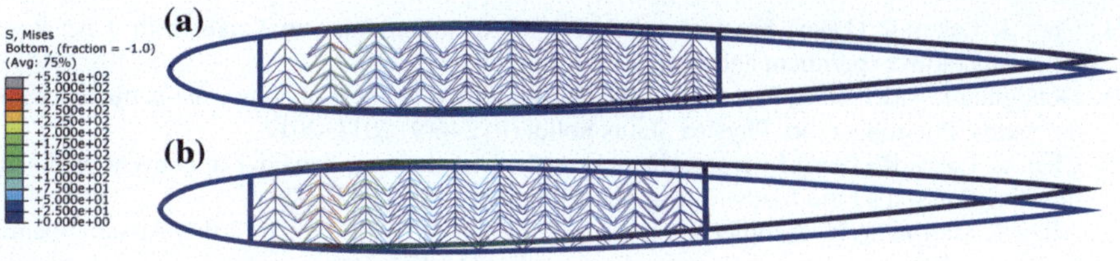


**Fig. 11** Deformation and stress distribution under: **a** 3 mm; **b** 5 mm

## 4 Conclusion

In this paper, we investigated the in-plane mechanical properties of re-entrant quadrangular lattices and its application to the morphing airfoil with flexible cores. The enhanced re-entrant lattice shows remarkable superior to the original one in stiffness and has similar flexibility. The morphing airfoil with auxetic lattice was studied under an aerostatic load. The re-entrant lattice core shows high shear flexibility. Besides the passive morphing, the active morphing also has been investigated in this paper. The actuator is readily arranged in a re-entrant quadrangular lattice to implement active control for morphing airfoil. By lengthening actuator in the right way, the airfoil can achieve deformation like the passive morphing generated by the concentrated force. The elementary result of this paper provides ideas for the design of passive and active morphing airfoil.

**Acknowledgements** The authors are grateful for the support provided by National Key R&D Program of China (No. 2017YFB0103204), Natural Science Foundation of Shanghai (No. 18ZR1440900) and Chinese Scholarship Council (No. 201706260056).

## References

1. Barbarino S, Bilgen O, Ajaj RM, Friswell MI, Inman DJ (2011) A review of morphing aircraft. J Intell Mater Syst Struct 22(9):823–877
2. Rodriguez A (2007) Morphing aircraft technology survey. In: 45th AIAA aerospace sciences meeting and exhibit 2007, p 1258
3. Thill C, Etches J, Bond I, Potter K, Weaver PJ (2008) Morphing skins. Aeronaut J 112(1129):117–139
4. Weisshaar TA (2013) Morphing aircraft systems: historical perspectives and future challenges. J Aircraft 50(2):337–353
5. Liu Y, Hu H (2010) Essays. A review on auxetic structures and polymeric materials. Scientif Res Essays 5(10): 1052–1063
6. Lakes R (1987) Foam structures with a negative Poisson's ratio. Science 235:1038–1041
7. Prawoto Y (2012) Seeing auxetic materials from the mechanics point of view: a structural review on the negative Poisson's ratio. Comput Mater Sci 58:140–153
8. Liu Q (2006) Literature review: materials with negative Poisson's ratios and potential applications to aerospace and defence. Defence science and technology organisation victoria (Australia) air vehicles div

9. Choi J, Lakes R (1996) Fracture toughness of re-entrant foam materials with a negative Poisson's ratio: experiment and analysis. Int J Fract 80(1):73–83
10. Donoghue J, Alderson K, Evans K (2009) The fracture toughness of composite laminates with a negative Poisson's ratio. Physica Status Solidi (b) 246(9):2011–2017
11. Chen C, Lakes R (1996) Micromechanical analysis of dynamic behavior of conventional and negative Poisson's ratio foams. J Eng Mater Technol 118(3):285–288
12. Allen T, Martinello N, Zampieri D, Hewage T, Senior T, Foster L et al (2015) Auxetic foams for sport safety applications. Procedia Engineering 112:104–109
13. Chen Z, Wang Z, Zhou S, Shao J, Wu X (2018) Novel negative Poisson's ratio lattice structures with enhanced stiffness and energy absorption capacity. Materials 11(7):1095
14. Peel LD, Mejia J, Narvaez B, Thompson K, Lingala MJ (2009) Development of a simple morphing wing using elastomeric composites as skins and actuators. J Mech Des 131(9):091003
15. Fleck N, Deshpande V, Ashby MJ (2010) Micro-architectured materials: past, present and future. Proc Royal Soc A: Math, Phys Eng Sci 466(2121):2495–2516
16. Olympio KR, Gandhi FJ (2010) Flexible skins for morphing aircraft using cellular honeycomb cores. J Intell Mater Syst Struct 21(17):1719–1735
17. Ramrakhyani DS, Lesieutre GA, Frecker MI, Bharti SJ (2005) Aircraft structural morphing using tendon-actuated compliant cellular trusses. J Aircraft 42(6):1614–1620
18. Bharti S, Frecker M, Lesieutre G, Browne J (2007) Tendon actuated cellular mechanisms for morphing aircraft wing. In: Modeling, signal processing, and control for smart structures, International Society for Optics and Photonics, p 652307
19. Olympio KR, Gandhi FJ (2010) Zero Poisson's ratio cellular honeycombs for flex skins undergoing one-dimensional morphing. J Intell Mater Syst Struct 21(17):1737–1753
20. Heo H, Ju J, Kim D-M (2013) Compliant cellular structures: application to a passive morphing airfoil. Compos Struct 106:560–569
21. Bettini P, Airoldi A, Sala G, Di Landro L, Ruzzene M, Spadoni A (2010) Composite chiral structures for morphing airfoils: numerical analyses and development of a manufacturing process. Compos B Eng 41(2):133–147

# An Improved Deep Convolutional Neural Network to Predict Airfoil Lift Coefficient

**Boping Yu, Liang Xie and Fuxin Wang**

**Abstract** Recently, significant developments in deep learning have made many possibilities in the field of fluid mechanics. This paper presents a new method of using improved convolutional neural network to learn airfoil lift coefficient calculated by OpenFOAM simulation tool. We propose a "feature-enhanced-image" data preprocessing method to prepare the training and testing data set. A novel convolutional neural network is designed which uses deeper convolution and pooling layers coupled with batch normalization technique. In addition, before linear regression, in fully connected layers, we use dropout method to reduce the risk of over-fitting. Mini-batch stochastic gradient descent (SGD) optimization algorithm is chosen, and mean square error (MSE) is used to do the model evaluation when training and testing the model. It is demonstrated that this improved deep convolutional neural network (IDCNN) provides more accurate lift coefficient prediction compared to other state-of-the-art neural networks. We also test the effect of batch size and full batch normalization implementation on the performance of the whole convolutional neural network. Finally, it is concluded that the best predicting performance is achieved in the condition of 10 batch size and the mean square error of blind test can reach $3.1 \times 10^{-4}$. Furthermore, the "feature-enhanced-image" method we proposed can achieve 85.2% decreasing of testing MSE.

**Keywords** IDCNN · Lift coefficient · Batch normalization · Dropout · Mini-batch SGD

B. Yu · L. Xie · F. Wang (✉)
Shanghai Jiao Tong University, Shanghai, China
e-mail: fuxinwang@sjtu.edu.cn

B. Yu
e-mail: boping_yu@sjtu.edu.cn

L. Xie
e-mail: xl2016@sjtu.edu.cn

 
Z. Jing (ed.), *Proceedings of the International Conference on Aerospace System Science and Engineering 2019*, Lecture Notes in Electrical Engineering 622,
https://doi.org/10.1007/978-981-15-1773-0_21

# 1 Introduction

The calculation of aerodynamic coefficients of airfoil has always been an important process of airfoil optimization and aircraft design [1]. At present, the main methods to obtain aerodynamic coefficients are wind tunnel test and computational fluid dynamics (CFD) simulation. In addition, there exist many effective surrogate models [2–4] to deal with these high dimensional computing problems, and many researchers also utilize artificial neural network (ANN) technique to perform the aerodynamic coefficients approximation [5, 6]. Recently, great advances in neural networks and machine learning have been made due to the booming development of computational power. The development of deep neural network (DNN) technique has opened the new doors to improve the robustness, generalization and precision of the aerodynamic coefficients approximation model.

More recently, more and more researchers have applied the deep learning techniques in the field of aerospace and achieve some positive results. Ling et al. [7] proposed a novel deep neural network architecture which contains a multiplicative layer with embedded invariance for the Reynolds stress anisotropy prediction, and this model shows improved prediction precision compared with a commonly used neural network. Suresh et al. [8] used a RNN method to predict lift coefficient at high angle of attack, and it demonstrates that RNN can have high predicted performance than memory neuron networks (MNN). Zhang et al. [9] used a CNN approach to predict airfoil aerodynamic coefficients for unseen airfoil shapes. However, there exists a MSE gap between the testing data set and training data set, which means the model they proposed has a risk of over-fitting.

Inspired by the work by Zhang et al. [9], we design an improved deep convolutional neural network to do the aerodynamic coefficients approximation, and this paper will demonstrate that the prediction model can provide more accurate and general coefficients approximation. For a CNN model, the most important thing is the preprocessing of the training data. Enough and extensive data can contribute to a good learning ability and generalization performance model [10]. Within our known knowledge, we made the following innovations in data preparation.

- We use user-defined OpenFOAM simulation tool to guarantee the accuracy of training data.
- We propose a "feature-enhanced-image" method to prepare the training data, thus broadening the width of the airfoil data image and changing the brightness of the image to do the data augmentation.
- Feed some bad shapes of airfoil data image into the IDCNN model we proposed, in order to enhance its generalization ability.

## 2 Improved Deep CNN for Lift Coefficient Prediction

Zhang et al. [9] used a highly similar architecture to LeNet-5 [11], to do the airfoil lift coefficients approximation. It has demonstrated that CNN technique has been a useful perspective in engineering meta-modeling task, but the CNN model they proposed did not get a satisfactory accuracy. We harness a deeper CNN architecture to predict lift coefficient, including three convolution layers, three pooling layers and four fully connected layers.

### *2.1 IDCNN Architecture for Lift Coefficient Prediction Model*

The improved CNN architecture we proposed is based on the traditional LeNet-5 architecture, but we have made the following changes and improvements. We deepening the whole neural network, including the convolutional layer, pooling layer and fully connected layer. Besides, batch normalization [12] is used after each pooling operation, and it can accelerate training speed and reduce internal covariate shift between previous pooling layer and next convolutional layer. In order to avoid the over-fitting problems effectively which may affect the prediction accuracy, we adopt the dropout [13] technique and set the dropout rate to 0.7. ReLU activation function and mini-batch SGD [14] optimized method are chosen in the training of the CNN model.

The data input layer is a mini-batch of airfoil images, and the resolution of the image is $100 \times 100$. The size of batch can affect the training results of model, and we will compare testing accuracy with different given batch sizes. In convolutional layer 1, we used 40 convolution kernels which have the size of $5 \times 5$. After feeding the output data into pooling layer 1, we add a BN layer and the same operation occurs between every convolutional layer and pooling layer. In pooling layer 1, we select max-pooling method and the pooling filter size is $2 \times 2$. In addition, zero-padding method and moving step size which is equal to filter size in each direction are chosen. In convolutional layer 2, 60 convolutional filters are used during the convolution operation and the filter size is $5 \times 5$. The settings of pooling 2 are the same as pooling layer 1, and a BN layer is added before the pooling 2. In the last convolutional layer, we use 100 kernels which have the same size as previous kernels. Then, the max-pooling method is chosen in the pooling layer 3. After the convolutional and pooling layers, four fully connected layers are used to do the linear regression, in other words, airfoil lift coefficient prediction. The details of the number of nodes in each fully connected layer are shown in Fig. 1.

To summarize, we have constructed an improved deep CNN model that contains 13 hidden layers, including three convolutional layers, three pooling layers, three BN layers and four fully connected layers. A visual illustration architecture of the IDCNN we proposed is shown in Fig. 1.

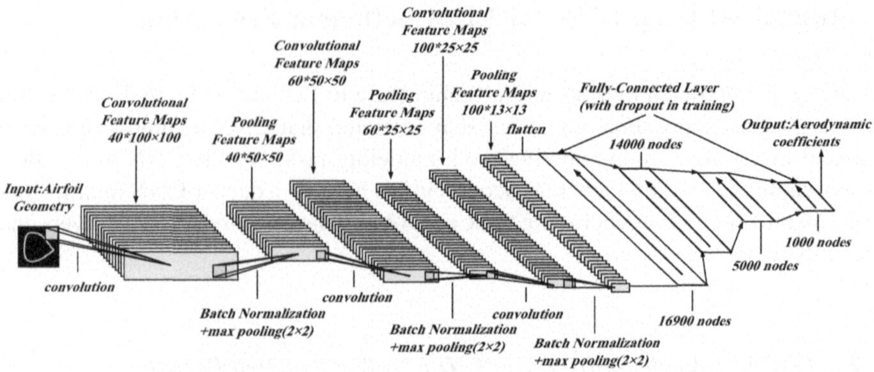


**Fig. 1** Improved deep CNN architecture for lift coefficient prediction model

## 2.2 Data Preparation

The CNN model we proposed is trained and tested on a database of airfoil data images and lift coefficients. According to the goal of airfoil optimization in the future, SC1095 airfoil is chosen, and Hicks-Henne fitting method [15] is used to generate the training and testing airfoil data sets. In this context, the user-defined OpenFOAM solver is chosen as the simulation tool to obtain the aerodynamic coefficients. The simulating Mach number is 0.4, Reynolds number is $3.63 \times 10^6$ and the angle of attack is 6.13°. Spalart-Allmaras turbulence model [16] and SIMPLE algorithm [17] are selected. In order to check the accuracy of CFD results, in this paper, we compare the CFD result with experimental data [18]. Figure 2 shows the comparing result and it shows that the CFD result is highly consistent with experimental data. In other words, Fig. 2 illustrates that the user-defined OpenFOAM solver and computing settings are feasible in this situation.

Data preprocessing is crucial to the training and testing of CNN model. Besides, we expect to get more training samples with less computational cost. In view of the above two situations, we propose a novel data preprocessing method "feature-enhanced-image" for the airfoil data image and the description of this method is given below. An initial SC1095 airfoil data image is shown at the top of Fig. 3. The maximum thickness of the SC1095 airfoil is only 9.50% of the chord length. So small changes of control variables could not lead to a visible shape changes in a low resolution data image. In order to maximize the shape changes of the given airfoil at low resolution, we broaden the width of airfoil data image by multiplying the thickness of airfoil by 10 times. Then, we do the data augmentation by changing the brightness of the data image into *three* different types. In other words, three different types of airfoil data image have the same lift coefficient label. We can get massive training and testing samples using only one-third of CFD computation cost through this method. Changing brightness means changing the number of airfoil data matrix, and the variance of all the data in the matrix can keep the same. Thus, the information

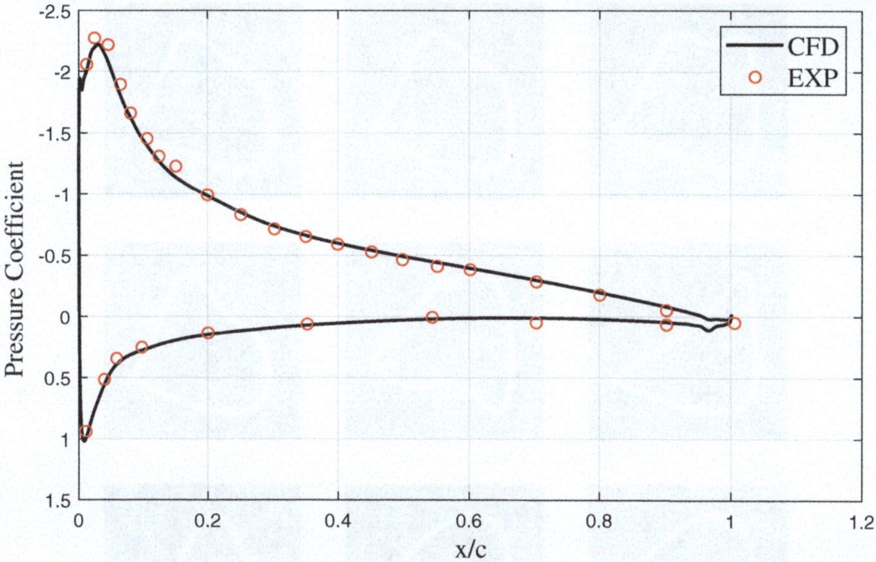


**Fig. 2** Comparing result of experimental and CFD data

**Fig. 3** "feature-enhanced-image" method

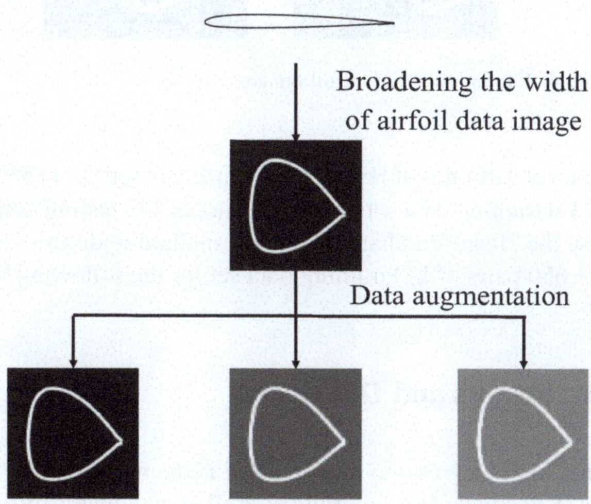


of the airfoil shape can be preserved and the cost of training data preparation can be reduced at the same time. The process of applying the "feature-enhanced-image" method in the IDCNN model is shown in Fig. 3.

In addition to the above data preprocessing methods, we also add some bad airfoil data images into the training data set. Some bad airfoil image examples are shown in Fig. 4. Bad images mean some airfoils that can not be regarded as a proper airfoil shape. We can consider this operation as the noise adding. Through changing the

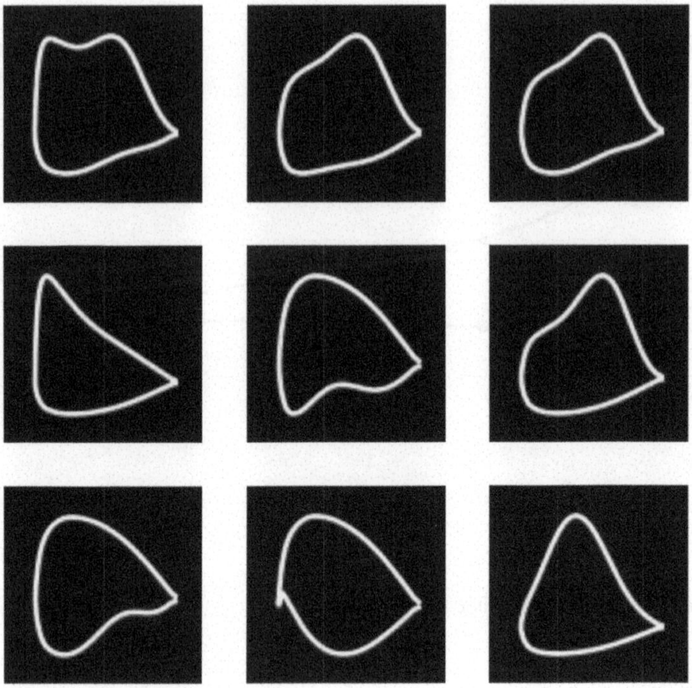

**Fig. 4** Examples of bad airfoil images

control variables of Hicks-Henne fitting function, 11550 pairs of normal input/output (I/O) training data set and 1050 pairs of I/O testing data set are generated. Then, we use the "feature-enhanced-image" method to do the data augmentation, thus we get 34,650 pairs of I/O training data set for the following hyper-parameter tuning.

## 3 Results and Discussion

In a linear regression model, MSE is the most popular used evaluation criteria. The MSE calculated by summing over all of the squared differences between the predicted and true values and then divided by the number of the batch size. In Eq. 1, $m_{batch}$ represents the number of batch size, $t_i$ is the true value and $y_i$ is the predicted value.

$$MSE = \sum_{i=1}^{m_{batch}} \frac{(t_i - y_i)^2}{m_{batch}} \quad (1)$$

In the training of the IDCNN model, we proposed. We utilize the early-stopping method to save the trained model at a specific epoch to get the minimum testing MSE.

## 3.1 "feature-enhanced-image" Method Influence on the Testing MSE

We select 11,550 training data randomly from the 34,650 training data set to check the influence of the "feature-enhanced-image" method on the testing MSE compared with the normal training data set mentioned above. The other parameters in the CNN model are the same and the comparing results of testing MSE are showing in Fig. 5. In these two CNN models, dropout technique is used to fight the over-fitting problem. The testing MSE of the CNN with normal training data set can only reach $2.5 \times 10^{-3}$. However, the testing MSE of the randomly selected training data that using the "feature-enhanced-image" method can reach $3.7 \times 10^{-4}$. The decreasing rate of the testing MSE is 85.2% which shows the "feature-enhanced-image" is a very effective method for reducing the testing MSE. We can also figure out that the amplitude of the MSE fluctuation has increased after using the "feature-enhanced-image" method. This is because we have added some noise airfoil data images into the training data set.

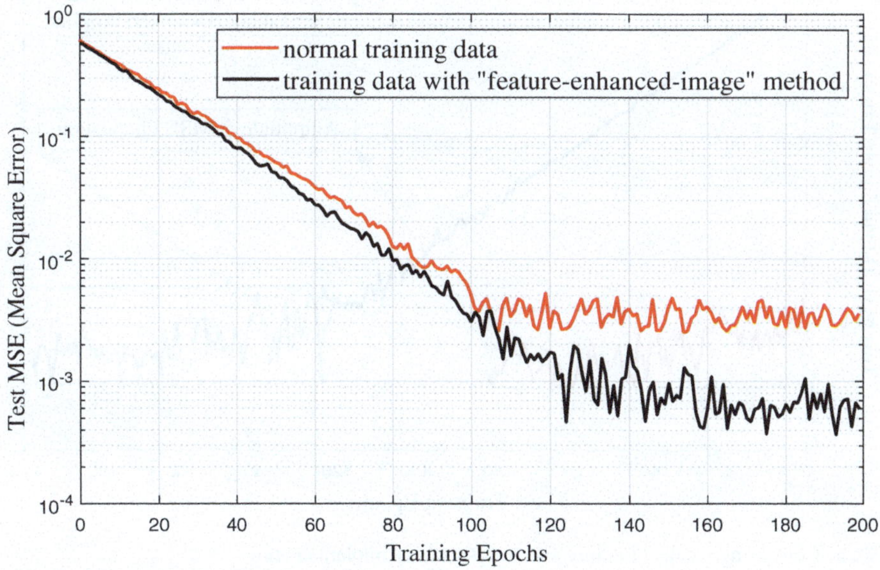


**Fig. 5** "feature-enhanced-image" method influence on the testing MSE

### 3.2 Comparing Results of Empty and Full BN Layers Implementation

Batch normalization is always be implemented after the data input layer, but multiple BN layers are added after each convolutional layer in order to reduce internal covariate shift in this paper. The comparing result is shown in Fig. 6, it is easy to conclude that full BN layers implementation can dramatically reduce the training time. One training epoch means feeding the whole training data to the CNN model. When we use a normal CNN architecture without full BN layers implementation, the MSE does not reach an acceptable level until about 120 epochs. However, when applying full BN layers in the previous CNN architecture, the testing MSE reaches $1 \times 10^{-3}$ at about 2 epochs. Moreover, the fluctuation of MSE during the model training has also been affected by the full BN layers implementation. In Fig. 6, the amplitude of the MSE fluctuation is enclosed by two black lines for easily comparing. When applying full BN layers in the CNN model, the amplitude of the MSE fluctuation reduces a little bit than that of CNN model without full BN layers applied.

The given comparing result shows that full BN implementation can highly reduce the training time, thus being good to the following hyper-parameters tunning and model constructing. Smaller amplitude of the MSE fluctuation can lead to easily early stopping at an appropriate epoch to save the CNN model.

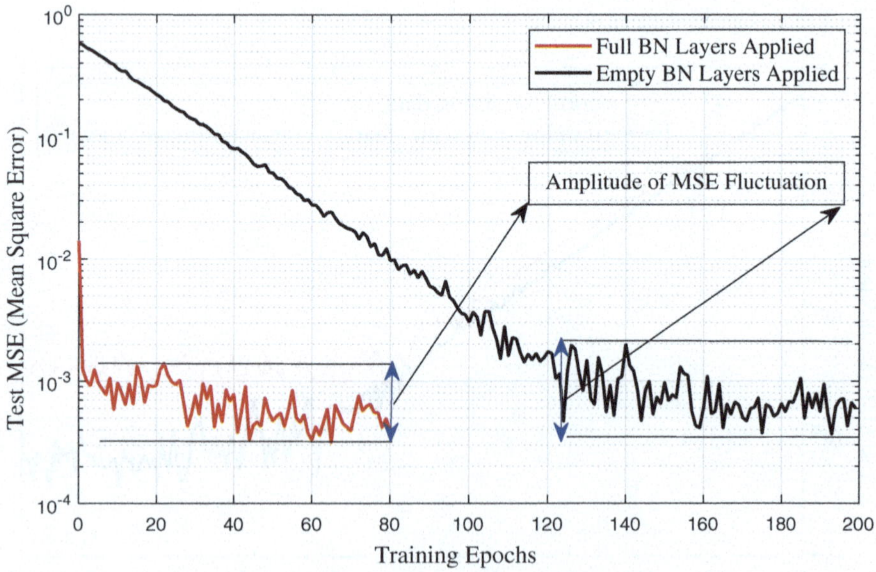


**Fig. 6** Comparing results of empty and full BN layers implementation

### 3.3 Batch Size Influence on the Testing Results

Mini-batch SGD optimized algorithm is used in the neural network training. Batch size is a highly influential factor on the performance of the CNN model and an intuitive judgment is that the larger the batch size, the slower the convergence. From the CNN model tunning experience, using about 1% of the training data set as the size of mini-batch is a proper choice. According to the goal of faster training and best MSE performance, we set the top level of comparing batch sizes to be 100. Then, five different batch sizes are compared when we do the mini-batch SGD optimization, and the results are presented in Fig. 7. It is easy to figure out that 10 batch size is an optimal choice for the improved CNN model. Because under this condition, both fast convergence and minimum MSE are guaranteed.

### 3.4 Over-Fitting Checking and MSE Quantitative Analysis

A satisfactory testing accuracy is not the end of training a neural network. Over-fitting is the most likely occur during neural network training. For over-fitting phenomenon, a simple judgment method is checking the difference between the training and testing accuracy of the neural network. In the improved CNN airfoil lift coefficient prediction model, dropout technique is used to avoid this problem and the comparing result is

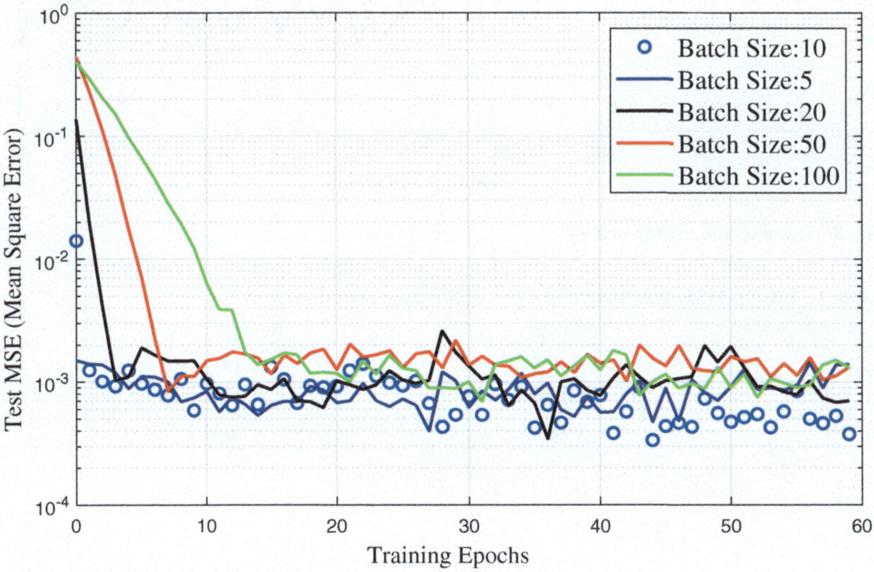


**Fig. 7** Comparing result of batch size on the performance of model

shown in Fig. 8. The testing MSE of CNN without dropout is higher than that of CNN with dropout in most training epochs, except for a few epochs in the beginning of the training. It is shown that dropout implementation can reduce the MSE gap between the training and testing procedure.

Three different neural networks have been compared with the same evaluation criteria MSE and the comparing results are shown in Table 1. The list two other neural networks, CNN and recurrent multilayer perceptron (RMLP), are also used to predict the lift coefficients. The improved deep CNN model we proposed has a much smaller MSE than other two neural networks. It demonstrates that the improved techniques we employed, including BN, dropout and mini-batch SGD, are feasible and of great value in predicting the airfoil lift coefficient.

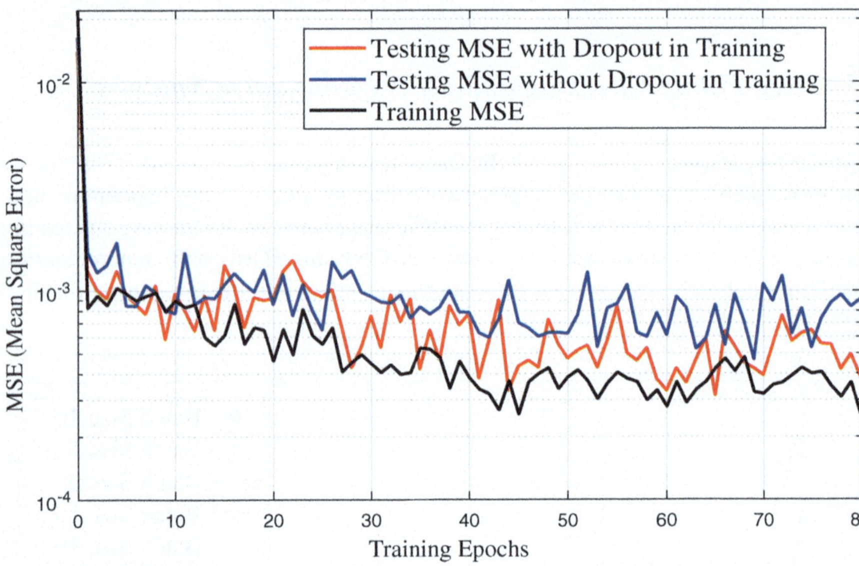


**Fig. 8** Training and testing history

**Table 1** Testing MSE quantitative analysis of different neural networks

| Network type | MSE | Remark |
|---|---|---|
| CNN | 0.005 | Ref. [9] |
| RMLP | 0.0009 | Ref. [8] |
| IDCNN | 0.0003 | Our model |

## 4 Conclusion

An improved deep CNN model for lift coefficient prediction is presented which uses full BN layers implementation, dropout and mini-batch SGD techniques. In the preprocessing of the training and testing data, the "feature-enhanced-image" method we proposed is utilized and it can gain a effective 85.2% declining of testing MSE in the model training. Besides, we also add some bad airfoil data images into the model training data to enhance the model generalization ability.

As for the improvements of the "feature-enhanced-image," full BN and dropout methods on the initial CNN model, the comparing results are given to show that these three techniques are effective in reducing the training time and increasing the prediction accuracy. The training and testing history is presented to show that the model we established has a good generalization ability than other state-of-the-art neural networks. Furthermore, the IDCNN model we constructed presents a more accurate prediction than other two neural networks. In conclusion, the IDCNN model we proposed can guarantee a much less training time and a more precise testing accuracy at the same time.

## References


1. Buckley HP, Zhou BY, Zingg DW (2009) Airfoil optimization using practical aerodynamic design requirements. J Aircr 47(5):1707–1719
2. Iuliano E (2017) Global optimization of benchmark aerodynamic cases using physics-based surrogate models. Aerosp Sci Technol 67:273–286
3. Jouhaud JC, Sagaut P, Montagnac M, Laurenceau J (2007) A surrogate-model based multi-disciplinary shape optimization method with application to a 2d subsonic airfoil. Computers Fluids 36(3):520–529
4. Laurenceau J, Sagaut P (2008) Building efficient response surfaces of aerodynamic functions with kriging and cokriging. AIAA Journal 46(2):498–507
5. Ignatyev DI, Khrabrov AN (2015) Neural network modeling of unsteady aerodynamic characteristics at high angles of attack. Aerosp Sci Technol 41:106–115
6. Linse DJ, Stenge RF (1993) Identification of aerodynamic coefficients using computational neural networks. J Guidance Control Dyn 16(6):1018–1025
7. Ling J, Andrew K, Jeremy T (2016) Reynolds averaged turbulence modelling using deep neural networks with embedded invariance. J Fluid Mech 807:155–166
8. Suresh S, Omkar SN, Mani V, Prakash TNG (2003) Lift coefficient prediction at high angle of attack using recurrent neural network. Aerosp Sci Technol 7(8):595–602
9. Zhang Y, Sung WJ, Mavris DN (2018) Application of convolutional neural network to predict airfoil lift coefficient. In: 2018 AIAA/ASCE/AHS/ASC structures, structural dynamics, and materials conference, p 1903
10. LeCun Y, Bengio Y, Hinton G (2015) Deep learning. Nature 521(7553):436
11. Lecun Y, Bottou L, Bengio Y, Haffner P (1998) Gradient-based learning applied to document recognition. Proc IEEE 86(11):2278–2324
12. Ioffe S, Szegedy C (2015) Batch normalization: accelerating deep network training by reducing internal covariate shift. arXiv preprint arXiv:1502.03167
13. Srivastava N, Hinton G, Krizhevsky A, Sutskever I, Salakhutdinov R (2014) Dropout: a simple way to prevent neural networks from overfitting. J Mach Learn Res 15(1):1929–1958

14. Bottou L (2010) Large-scale machine learning with stochastic gradient descent. In: Proceedings of COMPSTAT'2010. Springer, pp 177–186
15. Hicks RM, Henne PA (1978) Wing design by numerical optimization. J Aircr 15(7):407–412
16. Spalart P, Allmaras S (1992) A one-equation turbulence model for aerodynamic flows. In: 30th Aerospace sciences meeting and exhibit, p 439
17. Patankar S (1980) Numerical heat transfer and fluid flow. CRC Press
18. Qing W, Qi-jun Z (2016) Synthectical optimization design of rotor airfoil by genetic algorithm. J Aerosp Power 31(6):1486–1495

# Remote Sensing Image Change Detection and Location Based on Dynamic Level Set Model

**Yunkai Liu, Yuanxiang Li, Yongshuai Lu and Jiawei Liu**

**Abstract** Image change detection is established for extracting changed regions in multiple images of the same scene captured at different times. Recent research has demonstrated that the change detection methodologies using satellite images, such as multi-temporal visible light remote sensing image and synthetic aperture radar (SAR) image, are particularly useful for damage assessment after various disasters, e.g., earthquakes, fires, floods, and landslides. The level set method, because of its implicit handling of topological changes and low sensitivity to noise, is one of the most effective unsupervised change detection techniques for satellite images. The signed pressure force function (SPF) improved the performances of conventional level set methods through including two grayscale parameters, i.e., the average pixel intensity inside and outside the contour, respectively. However, the mean of region pixel intensity is not a good indicator in case that the images are inhomogeneities grayscale, e.g., confused-edge objects in satellite images. In order to address this problem, we propose a novel model, denoted as dynamic SPF (D-SPF) model, which can dynamically learn a discriminative indicator for distinguishing the pixels inside or outside the contour. Specifically, the principle of maximizing entropy between the regions inside and outside the contour is used to learn the K distinguish parameters, which help to guide the segmentation contour to end on the object's edge. The experiments are conducted on a public satellite image dataset, i.e., ERS, which contain 670 SAR and 670 optical images; each image covers approximately 150 $m^2$

Y. Liu · Y. Li (✉) · J. Liu
School of Aeronautics and Astronautics, Shanghai Jiao Tong University, Minhang 200240, Shanghai, China
e-mail: yuanxli@sjtu.edu.cn

Y. Liu
e-mail: sjtu10018lyk@sjtu.edu.cn

J. Liu
e-mail: eiwqueqie@sjtu.edu.cn

Y. Lu
Baidu Netcom Science and Technology Co, Beijing 201210, Shanghai, China
e-mail: lys_2008@163.com



Z. Jing (ed.), *Proceedings of the International Conference on Aerospace System Science and Engineering 2019*, Lecture Notes in Electrical Engineering 622,
https://doi.org/10.1007/978-981-15-1773-0_22

area including forests, lakes, and cities, etc. These images are challenging due to the inhomogeneities of landforms and unknown natural disasters. The experimental results demonstrate that D-SPF model reduces almost 30.4% missed detection rate on the optical images and 41.2% missed detection rate on the SAR images in comparison with SPF and obtains the best detection performances in ERS dataset.

**Keywords** Remote sensing image · Change detection · Level set method · Dynamic signed pressure force model

## 1 Introduction

In the analysis and processing of remote sensing images, change detection is a very important field. Multi-temporal remote sensing image change detection is a process of quantitatively analyzing the changes of remote sensing images at the same place at different times. Change detection technology is widely used in various aspects: monitoring crop growth and changes, land type changes; monitoring changes before and after various types of disasters, such as location and loss assessment of strong earthquake areas; monitoring forest vegetation, snow cover, soil changes in moisture; monitoring the movement of ice floes on the sea, landslide movements on land glaciers; dynamic monitoring of objects of military interest and assessment of the effects of destruction after battlefield strikes.

Image segmentation is a very important part of traditional change detection methods. The method based on the level set method is a relatively advanced algorithm in image segmentation. It has the characteristics of high segmentation precision and insensitivity to noise and is good at processing images with complex shapes and topological changes. The results of change detection often present a variety of complex shapes, so applying the level set segmentation method to change detection is theoretically feasible.

At present, there are many methods for multi-temporal remote sensing image change detection. Different scholars have established a large number of change detection methods and various models for different image characteristics. Singh concludes in the [1] that the common methods of change detection can be divided into direct comparison method and post-classification comparison method. The direct comparison method is to directly generate difference images from two remote sensing images (simple difference, ratio, etc.), uses various methods to process the difference images, and then obtains the change detection results. This method is simple and intuitive which the implementation method is also very easy, but it is sensitive to the sensor itself and the noise in the transmission. The post-classification comparison method classifies the pixel-by-pixel points in the two remote sensing images and then performs the difference detection on the two categories of interest to obtain the change detection result. This method requires high classification accuracy, and the classification error will directly lead to the error of the detection, such as Bayesian method and so on. Whether it is direct comparison or post-classification comparison, image segmentation (category of categories, division of "change" and "invariant" regions) is inevitable. The level set-based segmentation method is a more advanced technique in image segmentation. Most of the change detection studies use traditional segmentation methods such as threshold segmentation, and the study on the use of level set methods in the detection of remote sensing images is relatively rare. Bazi et al. in the paper [2] introduced a single-phase image classification method based on the multi-resolution level set (MLS) model, and Celik et al. in 2011 [3] proposed the C-V model to divide the "variable" and "invariant" regions of the difference image after the non-sampling discrete wavelet transform.

Among them, the problems of remote sensing image change detection mainly conclude:

(1) How to improve the quality of the input image
Change detection is a method that is accurate to the pixel level, and any small error will seriously affect the detection result. Therefore, the prior image registration and denoising process are particularly important. Only the registration error of the two images is less than 1 pixel, so we can carry out the following detection steps very well [4].
(2) How to construct a satisfactory difference region
How to construct a difference image is a direct difference, a direct ratio or a filter to reduce the interference caused by noise.
(3) How to determine an appropriate change area
The key to change detection is actually to quantitatively find a change area. The determination of the change area is generally obtained by some threshold or classification methods in which the threshold setting and category selection itself have certain uncertainty.
(4) How to improve the degree of automation
In the change detection method with supervision, many parameters need to be adjusted for different experiments. What is more, non-supervised change detection methods often have some threshold and other parameters and require manual intervention. This has affected the accuracy and speed of our detection.

For the above problems, the method of this paper made mainly progress in (2) (3). Multi-temporal remote sensing image change detection has always been a research focus in the field of remote sensing applications; in addition, image segmentation based on level set is also an emerging method in the field of image segmentation. This paper attempts to combine the two directions to improve the precision and accuracy of remote sensing image change detection.

## 2 The Proposed Method

### 2.1 Detection Algorithm Structure

The direct comparison is a method of obtaining a difference image of two images and then performing processing analysis of the differential phase image. The main idea of the change detection algorithm proposed in this paper is that the level set algorithm (mainly D-SPF model) [5] is used to segment and finally obtain the change region after the difference image is obtained by the direct comparison. The main algorithm flowchart is as follows in Fig. 1.

The difference image acquisition method used in this paper is the average log-ratio method, which is similar to the image ratio method. In order to reduce the noise interference and improve the discrimination of the foreground-background, we have added the process of windowing and the natural logarithm of the result.

**Fig. 1** Combination of direct comparison and level set

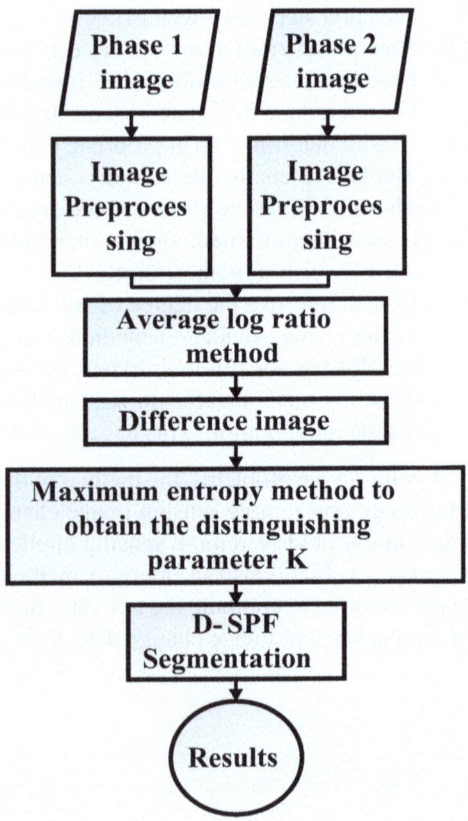

$$r = \text{abs}\left(\ln\left(\frac{\frac{1}{n}\sum_{i=1}^{n} I_1 + 1}{\frac{1}{n}\sum_{i=1}^{n} I_2 + 1}\right)\right) \tag{1}$$

wherein $n$ is the number of pixels in the window. In the actual program, the window of the size is taken, so $n = 9$. $I_1$, $I_2$ represent the gray values of the corresponding pixels on the phase 1 image and the time phase 2 image, respectively. The use of absolute values is used to prevent the occurrence of negative signs. In order to facilitate the comparison of various algorithms, all the algorithms mentioned below use the method of this section to obtain the difference image.

For the difference image, the part with the large gray value is the change area, and the part with the small gray value indicates that the change is not obvious. We use the level set segmentation algorithm to segment the difference image. The two parts that are separated are the areas of "change" and "unchanged." The "change" area is the purpose of our change detection.

There are many models of the level set segmentation algorithm, and a different potential energy function can construct a new segmentation model. I mainly use the C-V [6] model, the SPF model, and my D-SPF model to segment. The first two level set segmentation models can achieve no-parameter segmentation. However, our D-SPF model needs to adjust the distinguish parameters $k$ according to different images. We can use the maximum entropy algorithm to automatically obtain the distinguish parameter $k$ for the difference image obtained by the average logarithmic ratio method for images such as satellite remote sensing images.

## 2.2 Dynamic Signed Pressure Force

Constructing the driving force of a curve and letting the curve gradually evolve into the shape of the target we are interested in are a very important purpose in the active contour model. Inside and outside the target, the opposite sign of the driving force function is the main feature of this driving force. After all, the only way to get the curve to receive the contraction force outside the target is to have an expansion force inside and finally stay on the edge of the segmentation target. Based on the above characteristics, we call this driving force called signed pressure force (SPF).

Zhang et al. proposed a symbol pressure function based on the internal and external grayscale statistical information of the curve and combined it with the level set to achieve image segmentation. This method of segmentation is called the SPF model. A brief introduction to SPF is as follows:

$$\text{SPF}(I(x)) = \frac{I(x) - \frac{c_1 + c_2}{2}}{\max\left(\left|I(x) - \frac{c_1 + c_2}{2}\right|\right)} \tag{2}$$

wherein the parameters $c_1$ and $c_2$ are consistent with the parameters in the C-V model, respectively, representing the average gray value inside and outside in which

$$c_1 = \frac{\iint_{\Omega_1} I \mathrm{d}x\mathrm{d}y}{\iint_{\Omega_1} \mathrm{d}x\mathrm{d}y}, \quad c_2 = \frac{\iint_{\Omega_2} I \mathrm{d}x\mathrm{d}y}{\iint_{\Omega_2} \mathrm{d}x\mathrm{d}y} \tag{3}$$

The evolution equation of the level set function is obtained after calculation.

$$\frac{\partial \phi}{\partial t} = \mathrm{SPF}(I(x)) \cdot \alpha |\nabla \phi| \tag{4}$$

Although the SPF model has been greatly improved and simplified based on the C-V [7] model and the GAC [8] model, the correctness of the algorithm is not satisfactory when the object grayscale features are not obvious.

Based on the SPF model introduced above, the main parameters used to distinguish the edges of the object are fixed $\frac{c_1+c_2}{2}$; as a result, the model is not particularly effective in dealing with images with some grayscale features that are not obvious. Therefore, the proposed method in this paper makes some improvements on the basis of the above model, adding a distinguish parameter $k \in [0, 1]$, which is used to adjust the strength of the segmentation.

Then, the D-SPF function is defined as:

$$\mathrm{SPF}_k(I(x)) = \frac{I(x) - c_1 \cdot \left(\frac{c_2}{c_1}\right)^k}{\max\left(\left|I(x) - c_1 \cdot \left(\frac{c_2}{c_1}\right)^k\right|\right)} \tag{5}$$

Because of the definition of $c_1$ and $c_2$, $\mathrm{Min}(I(x)) \leq c_1, c_2 \leq \mathrm{Max}(I(x))$, so we can get the

$$\min(c_1, c_2) \leq c_1 \cdot \left(\frac{c_2}{c_1}\right)^k \leq \max(c_1, c_2). \tag{6}$$

For a more vivid description, we simply assume the object area and the background area of the image to be evenly gray, and the grayscale of the object is smaller than the background. Therefore, the object gradation is $\mathrm{Min}(I(x))$, and the background gradation is $\mathrm{Max}(I(x))$. Then for the points in the background area conform to Formula (7). Otherwise, it is less than 0.

$$\mathrm{SPF}_{\mathrm{sign}}(I(x)) = \max(I(x)) - c_1 \cdot \left(\frac{c_2}{c_1}\right)^k > 0 \tag{7}$$

The SPF symbol [9] has the characteristics of internal and external signs. It is this law of force that causes the curve to eventually stop at the edge of the object. By replacing the original SPF function with the improved SPF function described above, we obtain the corresponding level and evolution equation as

$$\frac{\partial \phi}{\partial t} = \mathrm{SPF}_k(I(x))\left(\mathrm{div}\left(\frac{\nabla \phi}{|\nabla \phi|}\right) + \alpha\right)|\nabla \phi| + \nabla \mathrm{SPF}_k(I(x)) \cdot \nabla \phi \tag{8}$$

Zhang [10] et al. mentioned in the paper that $\mathrm{div}\left(\frac{\nabla \phi}{|\nabla \phi|}\right)|\nabla \phi|$ is an adjustment in the level set function. Since it is satisfied $|\nabla \phi| = 1$ condition, the adjustment can be written as a Laplacian $\Delta \phi$. It can be known from the scale space theory [11] that the evolution of the Laplacian form of the equation is equivalent to initializing the initial condition of the equation with a Gaussian kernel filter. Therefore, we can add a Gaussian filter step to omit the $\Delta \phi$ calculation of this adjustment before the level set function evolves. By adjusting the standard error of the Gaussian filter, the depth of the curve evolution can be controlled. In addition, the SPF model utilizes regional statistical information, which has a larger search range and edge transition capability, so we can also omit $\nabla \mathrm{SPF}_k(I(x)) \cdot \nabla \phi$. Finally, our level set function became

$$\frac{\partial \phi}{\partial t} = \mathrm{SPF}_k(I(x)) \cdot \alpha|\nabla \phi| \tag{9}$$

Finally, we programmatically the above steps, and we can get our algorithm flow as follows:

(1) Initialize level set function

$$\phi(x, t = 0) = \begin{cases} 1, & x \in \Omega_0 \\ -1, & x \in \Omega - \Omega_0 \end{cases} \tag{10}$$

where $\Omega$ is the area of the entire picture and $\Omega_0$ is the inner area of initialization curve.

(2) Calculate $c_1$ and $c_2$ using Eq. (3).
(3) Evolve the level set function curve using Eq. (9).
(4) If $\phi > 0$, we made $\phi = 1$; otherwise, $\phi = -1$.
(5) Filter the entire level set function with a Gaussian filter.
(6) Check whether the evolution of the level set function converges, if not, repeat to the step (2). If it converges, the algorithm ends and the output is outputted.

Through the above algorithm, image segmentation can be achieved. As for how to distinguish the parameter $k$ in our D-SPF model, we will discuss it in the next chapter with the change detection algorithm.

## 3 Experimental Results

### 3.1 Datasets and Experimental Setting

The first set of visible light images in our ERS, an optical remote sensing image taken by the thematic mapping instrument of the remote sensing satellite Landsat-5, reflects the changes in the waters of Sardinia in Italy. At the end of 1995, the area of a lake in the center of the island suddenly expanded rapidly and it is firstly necessary to test the area where the lake is enlarged. Because the lake is surrounded by mountains and mountains, it can be solved by remote sensing image change detection. Figure 2a, b shows remote sensing images for September in 1995 and July in 1996, respectively, and Fig. 2c represents the true value of water expansion with resolution 412 × 300 and gray level 256.

The second set of SAR images in our ERS is a synthetic aperture radar (SAR) image taken by the RADARSAT satellite, reflecting the surface changes in the Ottawa region of Canada affected by the rainy season. It can be seen from Fig. 3a, b photographed at May in 1997 and August in 1997, respectively, that the image data of this group is mainly composed of land and water, and the change information mainly

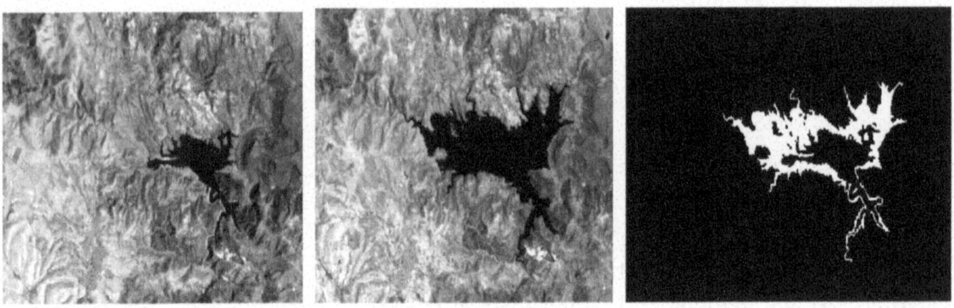

(a) September in 1995 (b) July in 1996 (c) True Value

**Fig. 2** Visible light remote sensing image in Sardinia

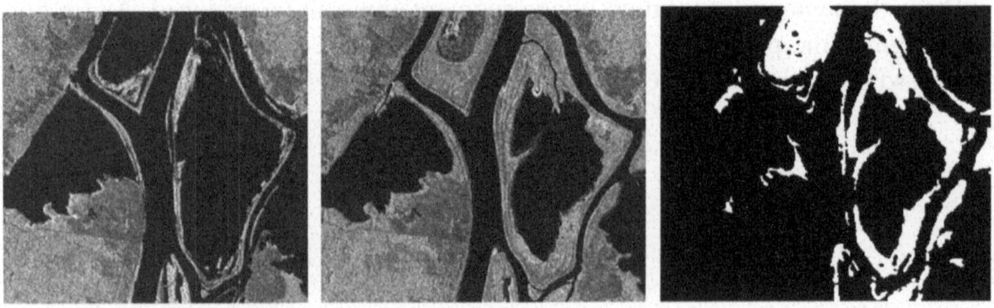

(a) May in 1997 (b) August in 1997 (c) True Value

**Fig. 3** SAR remote sensing image in Ottawa, Canada

comes from the surface changes caused by floods. Figure 3c shows the true value of the water change with resolution 290 × 350 and gray level 256.

## 3.2 Experimental Analysis

The overall appearance of the image can be represented by information entropy. If an image contains a target (a "variable" region in the difference image), then when we divide the image into target and non-target regions, the sum of information entropy on the two parts is the biggest. With the maximum entropy algorithm, we can get a segmentation threshold $T$. The traditional threshold segmentation is to directly use the threshold $T$ to binarize the image and get the final result.

In the selection of the distinguishing parameter $k$ of D-SPF model, we can learn from T algorithm. In the D-SPF of Eq. (5), we need to segment a type of image such as a different image. In general, our variation area has a grayscale greater than the non-target area. The internal average gray level of the curve is greater than the average gray level outside the curve. And as the discrimination parameter $k$ increases, the curve tends to move to a place where the gradation is large (internal change region).

The larger the threshold $T$ obtained by the difference image maximum entropy algorithm, the larger the interference of the non-changing region of the image. Therefore, a larger distinction between the distinguishing parameter $k$ is required to allow the curve to "over" the interference areas, rather than detecting the interference areas together. As a result, the distinction between the parameter $k$ and the maximum entropy threshold $T$ should be a positive correlation.

For the difference image, the gray value of the edge of the final curve should be much larger than the background gray value $c_1$ and close to the target gray value $c_2$. We can limit the difference items $c_1 \cdot \left(\frac{c_2}{c_1}\right)^k$ should be above is $\sqrt{c_1 c_2}$, be equivalent to $k > 0.5$.

We first select five satellite remote sensing difference images obtained by the average log-ratio method and then try to obtain the optimal distinguishing parameter $k$ in the D-SPF model. We also calculate the maximum entropy threshold of each difference image. We can see the quantitative relationship between the distinguishing parameter $k$ and threshold $T$ in Fig. 4.

We can roughly see that as the maximum entropy threshold $T$ increases, the distinguishing parameter $k$ of the D-SPF model also increases. And, as can be seen in Fig. 4, the two parameters are roughly an exponentially increasing relationship. Therefore, we first define a Formula (11) with parameters to describe the relationship between them.

$$k = a\mathrm{e}^{bT} + c \tag{11}$$

where $T$ represents the maximum entropy threshold, and parameter $k$ represents the optimal $k$ value of our D-SPF model. We can use the "curve fitting tool" in MATLAB

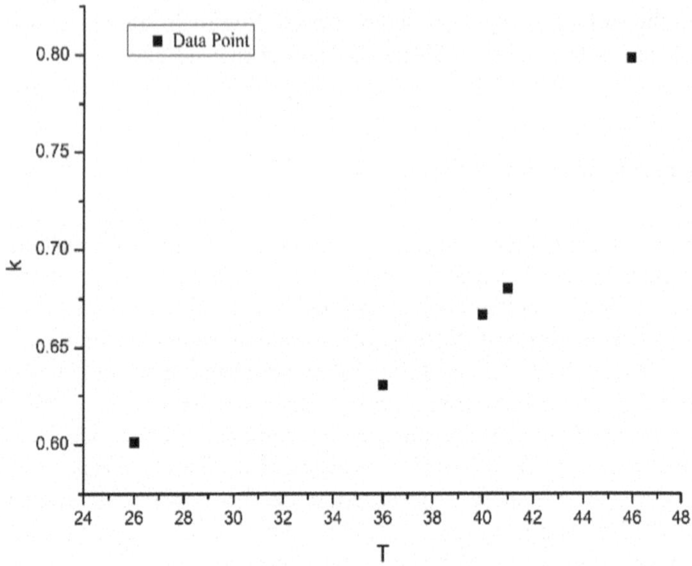


**Fig. 4** Scatter plot between the best distinguishing parameter $k$ and $T$

to get the values of the constants $a$, $b$, and $c$. The final output fitting results are $a =$ 0.000,068, $b = 0.174$, and $c = 0.595$.

By substituting this result into Eq. (11), we can construct a relation between the parameter $k$ in the D-SPF algorithm and the threshold $T$ in the maximum entropy threshold algorithm:

$$k = 6.8 \times 10^{-5} \times e^{0.174T} + 0.595 \quad (12)$$

## 3.3 Comparison with Other Methods

Five methods of maximum inter-class variance, maximum entropy, and level set method of C-V model, SPF model, and automatic D-SPF model were used to carry out change detection experiments, and the results of change detection were analyzed in our ERS (Figs. 5 and 6).

The most important parameter for change detection is the error rate in Table 1. In the two sets of data, the maximum entropy method has a higher error rate (the percentage of error pixels in the total pixels). From the segmentation results, it can be seen that there are many noises around the target, and the level set method can reduce the noise interference to the result to some extent. The reason why the missed detection rate (the number of missed detections as a percentage of the actual changed pixels) is larger than the false alarm rate (the number of false alarms as a percentage

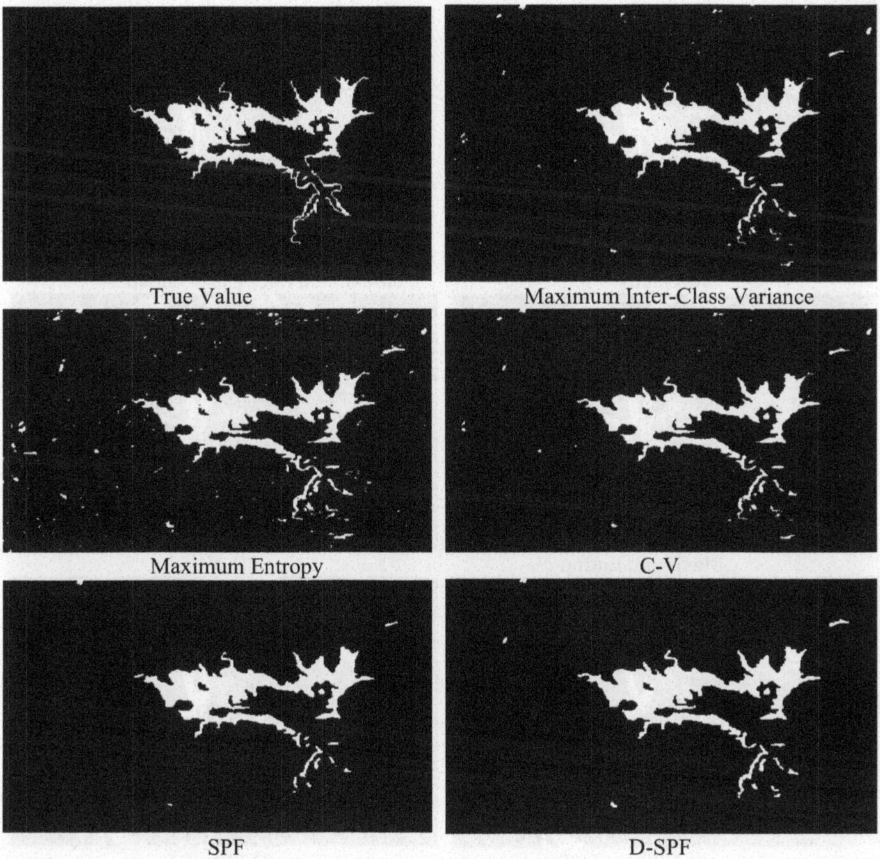


**Fig. 5** Visible light image detection results in Sardinia

of the actual unchanging pixels) is that change region generally only occupied a small part of the whole image, so the actual number of changed pixel points is far less than the actual number of pixel points. In general, the missed detection rate should not be greater than 10%. In summary, our D-SPF change detection algorithm can balance the false alarm rate and the missed detection rate under the premise of ensuring excellent error rate.

## 4 Conclusion

This paper proposes a new level set segmentation model—D-SPF model. The D-SPF model is based on the SPF model proposed by Zhang in this paper. It adds an improved model that distinguishes the parameter $k$ and makes the segmentation

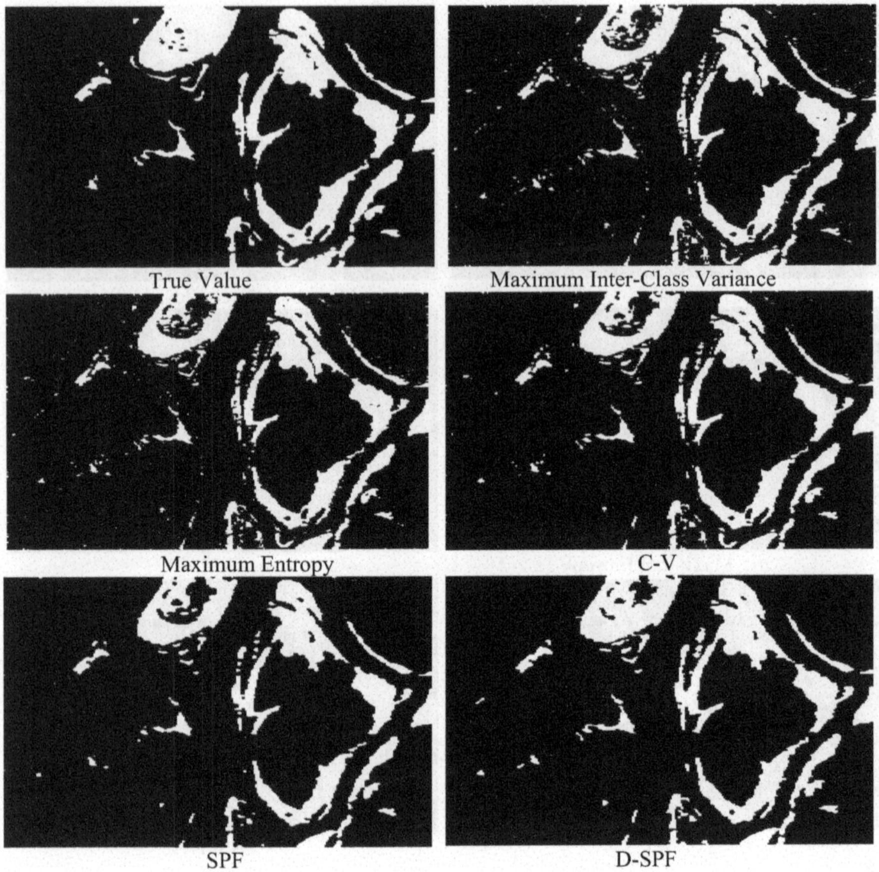


**Fig. 6** SAR remote sensing image detection results in Ottawa, Canada

result become controllable. For the change detection of the difference image based on the average logarithmic ratio method, this paper proposes an empirical formula that can automatically obtain the optimal distinguishing parameters k to make the detection process automatic. Finally, the change detection algorithm based on the D-SPF model has achieved good results in the experiment of change detection.

This paper attempts to combine the level set segmentation method with the change detection. From the experimental results of the change, the detection accuracy and accuracy are still very good. However, as a relatively new segmentation method, the level set segmentation method still has some limitations, such as the running speed and the distinguishing parameter $k$ selection problem in the D-SPF model. These problems also appear in our level-based change detection algorithm. If these problems can be solved effectively, our algorithm can be further improved.

**Table 1** Remote sensing image detection evaluation index

| Dataset | Evaluation index | Maximum inter-class variance (%) | Maximum entropy (%) | C-V (%) | SPF (%) | D-SPF (%) |
|---|---|---|---|---|---|---|
| Visible light image | Error rate | 1.6 | 2.4 | 1.7 | 1.5 | 1.5 |
| | False alarm rate | 1.1 | 2.2 | 1.4 | 0.9 | 1.3 |
| | Missed detection rate | 9.7 | 5.9 | 7.5 | 10.2 | 7.1 |
| SAR | Error rate | 3.7 | 3.8 | 3.2 | 3.3 | 2.8 |
| | False alarm rate | 1.2 | 1.1 | 0.9 | 0.6 | 1.4 |
| | Missed detection rate | 17.1 | 18.3 | 15.5 | 17.7 | 10.4 |

## References


1. Singh A (1989) Review article digital change detection techniques using remotely-sensed data. Int J Remote Sens 10(6):989–1003
2. Bazi Y, Melgani F, Al-Sharari HD (2010) Unsupervised change detection in multispectral remotely sensed imagery with level set methods. IEEE Trans Geosci Remote Sens 48(8):3178–3187
3. Celik T, Ma KK (2011) Multitemporal image change detection using undecimated discrete wavelet transform and active contours. IEEE Trans Geosci Remote Sens 49(2):706–716
4. Otsu N (1975) A threshold selection method from gray-level histograms. Automatica 11(285–296):23–27
5. Pun T (1981) Entropic thresholding, a new approach. Comput Graph Image Proc 16(3):210–239
6. Li C, Xu C, Gui C, Fox MD (2005) Level set evolution without re-initialization: a new variational formulation. In: 2005 IEEE computer society conference on Computer Vision and Pattern Recognition CVPR 2005, IEEE, vol 1, pp 430–436
7. Chan Tony F, Luminita A (2001) Active contours without edges. IEEE Trans Pattern Anal Mach Intell 10(2):266–277
8. Caselles V, Morel J, Sapiro G (1997) Geodesic active contours. Int J Comput Vision 22:61–79
9. Kass M, Witkin A, Terzopoulos D (1988) Snakes: Active contour models. Int J Comput Vision 1(4):321–331
10. Zhang K, Song H, Zhang L, Zhou W (2010) Active contours with selective local or global segmentation: A new formulation and level set method. J Image Vis Comput 28(4):668–676
11. Perona P, Malik J (1990) Scale-space and edge detection using anisotropic diffusion. IEEE Trans Pattern Anal Mach Intell 12(7):629–639

# Virtual Simulation Experiment System for Spacecraft Orbital Principle and Its Teaching Application

**Jiu-tian Li, Yan-gang Liang, Zheng Qin and Jian-yong Zhou**

**Abstract** Generally, space is far away from people's daily life. The motion characteristics of objects in space are very different from those of ground objects, which is easy to cause difficulties in understanding. In order to make everyone have an intuitive understanding of the basic principles of spacecraft orbital motion and the space targets distribution, a virtual simulation experiment system for spacecraft orbital principles was designed and developed. The system mainly includes two modules: space situation analysis module and spacecraft orbital principle module. Specifically, the space situation analysis module is supported by the space target database and can be used to perform classified query and multi-dimensional display and analysis of on-orbit space targets such as satellites and space debris. The spacecraft orbital principle module provides an intuitive and dynamic analysis of the satellite orbital operation principle and typical orbital characteristics under the influence of perturbations. To facilitate the application in teaching, the experimental system also comprehensively considers the students' attention to the space situation and the difficulty of teaching the spacecraft orbital principle. There are also a number of typical teaching cases in the system, through which students can cultivate interest, understand knowledge, train skills and quickly form the ability to use knowledge.

**Keywords** Virtual simulation · Experiment system · Spacecraft orbital principle

## 1 Introduction

With the development of human space activities, space has become more and more closely related to our lives. For example, we acquire the topography of the Earth through remote sensing satellites, live television broadcasts via communication satellites and provide ground navigation services through navigation satellites. Many

J. Li · Y. Liang (✉) · Z. Qin · J. Zhou
College of Aerospace Science and Engineering, National University of Defense Technology, Changsha 410073, People's Republic of China
e-mail: liangyg@nudt.edu.cn

 
Z. Jing (ed.), *Proceedings of the International Conference on Aerospace System Science and Engineering 2019*, Lecture Notes in Electrical Engineering 622,
https://doi.org/10.1007/978-981-15-1773-0_23

people have been curious about space. What artificial objects are in space? How is the satellite in space flying around the Earth? Will the satellite fall to the ground? These questions are both the puzzle of most ordinary people and the initial confusion of those who are preparing to enter the aerospace field for study and work. To answer these questions, it is not enough to explain by theory and principle alone. It is necessary to use certain experimental means to make these abstract knowledge more clearly. However, due to the high cost of space activities and the complexity of the space environment, it is almost impossible to use spacecraft objects or models in classroom teaching to simulate the orbital motion of spacecraft in the real space environment. Therefore, for the experimental needs of aerospace courses, simulation experiments can be said to be the only effective way [1]. With the development of information technology, the application of virtual simulation experiment in teaching is more and more extensive [2–4]. Satellite Tool Kit (or Systems Tool Kit), often referred to by its initials STK, is a physics-based software package from Analytical Graphics Inc. that allows engineers and scientists to perform complex analyses of space situation. With STK, users can customize spacecraft and ground system parameters, including curve tables, 2D and 3D, to support the entire process of aerospace missions. Unfortunately, the operation of the software is slightly complicated, and it requires a certain amount of training to be proficient in operation. Thus, it is not suitable for students who have no aerospace background to understand the operation principle of space and spacecraft.

In response to the above problems, a vivid and easy-to-use virtual simulation experiment teaching aid platform was designed and developed. The developed system consists of two modules, space situation analysis module and spacecraft orbital principle module. By using the system, students first understand the space targets and their distribution characteristics in space, and then learn more about the motion characteristics of these space targets and the differences in the application of various types of orbits.

The rest of the paper is organized as follows. In Sect. 2, we describe the virtual simulation experiment system, including the composition of the system and the functions of each part. Section 3 introduces the experimental teaching process, which is combined with classroom centralized teaching and independent appointment experiments. Summary and future expansion of the virtual simulation experiment system are made in Sect. 4.

## 2 Virtual Simulation Experiment System

The developed virtual simulation experiment system mainly includes two modules: space situation analysis module and spacecraft orbital principle module. Specifically, the space situation analysis module is supported by the space objects database and can be used to show what is included in the space. The spacecraft orbital principle module provides an intuitive and dynamic analysis of the satellite orbital operation principle and typical orbital characteristics under the influence of perturbations.

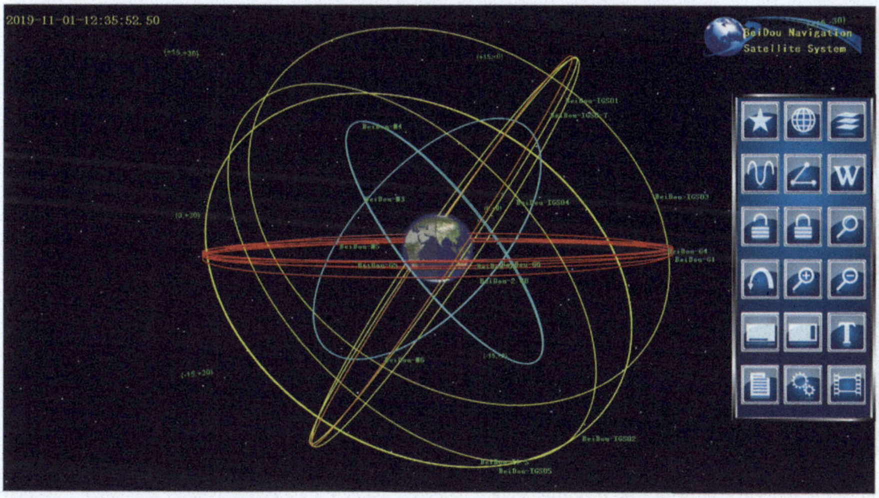


**Fig. 1** Space situation analysis module

Just like the STK software, the more powerful the software is, the more difficult it is to learn and operate, thus it is not easy to use in teaching. We comprehensively consider students' concerns and points of interest in space situation, difficult teaching points in spacecraft mission planning, etc., sort out typical teaching cases, and pre-set several experimental projects in system design. The combination of the project and the typical teaching case enables students to get started quickly in the experiment, open a typical project with simple operation, and learn and understand knowledge in case study. Moreover, for students who have strong learning ability and want to know more about the knowledge, the system reserves the function of self-selection, and students can design experimental projects to expand their learning according to their needs (Fig. 1).

## *2.1 Space Situation Analysis Module*

With the support of the space objects database and visualization technology, the space situation analysis module can be used to perform classified query and multi-dimensional display and analysis of on-orbit space objects such as satellites and space debris. Based on the space situation analysis module, the three types of experiments, including the overall situation analysis of space objects, the typical satellite constellation operation situation analysis and the satellite constellation design and analysis can be carried out.

***Overall Situation Analysis of Space Objects***
Students can view the distribution and operation status of space objects of global, the USA, Russia, and China through experimental case selection.

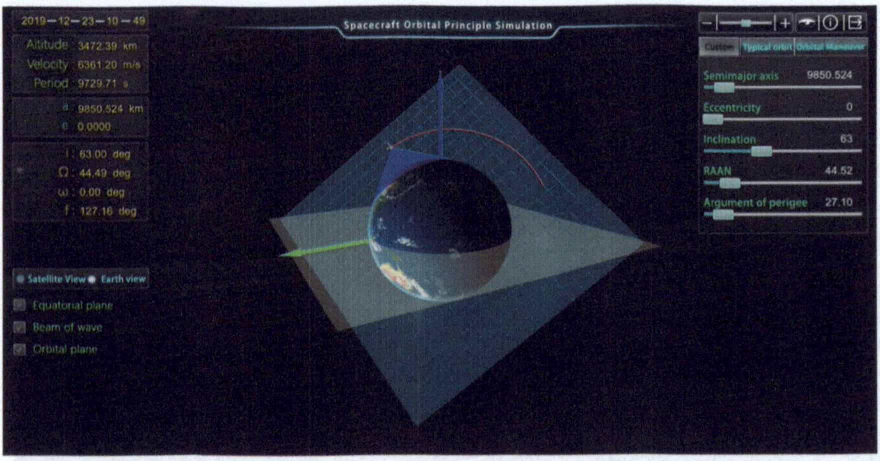


**Fig. 2** Space orbital principle module

***Typical Satellite Constellation Operation Situation Analysis***
Through the experimental case selection, students can view the distribution and operation status of China Beidou satellites, US Global Position System, China communication satellites, and US Iridium satellite constellation, and master the characteristics of typical satellite constellations (Fig. 2).

***Satellite Constellation Design and Analysis***
Students can independently design the satellite constellation through parameter configuration and analyze the characteristics of the satellite constellation.

## 2.2 Space Orbital Principle Module

In this module, intuitive dynamic analysis of satellite orbital principles and typical orbital characteristics are carried out in the form of cross-display of multi-dimensional views to help students understand the basic principles of spacecraft motion.

Based on the spacecraft orbit principle simulation module, three types of experiments, including the orbital elements analysis, the track of subsatellite point analysis and the satellite on-orbit acceleration and deceleration principle analysis are carried out.

***Orbital Elements Analysis***
By adjusting the orbital elements such as the semi-major axis of the orbit, the eccentricity, the inclination, the longitude of the ascending node, and the argument of periapsis, students observe and analyze the shape and characteristics of satellite orbits, the changes of orbital positions, and understand the difficulties such as orbital precession.

***Track of Subsatellite Point Analysis***
By adjusting the orbital elements, students analyze the trajectory changes of subsatellite points and the shape characteristics of the track of subsatellite point for several typical orbits.

***Satellite on-Orbit Acceleration and Deceleration Principle Analysis***
Students provide a momentary speed increment (forward or reverse) to the satellite at the current moment through parameter setting, analyze the changes of the satellite's orbit, and consider further research on how to improve satellite orbits and how to make satellites fall back into the Earth's atmosphere.

## 3 Experimental Teaching Process

The experimental teaching of the project is combined with classroom centralized teaching and independent appointment experiments. The classroom centralized teaching is carried out in the student computer room in a unified manner, and the experimental instructor guides the students in the classroom. The independent appointment experiments can be completed by logging into the experimental teaching management platform via the Internet.

In the classroom centralized teaching, the teaching is implemented in the way of principle explanation and offline design, practical operation and program verification, review and summary, and knowledge promotion. Before the experiment, the teacher briefly reviewed the relevant knowledge and principles of the experiment, and explained the teaching objectives of the experiment, the main functions of the experimental software, the experimental operation methods and precautions, etc. The students conduct experiments such as satellite constellation and spacecraft orbital parameters design according to the purpose of the experiment. Then, the students complete the experimental tasks and record the experimental results by setting different experimental parameters. The instructor will master the experimental design and implement the necessary guidance. After the experiment, the students summarize the experimental results, self-assess whether the experimental design content is completed, whether the experimental purpose requirements are met, and write the experimental report. Based on the student's classroom performance and experimental report, the teacher summarizes whether or not to achieve the teaching objectives and gives corresponding student experimental grade or experimental improvement suggestions.

In the independent appointment experiments, students can log into the experimental teaching management platform through the Internet according to the requirements of the task and the need for self-learning, and the experiment projects can be completed in sequence according to the experimental operation steps in help document. In the experimental operation, students conduct experiments and record the results by setting different experimental parameters and interactive operations. After the experiment, the students summarize the experimental results and write the experimental reports. The teacher evaluates whether the teaching goal is achieved according to the experimental reports and gives the corresponding experimental grade or experimental improvement suggestions.

## 4 Summary

The motion characteristics of objects in space are quite different from those of ground objects, and it is difficult to conduct physical operation experiments. The developed project makes full use of teaching aid software tools to combine knowledge learning with experimental design and operation, to integrate learning and practicing, to deepen students' understanding of the spacecraft orbit principle, and to improve the application of relevant knowledge. The virtual experiment system can not only visualize the complex space motion characteristics of spacecraft, but also stimulate student's learning and exploration interest through their interactive participation. The networked virtual simulation experiment project breaks the space and time limit of traditional teaching. Students can flexibly arrange their own experimental time through the network platform and can also flexibly select experimental projects according to their own professions and interests, thus effectively expanding the time and space of traditional teaching.

The current experimental system is still focused on the principle cognitive experiment, and the subsequent comprehensive experimental content can be added on this basis. It is anticipated that the future system will include a number of experiments, such as the coverage of the ground targets, the revisiting cycle, the ground stations, and the space and ground coordination system, etc.

**Acknowledgements** This work was supported by Research Project of Teaching Reform in Ordinary Colleges and Universities in Hunan Province.

## References

1. Qinghua Z, Weihua Z, Zhezhi H, Ronghua D (2014) Improving aerospace engineering students' achievements by an open aero control experiment apparatus. IEEE Trans Educ 57(4):229–234
2. Guangyang X, Haiyan Y, Li Z, Mei Z, Delan W (2018) The exploration of the application of virtual simulation in physics experiment teaching. In: Proceedings of the 2018 8th international conference on education and management (ICEM 2018)
3. Liucheng Z (2018) Design and practice of virtual simulation experiment teaching center for modern enterprise business operation. In: Proceedings of the 2018 9th international symposium on advanced education and management (ISAEM 2018)
4. Jin L, Chunqing Z, Huixiang L, Weiguang Z, Lingguang K (2019) The application of virtual simulation technology in experimental and practical teaching in colleges and universities. China Mod Educ Equip 309(3):6–8

# Verification of GPS III and QZSS L1C Signal Joint Positioning Performance with Software-Defined Receiver

**Xiang Huo, Xue Wang, Sen Wang, Xiaofei Chen and Jing Ke**

**Abstract** Satellite navigation and positioning have become an indispensable component of smart devices and applications. In order to avoid signal interference in L1 frequency and provide novel services, Global Positioning System III (GPS III) updated a modern signal named L1C and tested in three Quasi-Zenith Satellite System (QZSS) satellites launched in 2017. In December 2018, the first GPS III satellite was launched, which implies the joint positioning using QZSS and GPS L1C signal is improved. L1C signal offers a series of advanced designs in signal modulation, message structure and coding, etc. In this regard, this paper presents complete methodologies for L1C signal joint receiving processing. For the on-air signals, this paper presents a methodology and results from collecting and assessing binary offset carrier (BOC) modulation and time-multiplexed BOC (TMBOC) modulation used in L1C signal. Using the same omnidirectional antenna and the test-equipment, the L1C signal was collected in Xi'an and Sanya, China, respectively. Experiments in Xi'an verified the joint positioning method to complement the GPS III and QZSS satellite constellations. The methodology in this paper evaluates the positioning error of BOC(1,1) and TMBOC(6,1,4/33) under the same environment and satellite constellation in Sanya. It is also verified that the joint positioning error is less than the QZSS-only positioning due to the optimization of the satellite constellation.

**Keywords** L1C · GPS III · QZSS · Software-defined receiver · Joint positioning

X. Huo · X. Wang (✉) · S. Wang · X. Chen · J. Ke
National Time Service Center,
Chinese Academy of Sciences, Xi'an 710600, China
e-mail: wangxue@ntsc.ac.cn

X. Huo
e-mail: 307913917@qq.com

School of Electronic and Communication Engineering, University of Chinese Academy of Sciences, Beijing 101408, China

X. Wang
Key Laboratory of Precision Navigation, Positioning and Timing Technology, Chinese Academy of Sciences, Xi'an 710600, China



Z. Jing (ed.), *Proceedings of the International Conference on Aerospace System Science and Engineering 2019*, Lecture Notes in Electrical Engineering 622,
https://doi.org/10.1007/978-981-15-1773-0_24

# 1 Introduction

The Global Positioning System (GPS) uses radio waves to provide users with positioning, navigation, and timing (PNT) services. GPS navigation signals have numerous applications in research, industry, and everyday life [1, 2]. Civil GPS receivers could solely use legacy L1C/A signal for positioning before GPS modernization. In 2005 and 2010, GPS broadcasted two civilian signals, L2C and L5. In 2000, the US Congress approved a plan to establish a new GPS, called GPS III. GPS III deploys a signal compatible with other Global Navigation Satellite System (GNSS) such as Galileo and QZSS at the L1 frequency and is also the fourth-generation civil signal, called L1C. The first satellite of the GPS III was launched in December 2018 [3].

The Quasi-Zenith Satellite System (QZSS) is a Japanese regional satellite positioning system. Take advantage of the high elevation of highly elliptical orbit, QZSS has been developed as a GPS-complementary system to increase the availability, reliability, integrity, and accuracy of positioning performance in the Asia-Oceania region. QZS-1, the first satellite of QZSS, is in orbit since 2010. And in 2017, Japan launched three satellites to join QZSS, including QZS-2, QZS-3, and QZS-4. These four satellites transmit all GPS civil signals since the modernization of GPS, including L1C/A, L1C, L2C, and L5 [4].

IS-GPS-800 [5] illustrates the L1C signal consisting of data (L1CD) and pilot (L1CP). The novel binary offset carrier (BOC) modulation and time-multiplexed binary offset carrier (TMBOC) modulation are used in the GPS III L1C signal, which is virtually seamless interoperability with L1C from QZSS. Up to now, there are a total of five satellites broadcasting L1C signal in orbit. A GPS III satellite with a code number of 4 is called Vespucci. The other four satellites of QZSS called QZS-1, QZS-2, QZS-3, QZS-4 use the code number 193, 194, 199, and 195, respectively. It should be noted that QZS-1 differs from GPS III when it comes to modulation and phase. QZS-1 uses BOC(1,1) modulation for both data and pilot, and its pilot advances data by 90° (in phase quadrature). In GPS III and three other QZSS satellites, L1CD uses BOC(1,1) modulation, and L1CP uses TMBOC(6,1,4/33) modulation. And they are in the same phase.

Combined with GPS, Yu-Hsuan Chen and his team verified that the BOC modulated signal broadcast by QZS-1 has better performance than the BPSK modulated signal in the positioning error [6]. While the anti-interference, anti-noise, and other benefits of the TMBOC over the BOC have been theoretically assessed [7], no on-air, operational assessments have been published.

The research in this paper develops methods to make a direct comparison of the positioning error of TMBOC(6,1,4/33) and BOC(1,1) from GPS III and QZSS. In view of the fact that there is barely one satellite in the current GPS III constellation, this paper proposes a joint positioning method to use GPS III for positioning service and verified its feasibility by means of the experiments in Xi'an. The use of satellites of the same time period rather than the two periods of close proximity is the key to accurate comparison as the positioning results are sensitive to the slightest difference in the constellation geometry. Therefore, the experiments in Sanya made a detailed comparison of positioning error between the TMBOC(6,1,4/33) and BOC(1,1) modulated signals of GPS III + QZSS and QZSS-only.

## 2 Receiving and Evaluation Methodology

### 2.1 *Joint Positioning Method*

In this paper, a static test-equipment shown in Fig. 1a is used to collect signals in Xi'an and Sanya, China, which will be introduced later. The extremely high sampling

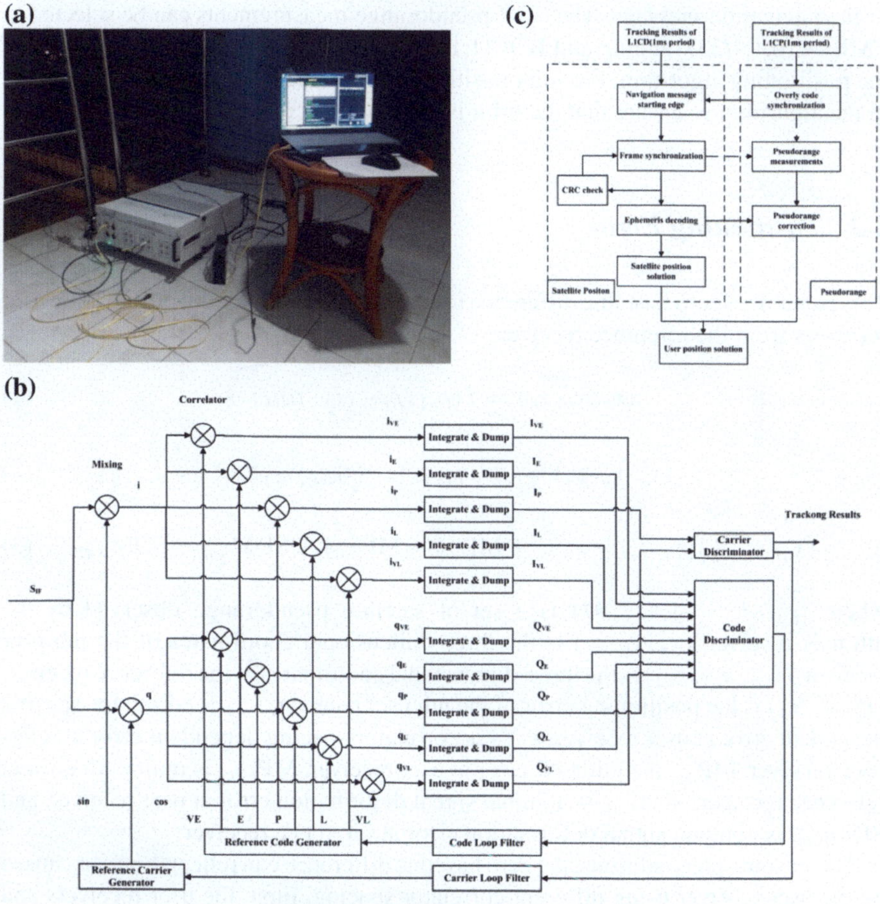


**Fig. 1** Test-equipment and the SDR: **a** the test-equipment for collecting signals, **b** a representation of DLL and PLL, **c** a representation of the positioning algorithm

rate of 250 and the 80 MHz bandwidth of front-end filter meet the ultra-narrow correlation used in this paper. The software-defined receiver (SDR) used in this paper uses the L1C signal transmitted on GPS III and QZSS to processing off-line and takes into account multiple differences in their signal structure [4, 5]. Figure 1b is a tracking loop diagram of the SDR used in this paper. Many of its receive-related parameters are flexible and adjustable, including correlator spacing, bandwidth, and damping factor for delay lock loop (DLL) and phase lock loop (PLL). The local signal generated by the SDR simulation for tracking is full bandwidth unfiltered. The SDR tracking method uses Bump-Jump [8], and the five pairs of correlators independently demodulate the data component and the pilot component by a coherent integration of 1 ms. The squaring operation of the non-coherent integration destroys the zero-mean of the thermal noise and brings about the squaring loss, so it is not used. Continuously tracking the carrier frequency and code phase of the L1CD demodulates the navigation massage bit stream. And the SDR demodulates the overlay code by tracking L1CP. The SDR uses the overlay code to complete the frame synchronization of the navigation message. The SDR pseudorange measurements can be selected in TMBOC(6,1,4/33) ranging and BOC(1,1) ranging. Figure 1c is a block diagram of the positioning algorithm. For joint positioning, the ephemeris needs to be modified in the algorithm to ensure that the orbit is correctly calculated.

## 2.2 *Positioning Error*

This paper uses zero baseline difference to illustrate the positioning errors of the user receivers from the reference receiver.

$$(x_n, y_n, z_n) = f(\rho_{n1}, \rho_{n2}, \ldots, \rho_{nM}) \tag{1}$$

$$\left(x_n', y_n', z_n'\right) = (x_n, y_n, z_n) - (x_{\text{ref}}, y_{\text{ref}}, z_{\text{ref}}) \tag{2}$$

$$\left(x_n', y_n', z_n'\right) = \varepsilon_{\rho,n} - \varepsilon_{\rho,\text{ref}} + \text{MP}_{\rho,n} - \text{MP}_{\rho,\text{ref}} + \text{SDM}_{\rho,n} - \text{SDM}_{\rho,\text{ref}} \tag{3}$$

where $\{\rho_{nj}, j = 1, \ldots, M\}$ is a set of satellite pseudorange observed by the nth user receiver, $(x_n, y_n, z_n)$ is the three-dimensional coordinates of the nth user receiver, $(x_{\text{ref}}, y_{\text{ref}}, z_{\text{ref}})$ is the three-dimensional coordinates of the reference receiver, $\left(x_n', y_n', z_n'\right)$ is the positioning error of the nth user receiver, $\varepsilon_{\rho,n}$ is correlator-spacing dependent error at user receiver, $\varepsilon_{\rho,\text{ref}}$ is correlator-spacing dependent error at reference receiver, $\text{MP}_{\rho,n}$ is multipath error at user receiver, $\text{MP}_{\rho,\text{ref}}$ is multipath error at reference receiver, $\text{SDM}_{\rho,n}$ is nominal signal deformation error at user receiver, and $\text{SDM}_{\rho,\text{ref}}$ is nominal signal deformation error at reference receiver.

Under some preconditions, the zero baseline difference can reflect the error caused by the user receiver using different correlator spacing. First, the user receivers and the reference receiver use the same satellite constellation. Second, they must use the same satellite signals. Third, their positioning algorithm must be the same. Provided that the above necessary conditions are met, the zero baseline difference can reflect the positioning error associated with the correlator spacing.

For each user in which the correlator spacing is fixed, the clock error, ionosphere, and tropospheric error are common mode in the positioning error equation and are therefore removed in the equation. The noise can be effectively removed with an average. In the experimental scenario below, the main residual multipath, signal nominal deformation, and DLL noise are all relevant to the correlation spacing. The user receivers in this paper use different correlator spacing, including 0.1 chips, 0.15 chips, 0.175 chips, and 0.4 chips, which are differentiated from reference receiver with a correlator spacing of 0.05 chips.

# 3 Measurement Campaign

## *3.1 Signal Collection in Xi'an, China*

QZSS satellites have elliptical geosynchronous orbit which allows them to spend a significant portion of their orbital period visible to East Asia. They provide ever present high elevation navigation signals for places such as Japan, Korea, Australia, or eastern China. At present, Vespucci is in medium earth orbit and its surrounding cycle is about 24 h, making it not visible at any time in the collection site. In southern China, Vespucci can be observed at higher elevation.

The skyplot for a user in Xi'an, China, is shown in Fig. 2a. In Xi'an, the elevations of QZS-1, QZS-2, and QZS-4 are sometimes less than 10° or even invisible. The elevation of each satellite is greater than 10° for some duration which is approximately 18 h. Meanwhile, QZS-3 is always at about 46°. Vespucci has a total of 8 h of elevation above 10° for two periods. The trend of the elevation of Vespucci is always

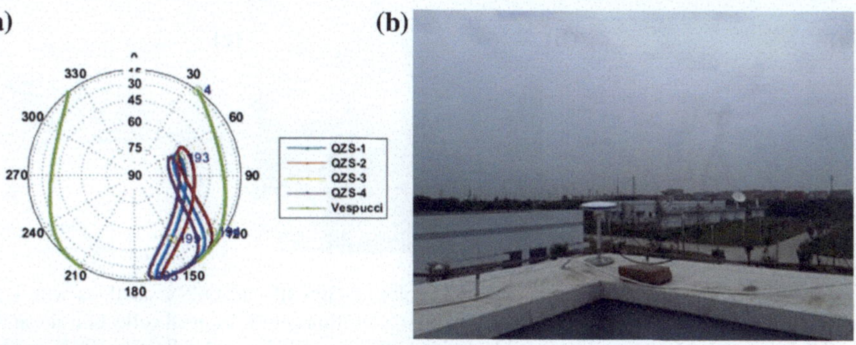


**Fig. 2** Skyplot and signal collection site: **a** skyplot of GPS III and QZSS satellites with L1C observed at Xi'an, China, on January 25, 2019, with 0° elevation cutoff, **b** signal collection site at NTSC roof

opposite to that of a satellite in QZSS. It should be noted that the SDR cannot use the constellation with less than four visible satellites for positioning measurement.

The signal collection site is located on the 1st floor roof of the National Time Center (NTSC) of the Chinese Academy of Sciences. The position where the omnidirectional antenna was placed is shown in Fig. 2b. Although only four satellites are visible in Xi'an, they can be used to verify the optimization of joint positioning on GPS III + QZSS constellations.

## *3.2 Signal Collection in Sanya, China*

Since Sanya is in southern China, the elevations of satellites of QZSS and Vespucci of GPS III are several degrees higher than those of Xi'an. The skyplot for a user in Sanya, China, observing GPS III and QZSS orbits over its 24 h periods is shown in Fig. 3a. At Sanya, the satellites of QZSS are always visible and always significantly above 10° elevation. Vespucci of GPS III is greater than 0° about 12 h. As shown in Fig. 3a, most of the dwell time, QZSS is at high elevation that greater than 45° (~15 h). But the trend of the elevation of Vespucci is always opposite to that of a satellite in QZSS, which is the same as in the Xi'an constellation. The signal collected in Sanya also using the test-equipment. If the constellation contains satellites with elevation below 10°, the positioning results will get tremendous deterioration. So satellites collected below 10° should be avoided.

The location at the hotel building by the sea was used for the signal collection. To receive GPS III and QZSS satellite signals, an omnidirectional antenna was placed on the third floor balcony, as shown in Fig. 3b. This paper analyzes the signal collected in Sanya at UTC 12:08:10, April 13, 2019. The skyplot of the satellites visible in the course of signal collection is shown in Fig. 3c. As seen in the skyplot, GPS III and QZSS satellites are at high elevation and their direct signals could be visible at the balcony location.

**Fig. 3** Skyplot and signal collection site: **a** skyplot of GPS III and QZSS satellites with L1C observed at Sanya, China, on April 13, 2019, with 0° elevation cutoff, **b** signal collection site at the hotel balcony, **c** skyplot of satellites visible during GPS III + QZSS signal collection (PRN number shown)

# 4 Results

## 4.1 Constellation Supplement Effect

In order to illustrate the influence of the geometry of the satellite constellation, the positioning results of Xi'an are shown in the topocentric coordinate system. However, the positioning error of Sanya is expressed in the WGS-84 coordinate system for the purpose of clearly comparing the error.

This section verifies the supplementary effect of joint positioning on constellations. Table 1 lists tow periods of QZSS and GPS III satellites visible in Xi'an. With the elevation of QZS-1 reduced to almost 10°, L1C of the QZSS-only can still be used for autonomous positioning at 8:34, as shown in Fig. 4. It can be seen from the skyplot that the QZSS satellites are generally concentrated in the southeast direction of the observation site. Influenced by the geometric configuration of the constellation, the positioning error is gigantic in the east (E) direction. Soon after, SDR using L1C signal lost its positioning ability because barely the remaining three satellites were visible in the QZSS-only constellation. The SDR using the GPS III + QZSS L1C joint positioning method restored the positioning capability at 11:02, although the PDOP value became worse than at 8:34, as shown in Fig. 5. Since the satellites are generally concentrated at a higher elevation, the positioning results show a large deviation in the up (U) direction.

**Table 1** Satellites visibility during two acquisition periods on January 25, 2019

| Period (UTC) | Visible | Invisible |
|---|---|---|
| 8:34:28–8:35:28 | QZS-1, QZS-2, QZS-3, QZS-4 | Vespucci |
| 11:02:33–11:03:33 | QZS-2, QZS-3, QZS-4, Vespucci | QZS-1 |

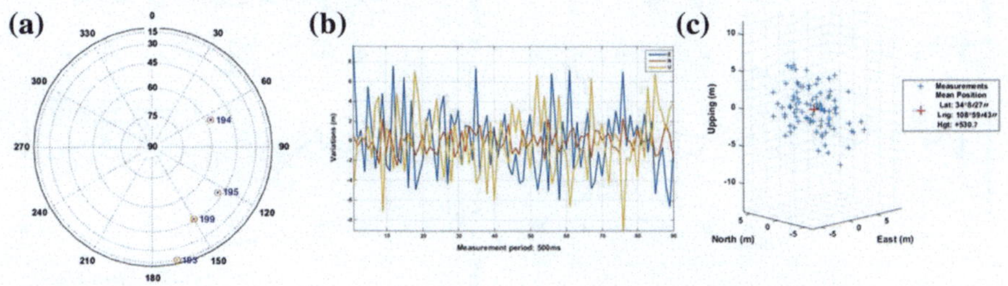


**Fig. 4** Positioning results of data collected in Xi'an at 8:34:28 (PDOP = 9.604): **a** the skyplot, **b** the variation in East (E), North (N) and Up (U), **c** the positioning result in the 3D map

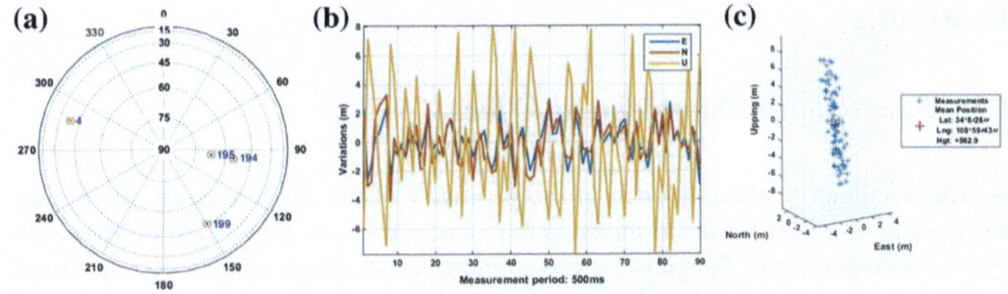


**Fig. 5** Positioning results of data collected in Xi'an at 11:02:33 (PDOP = 13.848): **a** the skyplot, **b** the variation in East (E), North (N) and Up (U), **c** the positioning result in the 3D map

## 4.2 Positioning Error

The signal collected in Sanya when five satellites are simultaneously visible is used to compare the positioning performance of GPS III + QZSS and QZSS-only. By use of the same correlator spacing (0.1 chips), the three-dimensional positioning error of TMBOC(6,1,4/33) and BOC(1,1) is given in Fig. 6, respectively, using the constellations of GPS III + QZSS or QZSS-only. The positioning error is averaged by 2000 positioning samples. Whether using joint positioning or QZSS-only positioning, the positioning error of TMBOC(6,1,4/33) is generally superior to BOC(1,1) in three dimensions. At 15 s, the positioning performance of the signal is affected by the environment which may include multipath and signal distortion. It can be seen that TMBOC(6,1,4/33) shows better resistance at this time. Regardless of whether TMBOC(6,1,4/33) or BOC(1,1) is used, the positioning error of joint positioning is

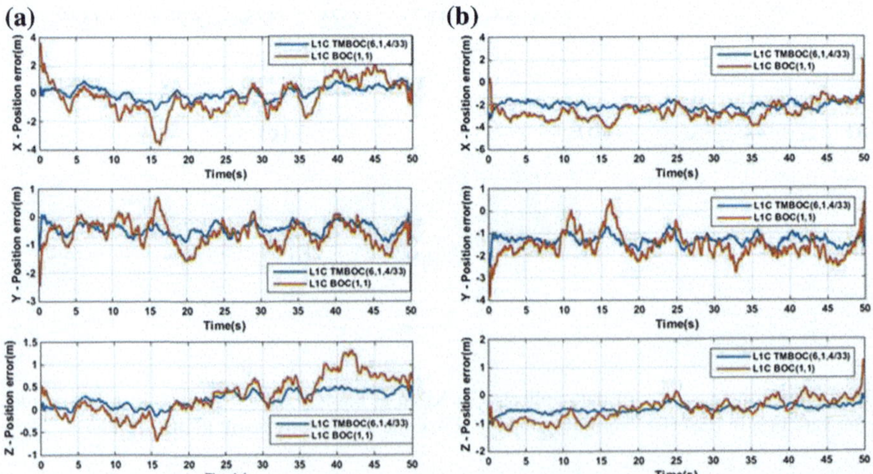


**Fig. 6** Position error (correlator spacing = 0.1 chips): **a** position error from GPS III + QZSS (PDOP = 6.45), **b** position error from QZSS-only (PDOP = 11.22)

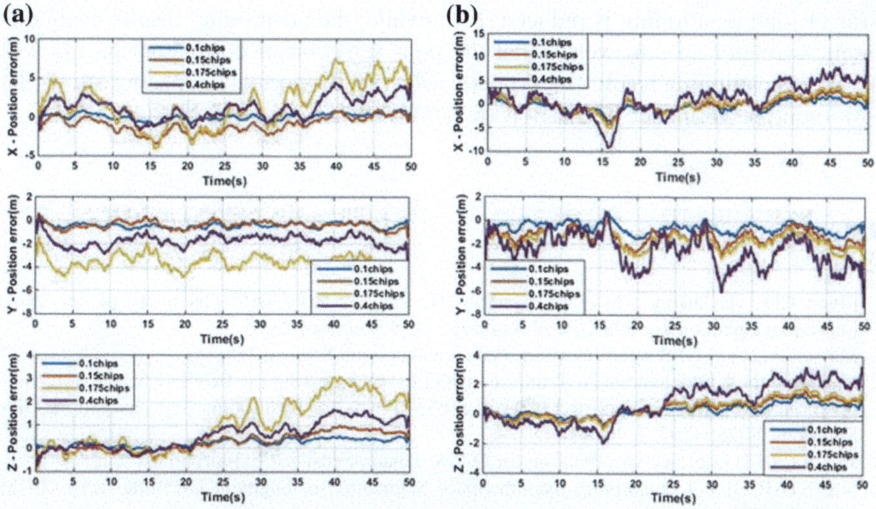


**Fig. 7** Position error (GPS III + QZSS): **a** position error from L1CP when using different correlator spacing, **b** position error from L1CD when using different correlator spacing

little than that of QZSS-only in three dimensions. This is the result of joint positioning optimizing the constellation composition while reducing the PDOP from 11.22 to 6.45. Since the SDR cannot obtain a priori position information, the position coordinates must be zeroed at the initial of regression, which results in a large positioning error in the initial stage.

Figure 7 shows the positioning error of several user receivers with different correlation spacing. Although the positioning error of TMBOC(6,1,4/33) deteriorated at 0.175 chips, the positioning error at 0.4 chips was significantly restored, as shown in Fig. 7a. As shown in Fig. 7b, the positioning error increases with the correlator spacing for BOC(1,1).

## 5 Summary and Conclusions

The transmission of the new navigation signal leads to the need to verify its positioning performance. This paper proposes a method for SDR to use L1C signal for joint positioning. And the evaluation methodology is proposed for the results using on-air signals. In this paper, the experiments of Xi'an adopted the L1C signal transmitted by the first GPS III satellite, Vespucci, to verify the positioning performance. The joint positioning results prove that the QZSS L1C signal has the claimed enhancement and supplement effect on GPS. This paper gives a direct comparison of GPS III + QZSS joint positioning versus QZSS-only using on-air signals collected in Sanya. Due to the optimization of the constellation configuration, the positioning error of joint positioning is reduced. Meanwhile, the positioning results using different correlator spacing verify that the poor selection of correlator spacing will affect the positioning result. But properly selecting the correlator spacing allows the positioning performance of TMBOC(6,1,4/33) outperforms that of BOC(1,1).

## References


1. Flynn CD, McCaffrey AM, Jayachandran PT, Langley RB (2019) Discovery of new code interference phenomenon in GPS observables. GPS Solutions 23:65
2. Voosen P (2017) GPS satellites yield space weather data. Science 355:443
3. Nadarajah N, Khodabandeh A, Teunissen PJG (2016) Assessing the IRNSS L5-signal in combination with GPS, Galileo, and QZSS L5/E5a-signals for positioning and navigation. GPS Solutions 20:289–297
4. JAXA (2018) Interface specification for QZSS. Japan Aerospace Exploration Agency
5. NAVSTAR Global Positioning System Space Segment/User Segment L1C Interfaces (2018) IS-GPS-800E
6. Lee, C, Chen, Y-H, Wong, G, Lo S, Enge, P (2013) Multipath benefits of BOC versus BPSK modulated signals using on-air measurements. In: Proceedings of the 2013 international technical meeting of The Institute of Navigation, pp 742–751
7. Rushanan JJ (2007) The spreading and overlay codes for the L1C signal. Navigation 54:43–51
8. Fine P, Wilson W (1999) Tracking algorithm for GPS offset carrier signals. In: Proceedings of the 1999 national technical meeting of The Institute of Navigation, pp 671–676

# Study on Impact Behavior of Composite Sandwich Structure with Different Interlayer Angles

**Peng Yuyang and Chen Xiuhua**

**Abstract** The paper investigates the influence of different interlayer angles on dynamic behavior of composite sandwich structure subjected to low-velocity impact. The composite sandwich structure is composed of CFRP/epoxy laminates and Nomex honeycomb core. Seven finite element models with seven different interlayer angles, from [0°/0°/0°/Honeycomb/0°/0°/0°] to [0°/90°/0°/Honeycomb/0°/90°/0°], of composite sandwich are implemented by ABAQUS/Explicit. The damage mechanism for composite laminate is applied with 3D Hashin's damage by VUMAT, while the honeycomb is based on traction separation laws. The modeling results show that as interlayer angle increases, the indent area is decreasing, but the rate of decrease becomes slower. Meanwhile, the energy absorption is increasing with the higher interlayer angle. From mentioned above, it can be concluded that the interlayer angle has an effect on the impact behavior of composite, and the bigger interlayer angle will result in better impact resistance.

**Keywords** Composite sandwich structure · Impact · Interlayer angle · ABAQUS

## 1 Introduction

Advanced composite structures offer many advantages compared to conventional materials, especially where high strength and stiffness to weight ratio is concerned. Thus, composites have been used widely in many applications such as aerospace, sport equipment, pressure vessels and automotive parts. For composite sandwich structures, they have good resistance of fatigue, corrosion and vibration, and the core, the panel can be chosen different to combine. However, composite sandwich structures perform not well in interlaminar properties and low-velocity impact resistance, when subjected to low-velocity impact of foreign objects during manufacturing, transportation, utilization and maintenance, such as tools dropping and hail impact, different types of damage such as matrix cracking, fiber breakage and interface debonding between the panel and the core will occur. It should be emphasized that

P. Yuyang (✉) · C. Xiuhua
Shanghai Jiaotong University, Dongchuan Road 800, Shanghai, China



Z. Jing (ed.), *Proceedings of the International Conference on Aerospace System Science and Engineering 2019*, Lecture Notes in Electrical Engineering 622,
https://doi.org/10.1007/978-981-15-1773-0_25

the damage caused by low-velocity impact is considered very detrimental, this damage will lead to internal defects, resulting in weakening of structural bearing capacity during service, especially in the compress and shear bearing capacity, which seriously restrict the development and utilization of composite sandwich structure [1]. Meanwhile, in most cases, these defects are not visually identified so the detection techniques are needed. Non-destructive techniques are widely used because it can detect the damage without destructing the structure. Ultrasonic C scanning, X-radiography and shearography also are applied for detecting the damage [2]. However, all these techniques are inadequate for in-service inspection of large components. As a new light structural material which widely used in aeronautic and astronautic, researches that related to impact damage of composite material are not complete and mature, no matter the dynamic mechanical properties of material or the theoretical analysis and numerical modeling of impact damage, which cannot satisfy the needs for military technical development [3]. Therefore, go in deep with the research of impact damage of composite material has important theoretical value and practical significance.

Many parameters will influence the impact response of the composite sandwich structure. Wang [4] did the research about the influence of impactor size, plate thickness and core thickness on the impact response. Amaro [5] studied the influence of the boundary conditions on the impact of composite, Körbelin [6] investigated how the temperature and impact energy would affect impact process and the effects of stacking sequence on impact was studied by Riccio and Di Felice [7]. In this paper, the influence of different interlayer angles was investigated by using ABAQUS simulation to create a 3D damage model of composite sandwich structures under low-velocity impact. The model comprehensively considered main damage modes such as matrix cracking, fiber breakage, matrix compression and delamination, using VUMAT to realize the definition of material constitutive relationship, and predicted the impact response of composite sandwich structures.

## 2 Failure Criterion

### 2.1 Composite

About the failure criterion of composite, many scholars have done plenty of researches, such as Max stress criterion, Tsai-Wu criterion and Chang-Chang criterion. In this paper, 3D Hashin's failure criterion was applied by combining user subroutine VUMAT.

The failure modes included in Hashin's criteria are as follows.

1. Tensile fiber failure for $\sigma_{11} \geq 0$

$$\left(\frac{\sigma_{11}}{X_T}\right)^2 + \frac{\sigma_{12}^2 + \sigma_{13}^2}{S_{12}^2} = \begin{cases} \geq 1 & \text{failure} \\ < 1 & \text{no failure} \end{cases}$$

2. Compressive fiber failure for $\sigma_{11} < 0$

$$\left(\frac{\sigma_{11}}{X_C}\right)^2 = \begin{cases} \geq 1 & \text{failure} \\ < 1 & \text{no failure} \end{cases}$$

3. Tensile matrix failure for $\sigma_{22} + \sigma_{33} > 0$

$$\frac{(\sigma_{22} + \sigma_{33})^2}{Y_T^2} + \frac{\sigma_{23}^2 - \sigma_{22}\sigma_{33}}{S_{23}^2} + \frac{\sigma_{12}^2 + \sigma_{13}^2}{S_{12}^2} = \begin{cases} \geq 1 & \text{failure} \\ < 1 & \text{no failure} \end{cases}$$

4. Compressive matrix failure for $\sigma_{22} + \sigma_{33} < 0$

$$\left[\left(\frac{Y_C}{2S_{23}}\right)^2 - 1\right]\left(\frac{\sigma_{22} + \sigma_{33}}{Y_C}\right) + \frac{(\sigma_{22} + \sigma_{33})^2}{4S_{23}^2} + \frac{\sigma_{23}^2 - \sigma_{22}\sigma_{33}}{S_{23}^2}$$
$$+ \frac{\sigma_{12}^2 + \sigma_{13}^2}{S_{12}^2} = \begin{cases} \geq 1 & \text{failure} \\ 1 & \text{no failure} \end{cases}$$

5. Interlaminar tensile failure for $\sigma_{33} > 0$

$$\left(\frac{\sigma_{33}}{Z_T}\right)^2 = \begin{cases} \geq 1 & \text{failure} \\ < 1 & \text{no failure} \end{cases}$$

6. Interlaminar compression failure for $\sigma_{33} < 0$

$$\left(\frac{\sigma_{33}}{Z_C}\right)^2 = \begin{cases} \geq 1 & \text{failure} \\ < 1 & \text{no failure} \end{cases}$$

where $\sigma_{ij}$ denote the stress components, and the tensile and compressive allowable strengths for lamina are denoted by subscripts $T$ and $C$, respectively. $X_T$, $Y_T$, $Z_T$ denote the allowable tensile strengths in three respective material directions. Similarly, $X_C$, $Y_C$, $Z_C$ denote the allowable compressive strengths in three respective material directions. Further, $S_{12}$, $S_{13}$ and $S_{23}$ denote allowable shear strengths in the respective principal material directions [8, 9].

### 2.2 Honeycomb

The damage mechanism for composite sandwich structure is quite complicated, so during the research following assumptions are often proposed: The plate is only subjected to interlaminar stress $\sigma_x$, $\sigma_y$, $\tau_{xy}$, the honeycomb core is only subjected to transverse shear stress, evenly distributed through the thickness, the state of stress is: $\sigma_x = \sigma_y = \tau_{xy} = 0$, $\tau_{xz} \neq 0$, $\tau_{yz} \neq 0$; when the sandwich structure deforms, $\varepsilon_z = 0$, and the effect of $\sigma_z$ is omitted. Z-direction is vertical to the plate, and $x$-direction and $y$-direction are the two directions on the plate. These assumptions can be considered that the interlaminar tensile, compression and shear are not taken into account, only the transverse shear failure and $z$-direction compression [10]. The failure criterion is presented as follows:

1. Z-direction compression failure: $\left\{\frac{\sigma_z}{Z}\right\}^2 = \begin{cases} \geq 1 \text{ failure} \\ < 1 \text{ no failure} \end{cases}$
2. Transverse ($xz$) shear failure: $\left\{\frac{\tau_{xz}}{S_{xz}}\right\}^2 = \begin{cases} \geq 1 \text{ failure} \\ < 1 \text{ no failure} \end{cases}$
3. Transverse ($xz$) shear failure: $\left\{\frac{\tau_{yz}}{S_{yz}}\right\}^2 = \begin{cases} \geq 1 \text{ failure} \\ < 1 \text{ no failure} \end{cases}$

where $Z$ presents the $z$-direction compression strength, and $S_{xz}$ and $S_{yz}$ are the transverse shear strength for the $xz$ and $yz$ direction, respectively.

## 2.3 Cohesive

To simulate the formation and evolution of the interlaminar damage of composite plate, cohesive behavior was introduced in interaction between each ply of composite. The stress criterion used is given by damage for traction-separation laws–Maximum nominal stress criterion, which damage is assumed to initiate when the maximum nominal stress ratio (as defined in the expression below) reaches a value of one. Figure 1 shows a typical traction-separation response with a failure mechanism [11].

This criterion can be represented as

$$\max\left\{\frac{\langle t_n\rangle}{t_n^0}, \frac{t_s}{t_s^0}, \frac{t_t}{t_t^0}\right\} = 1$$

where $t_n^0$, $t_s^0$, $t_t^0$ represent the peak values of the contact stress when the separation is either purely normal to the interface or purely in the first or the second shear

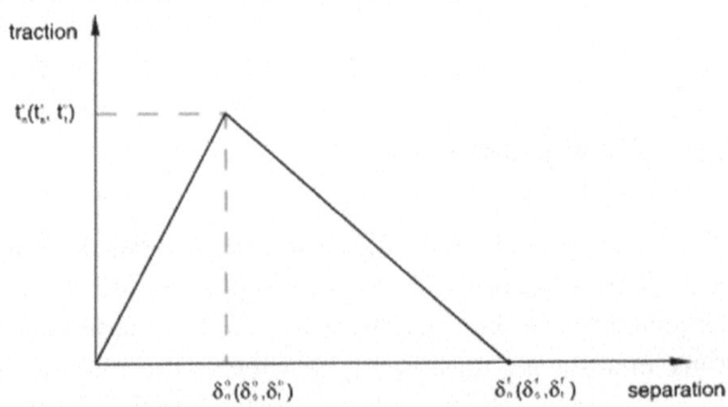


**Fig. 1** Typical traction-separation response

direction, respectively. $t_n$, $t_s$, and $t_t$ denote the traction stress vector in the normal and shear directions. The symbol $\langle\ \rangle$ represents the Macaulay bracket with the usual interpretation. The Macaulay brackets are used to signify that a purely compressive displacement (i.e., a contact penetration), or a purely compressive stress state does not initiate damage.

## 3 FEM Model

As shown in Figs. 2 and 3, the finite element model is built in ABAQUS/Explicit.

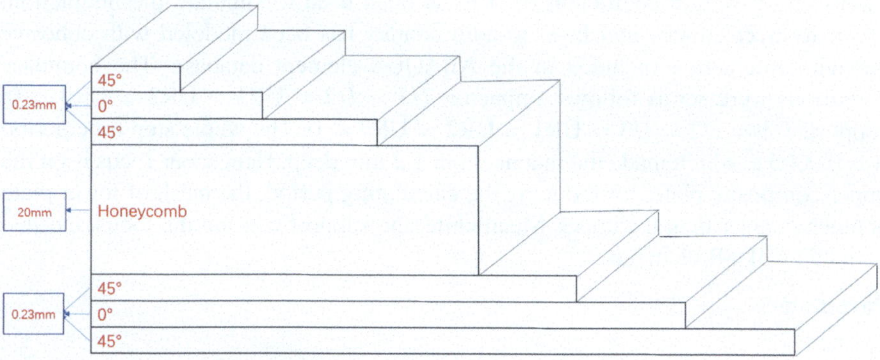


**Fig. 2** Composite sandwich structure

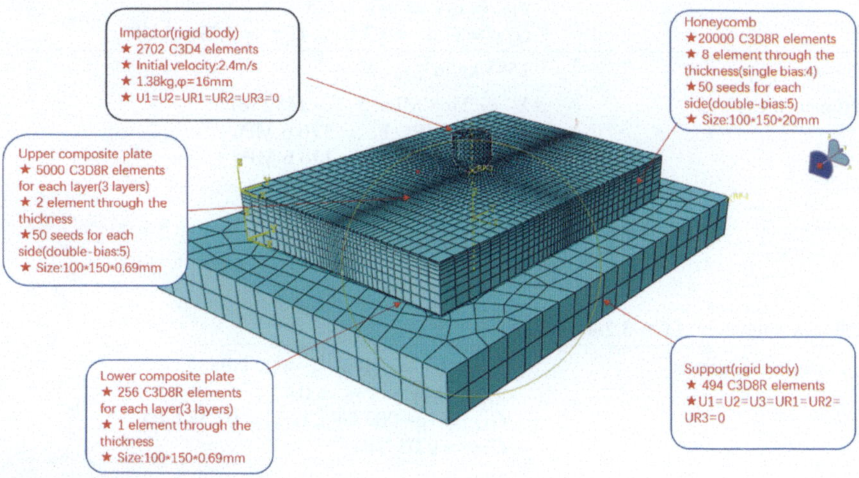


**Fig. 3** ABAQUS FEM model

The composite sandwich structure was composed of seven layers: [0°/x°/0°/Honeycomb/0°/x°/0°], including composite plate and honeycomb core. Each layer of composite plate was 0.23 mm, and each upper plate was modeled with 5000 three-dimensional eight-node solid finite elements (C3D8R element type), the mesh was implemented by double-bias seeds with ratio 5, each side 50 elements, while the lower plate was modeled with 256 C3D8R elements. The honeycomb was 20 mm thick, modeled with 20,000 C3D8R elements. To better observe the impact response, on the thickness direction, eight seeds with single-bias ratio 4 were implemented on the honeycomb core. The diameter of the impactor was 16 mm, and the mass was 1.38 kg. The impactor was assumed to be rigid with a velocity of 2.4 m/s and meshed as 2702 four-node linear tetrahedron solid finite elements (C3D4 element type). Between the impactor and the composite sandwich, a friction coefficient of 0.3 has been used to simulate the interaction. Layer-to-layer contact and layer-to-core contact has been modeled with cohesive behavior interaction included in the ABAQUS element database. The boundary conditions were set as follows: impactor: $U1 = U2 = \mathrm{UR1} = \mathrm{UR2} = \mathrm{UR3} = 0$, support: $U1 = U2 = U3 = \mathrm{UR1} = \mathrm{UR2} = \mathrm{UR3} = 0$. The whole step time period was 0.0005 s, which made the indent about 1.2 mm deep. Hence, our focus is on the upper composite plate, for reducing the calculating period, the mesh of lower plate is much coarser than the upper. Meanwhile, the support was not the focus, meshed with 494 C3D8R elements.

## Parameters

| 1. Composite plate: Cycom 381 IM7 UD Nominal | |
|---|---|
| Engineering constants | $E_1 = 156.5$ GPa, $E_2 = E_3 = 8.83$ GPa<br>$\nu_{12} = \nu_{13} = \nu_{23} = 0.3$<br>$G_{12} = G_{13} = 4.3$ GPa, $G_{23} = 3.39615$ GPa |
| Density | 1583 kg/m$^3$ |
| Ultimate stress | $X_t = 2468$ MPa, $X_c = 1482$ MPa<br>$Y_t = 38$ MPa, $Y_c = 176.6$ MPa<br>$Z_t = 38$ MPa, $Z_c = 176.6$ MPa<br>$S_{12} = S_{13} = 128$ MPa<br>$S_{23} = 50.9423$ MPa |

| 2. Honeycomb core: ECA 3.2-48-(51) | |
|---|---|
| Engineering constants | $E_1 = E_2 = E_3 = 1\mathrm{e}{-009}$ GPa<br>$\nu_{12} = \nu_{13} = \nu_{23} = 0.42$<br>$G_{12} = 1\mathrm{e}{-009}$ GPa, $G_{13} = 0.048$ GPa<br>$G_{23} = 0.03$ GPa |
| Density | 48 kg/m$^3$ |

(continued)

(continued)

| 2. Honeycomb core: ECA 3.2-48-(51) | |
|---|---|
| Ultimate stress | Z-direction compression: 2.1 MPa<br>Transverse direction(13):1.32 MPa<br>Transverse direction(23):0.72 MPa |

| 3. Cohesive | |
|---|---|
| Traction-separation behavior | $K_{\mathrm{nn}} = 1500$ MPa, $K_{\mathrm{ss}} = K_{\mathrm{tt}} = 1000$ MPa |
| Damage | Normal: 45 MPa<br>Shear-1, Shear-2:30 MPa |
| | Normal-only mode: 2.1 MPa<br>1-direction: 1.32 MPa<br>2-direction: 0.72 MPa |

## 4 Results

The numerical results were obtained from the FEM model.

From Fig. 4a, b, it is observed that energy absorption rate increases with the interlayer angle rising. $E_0$, $E_{15}$ and $E_{90}$ mean energy-time curve with different interlayer angles. For 15° and 75° interlayer angles, the results have some deviation, but the overall trend is certain.

Next, for indent depth, as is shown in Fig. 5a, b, it has the opposite tendency with energy absorption rate. As interlayer angle goes up, the indent depth presents an overall downward trend.

Meanwhile, it is consistently observed in Fig. 6a, b that residual velocity corresponds well with the relationship between residual energy and interlayer angle.

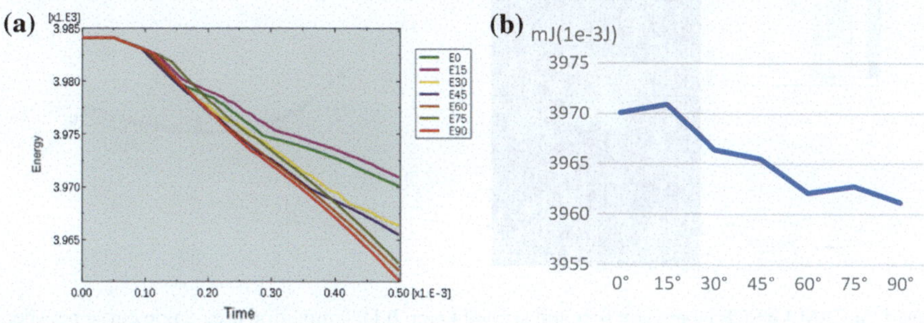


**Fig. 4** **a** Energy-time curve. **b** Residual energy-angle curve

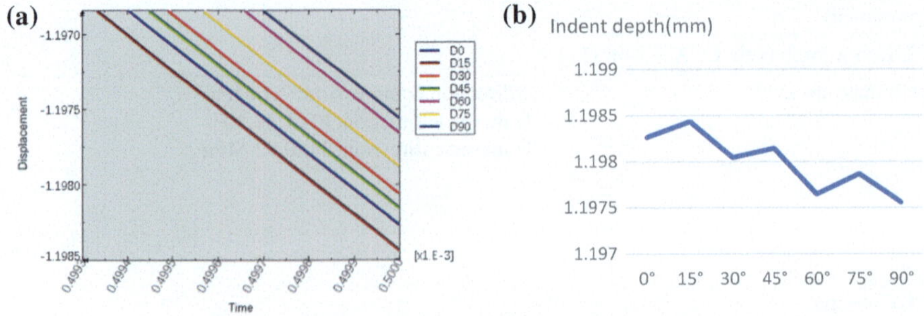


**Fig. 5** **a** Displacement-time curve. **b** Indent depth-angle curve

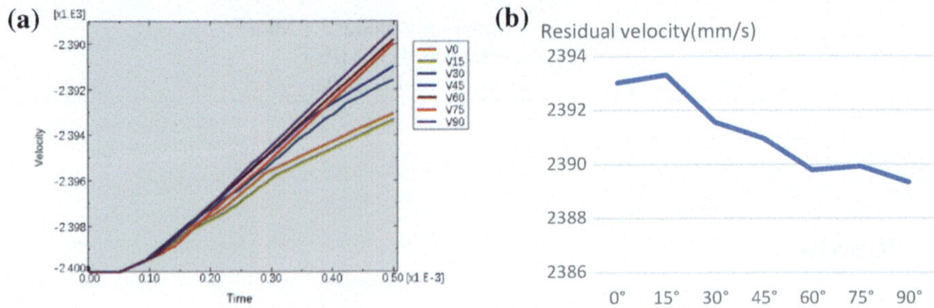


**Fig. 6** **a** Velocity-time curve. **b** Residual velocity-angle curve

Figures 7 and 8 show that the influence of interlayer angle on the delamination area of the upper composite plate and interface delamination is presented by maximum contact stress damage initiation criterion (CSMAXSCRT). We can observe that the delamination area between second and third layer is much bigger than the one between first and second layer. Because during the impact process, the plate was

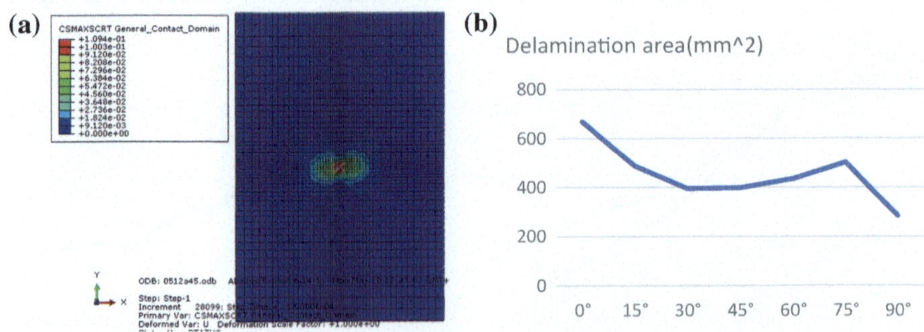


**Fig. 7** **a** CSMAXSCRT between first and second layer. **b** Delamination area-angle curve between first and second layer

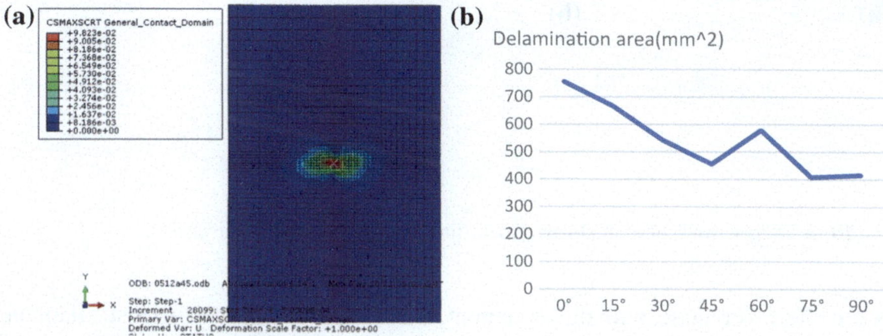


**Fig. 8** **a** CSMAXSCRT between second and third layer. **b** Delamination area-angle curve between second and third layer

mainly subjected to bending deformation, for the second layer, the upper surface was subjected to compress stress, while the lower surface was subjected to tensile stress, and the bending deformation of the lower surface was larger than the upper, so the stress on the lower surface was much higher, resulting in damage easilier. Meanwhile, for the delamination of both interfaces, as interlayer angle multiplies, the delamination area presents an overall decreasing trend. Except for 60° and 75°, the others are monotonically decreasing. This represents the composite sandwich with a larger interlayer angle has a better interlaminar performance.

The matrix tensile damage of each layer of the upper composite plate is represented by Fig. 9. It can be observed that how the matrix tensile damage is influenced

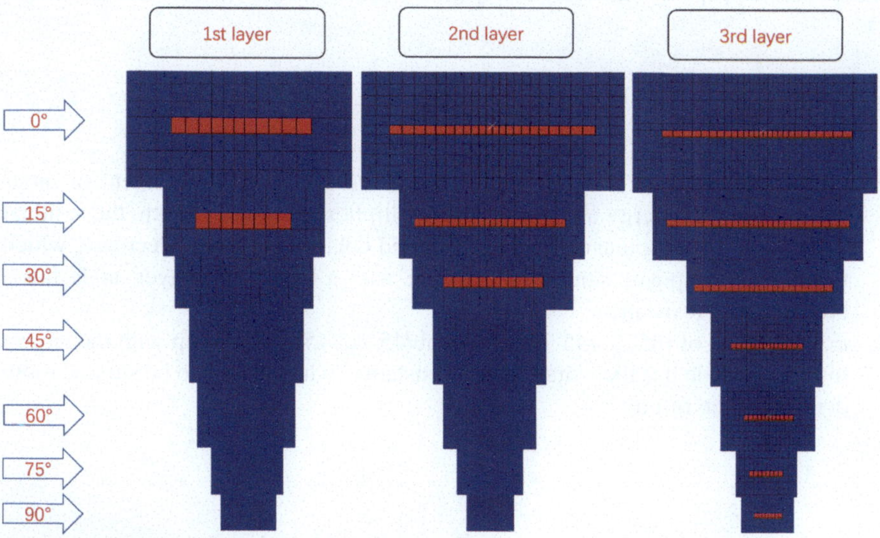


**Fig. 9** Matrix tensile damage of upper composite plate

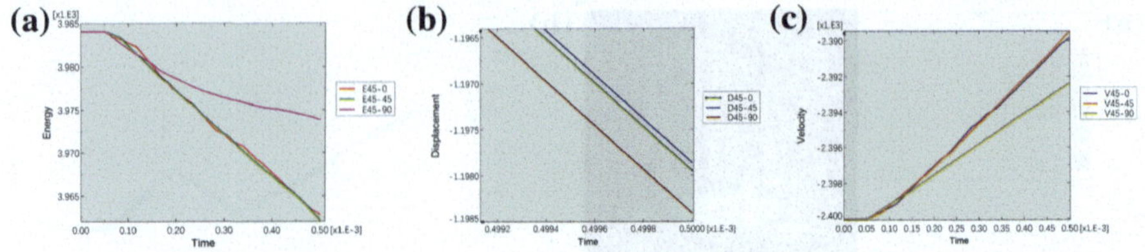


**Fig. 10** **a** Energy-time curve. **b** Displacement-time curve. **c** Velocity-time curve

by the interlayer angle, with the increment of interlayer angle, matrix tensile damage area is decreasing for all 3 layers. In summary, we can conclude that for the layup of [0°/*x*°/0°/Honeycomb/0°/*x*°/0°], as the interlayer angle increases, the energy absorption rate also goes upward, the indent depth decreases, the residual velocity becomes lower, the matrix tensile damage area and delamination are reduce, which means the composite sandwich structure has the better impact resistance.

For further research, the layup of [45°/*x*°/45°/Honeycomb/45°/*x*°/45°] with different interlayer angles was studied. Three models with the interlayer angles of 0°, 45° and 90° are created with ABAQUS/Explicit. All parameters and settings are the same as the previous model except the layup. And the results are as follows.

From Fig. 10a–c, the impact energy-time, displacement-time and velocity-time curves for the layup of [45°/*x*°/45°/Honeycomb/45°/*x*°/45°] with different interlayer angles are reported. As is shown in the figure, the interlayer angle of 90°, whose layup of [45°/−45°/45°/Honeycomb/45°/−45°/45°] has the worst impact resistance in terms of energy absorption, indent displacement and velocity. We presume that the stiffness gets weaker after the stiffness matrix transformation, and however, this is beyond the scope of this study and ought to be researched in detail in the future.

## 5 Conclusion

1. For the layup of [0°/*x*°/0°/Honeycomb/0°/*x*°/0°], with the increment of interlayer angle, the energy absorption rate multiplies, the indent depth, the residual velocity, the matrix tensile damage area and delamination area decrease, which means the composite sandwich structure with a higher interlayer angle has a better impact resistance.
2. For the layup of [45°/*x*°/45°/Honeycomb/45°/*x*°/45°], the layup with the largest interlayer angle has the worst impact resistance, which needs to be studied more detailed in the future.

## References


1. Xiaoxia Z, Xitao Z, Tianjiao Q (2010) Analysis of low velocity impact damage of sewing foam core sandwich structure. J Northwest Polytechnical Univ 05:139–144
2. Amaro AM, Santos JB, Cirne JS (2004) Comparative study of different non-destructive testing techniques in the characterisation and quantification of the damage effects in carbon-epoxy laminates. Insight 46:559–565
3. Bhuiyan A, Hosur MV, Jeelani S (2009) Low-velocity impact response of sandwich composites with nanophased foam core and biaxial braided face sheets. Compos B 40:561–571
4. Jie W, Baoxing C, Hai W (2008) The effects of impact variables and sandwich configuration parameters on low-velocity impact response. China Aviation Society. Completely build a well-off society and China's aviation development—Proceedings of the first China aviation science and technology conference 2013 Chinese Aviation Society: Chinese Aviation Society, p 8
5. Amaro AM, Reis PNB, Magalhaes AG, de Moura MFSF (2011) The influence of the boundary conditions on low-velocity impact composite damage. Strain 47:e220–e226
6. Körbelin J, Derra M, Fiedler B (2018) Influence of temperature and impact energy on low velocity impact damage severity in CFRP. Compos A 115:76–87
7. Riccio A, Di Felice G, Saputo S, Scaramuzzino F (2014) Stacking sequence effects on damage onset in composite laminate subjected to low velocity impact. Procedia Eng 88:222–229
8. Hashin Z (1980) Failure criteria for unidirectional fibercomposites. J Appl Mech 47(2):329–334
9. Hashin Z, Rotem A (1973) A fatigue failure criterion forfiber reinforced materials. J Compos Mater 7(4):448–464
10. Yangyang FA, Xiuhua C (2012) Finite element analysis of honeycomb sandwich in pull-off damage. Mech Eng Mater 36(10):86–91 + 96
11. Abaqus Analysis User's Manual 28.5.6 (2009) Defining the constitutive response of cohesive elements using a traction-separation description

# Positioning and Timing Test Campaign Based on China Area Positioning System (CAPS)

**Yang Zhang, Yu Su, Wenfang Jing and Xiaochun Lu**

**Abstract** With high-performance ground atomic clock and satellite retransferring feature, CAPS achieved higher positioning and timing precision than traditional GNSS. Positioning and timing high-performance verification test has been conducted by National Time Service Center (NTSC in Xi'an, China) in Xi'an city Shaanxi province. Test results show that (1) the accuracy of the code-phase timing and the carrier-phase timing with single GEO satellite is better than 5 ns and 0.5 ns, respectively, and (2) the accuracy of the code-phase positioning and the carrier-phase positioning with 1 IGSO and 3 GEO satellites is better than 1.5 m and 0.5 m, respectively. The results indicate the ability of high-precision positioning and timing of CAPS.

**Keywords** China Area Positioning System (CAPS) · Positioning and timing · Test campaign

## 1 Introduction

With the development of science and technology, many fields need more accurate real-time positioning and timing service, such as space target interception and strike need nanoseconds level timing and decimeter level positioning accuracy. Traditional GNSS cannot meet the requirement. With C-band radio frequency, China Area Positioning System (CAPS), a novel regional satellite navigation system proposed by Chinese Academy of Sciences (CAS), can provide the high-performance positioning and timing service based on carrier phase. GNSS uses satellite specially developed for navigation to broadcast navigation signal to users [1, 2], while in CAPS, the

Y. Zhang (✉) · Y. Su · W. Jing · X. Lu
National Time Service Center, CAS, Xi'an 710600, China
e-mail: zhangyang14@mails.ucas.ac.cn

Key Laboratory for Precision Navigation, Positioning and Timing of the CAS, Xi'an 710600, China

Y. Zhang · Y. Su · X. Lu
University of Chinese Academy of Sciences, Beijing, China



Z. Jing (ed.), *Proceedings of the International Conference on Aerospace System Science and Engineering 2019*, Lecture Notes in Electrical Engineering 622,
https://doi.org/10.1007/978-981-15-1773-0_26

**Table 1** Four GEO communication satellite resources

| NUM | Satellite | Location | Available frequency (MHz) | EIRP (dBW) |
|---|---|---|---|---|
| 1 | APSTAR 7 | 76.5° E | 3826 ± 18 | 39 |
| 2 | ChinaSat 12 | 87.5° E | 3826 ± 18 | 40 |
| 3 | ChinaSat 10 | 110.5° E | 3826 ± 18 | 44 |
| 4 | Measat2 | 148° E | 3700–4200 | 41 |

satellite transponder without onboard atomic clock is used as navigation satellite. It aims to transparently transfer the navigation signal generated on the ground.

For working on C-band, the wavelength of the CAPS and GPS L1 signal is 7.84 cm and 19.04 cm, respectively, and therefore, the precision of CAPS carrier-phase measurement would be theoretically higher than GNSS [3].

In this paper, the CAPS architecture, principle, and the features are introduced, the timing and positioning test is described, and test results will be shown.

## 2 System Architecture

CAPS consists of the space segment, the ground segment, and the user segment.

### 2.1 Space Segment

The space segment includes four GEO communication satellites (transponder) and an inclined geosynchronous orbit (IGSO) satellite (transponder) at present. Table 1 shows the orbit locations of the four communication satellites.

The ascending node of the IGSO satellite is 95° E, and the orbital inclination angle is 55°. All of the satellite transponders operate on C-band.

### 2.2 Ground Segment

The ground segment mainly realizes the generation, maintenance, and transmission of the system time, the precision orbit observation and determination of each satellite, and the navigation signal transmission and reception processing. It is mainly divided into three parts:

(1) Time and frequency reference generation and maintenance center are mainly composed of high-performance atomic clock group, time and frequency integrated equipment, two-way comparison equipment of high-precision time and

frequency signal, and time and frequency signal distribution network. It provides high stability and accuracy time and frequency signals and time code information for system.

(2) The satellite orbit observation and determination center are made up of the orbit observation and determination main station located in Xi'an and the secondary stations located in Changchun, Xuyi, Kashgar, Sanya, and Kunming. It mainly realizes precise forecasting and determination of the GEO satellites and IGSO satellite orbits and provides highly reliable precise ephemeris data.

(3) The navigation signal controlling center consists of four ground stations with 13-m aperture antenna, one ground station with 16-m aperture antenna, navigation message and virtual clock data processing, and management equipment and supporting equipment. It mainly realizes the generation of virtual atomic clock, the generation and transmission of satellite uplink navigation signals, the reception and processing of satellite downlink navigation signals and high-precision ranging.

## 2.3 *User Segment*

Multiple types of user receivers including pseudo-random code user receivers and carrier-phase user receivers have been developed by NTSC, which can provide high-precision positioning, timing, and velocity measurement (PVT) services for users. The carrier-phase high-precision PVT user receiver is shown in Fig. 1.

**Fig. 1** Carrier-phase high-precision PVT user receiver

# 3 Basic Principle

CAPS is different from GNSS in that the navigation signal is generated on the ground and uploaded to the communication satellites [4]. And then the navigation signal is broadcasted to users by the satellite transponder. The basic principle of CAPS based on communication satellite transfer mode is shown in Fig. 2.

Figure 2 shows that CAPS adopts commercial geostationary orbit (GEO) communication satellites to form a navigation constellation. The navigation signal and the navigation message are generated on the ground and retransferred to users via satellite. Therefore, the range which is measured by user receiver is that the signal path from the ground station to satellite including the time delay of satellite transponder and plus the path from the satellite to the receiver. The range can be expressed as:

$$\rho = c \cdot (\tau_{\text{uplink}} + \tau_{\text{transp}}) + \rho_{\text{down}} = c \cdot \tau_{\text{Vclk}} + \rho_{\text{down}} \tag{1}$$

where $\rho$ is the range of the whole loop, $\rho_{\text{down}}$ is the range from the satellite to the user receiver, $\tau_{\text{uplink}}$ is the time delay from the ground station to satellite, $\tau_{\text{transp}}$ is the time delay of the satellite transponder, and $c$ is the speed of light.

Compared with GNSS, there is an additional uplink path from the ground station to the satellite in the signal transmission path of CAPS. However, the pseudo-range used for positioning and timing is the measurement between the satellite and the users [4]. Hence, this additional uplink distance must be known to user for PVT services.

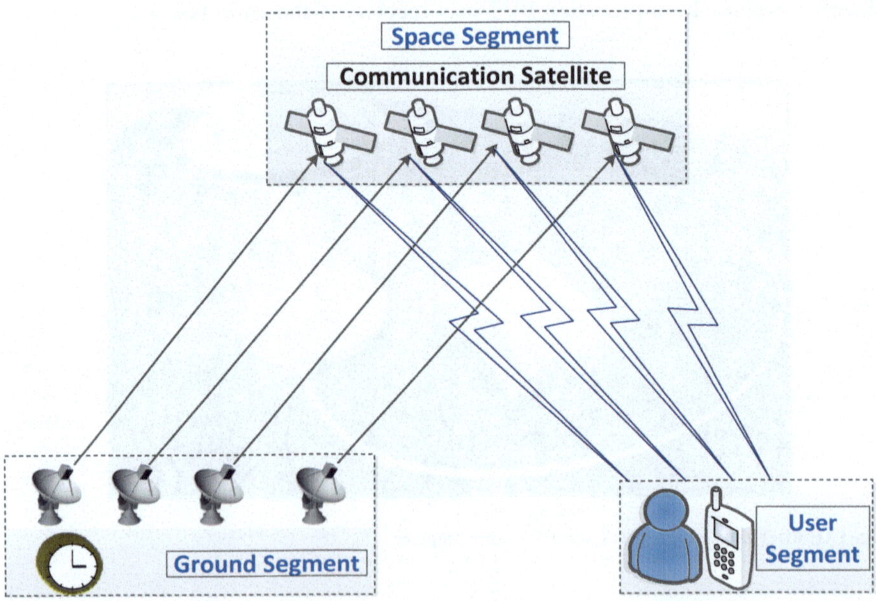


**Fig. 2** Basic principle of CAPS based on the communication satellite transponder

A method called "virtual atomic clock" is proposed to solve this additional uplink [5]. The modems on ground station transmit the ranging signal and receive it, and the real-time measurement ranging data is used to build the "virtual atomic clock" model to predict the uplink delay $\tau_{\text{uplink}}$ and the satellite transponder time delay $\tau_{\text{transp}}$. The virtual clock correction includes the uplink atmospheric delay, satellite receiving, and retransfer delay [4]. The model will be broadcasted to users in the navigation message. From the perspective of the user terminal, the CAPS users obtain PVT service by passively receiving navigation signals, which is almost as same as the way that of GNSS users. There is only a difference in calculation method of the pseudo-range from the satellite to the user receiver which will be explained in detail below.

## 3.1 Single-Satellite Time Service Principle

The precise position coordinates $(X_u, Y_u, Z_u)$ are known to a static user, precise time measurement, or synchronization can be achieved by observing only one satellite $(x_s, y_s, z_s)$ [6]. The principle of single-satellite time service of the CAPS is shown in Fig. 3.

The unknown amount $\Delta t_u$ is defined as the clock difference between the system clock and the user clock. $\rho$ is the pseudo-range from the ground station to the user which is measured by receiver directly, and the observation equation is:

$$\sqrt{(X_u - x_s)^2 + (Y_u - y_s)^2 + (Z_u - z_s)^2} + c \cdot \Delta t = \rho - c \cdot \tau_{\text{Vclk}} \tag{2}$$

The satellite position $(X_u, Y_u, Z_u)$ can be obtained from the broadcast ephemeris in the navigation message, and $\tau_{\text{Vclk}}$ is the time delay from the ground station to the

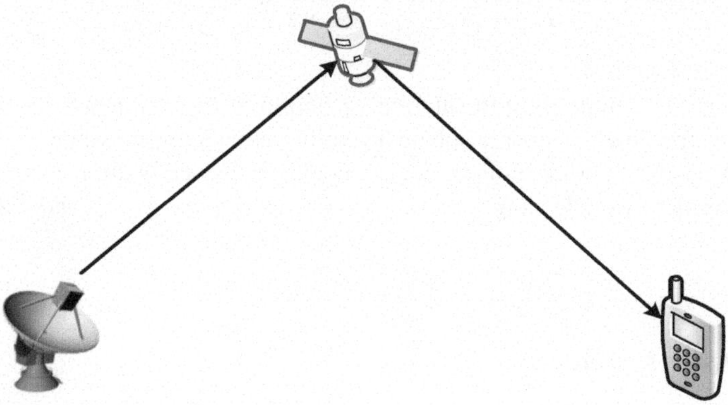

**Fig. 3** Single-satellite time service of the CAPS

satellite and the satellite transponder time delay. $\tau_{\text{Vclk}}$ can be calculated from the "virtual atomic clock" model.

## 3.2 Positioning Principle

Similar to GNSS, CAPS utilizes time of arrival (TOA) ranging to determine user position [2]. By making TOA measurement to at least four satellites, three-dimensional positioning can be achieved [2]. When using the CAPS system to determine user position, it is necessary to simultaneously observe four or more satellites to obtain the user coordinate and receiver's clock error. Pseudo-range measurements can be done by observing the tracking the satellite, and the observation equations are:

$$\left\| S_j - u \right\| + c \cdot \Delta t_u = \rho_j - c \cdot \tau_{\text{Vclk}}^{j} \tag{3}$$

where $j$ refers to different satellites and their corresponding signal transmitting ground stations, $S$ indicates the satellite coordinate, $u$ is the position of the user receiver, $\Delta t_u$ is the clock difference between the user clock and the system clock, $\rho_j$ is the pseudo-range from the ground station $j$ to the user, and $\tau_{\text{Vclk}}^{j}$ is the time delay from the ground station $j$ to satellite $j$ and its transponder time delay. When $j =$ 1,2,3,4, Eq. (3) can be expanded into the following set of equations in the unknowns $x_u$, $y_u$, $z_u$ and $\Delta t_u$:

$$\begin{aligned}
\sqrt{(x_u - x_s^1)^2 + (y_u - y_s^1)^2 + (z_u - z_s^1)^2} + c \cdot \Delta t &= \rho_1 - c \cdot \tau_{\text{Vclk}}^{1} \\
\sqrt{(x_u - x_s^2)^2 + (y_u - y_s^2)^2 + (z_u - z_s^2)^2} + c \cdot \Delta t &= \rho_2 - c \cdot \tau_{\text{Vclk}}^{2} \\
\sqrt{(x_u - x_s^3)^2 + (y_u - y_s^3)^2 + (z_u - z_s^3)^2} + c \cdot \Delta t &= \rho_3 - c \cdot \tau_{\text{Vclk}}^{3} \\
\sqrt{(x_u - x_s^4)^2 + (y_u - y_s^4)^2 + (z_u - z_s^4)^2} + c \cdot \Delta t &= \rho_4 - c \cdot \tau_{\text{Vclk}}^{4}
\end{aligned} \tag{4}$$

where $x_s^j$, $y_s^j$, $z_s^j$ denote the $j$-th satellite's position in three dimensions which can be obtained from the broadcast ephemeris in the navigation message. $\tau_{\text{Vclk}}^{j}$ can be calculated by the "virtual atomic clock" model in the navigation message. $\rho_j$ is the measurement pseudo-range of the receiver. Hence, $x_u$, $y_u$, $z_u$ and $\Delta t_u$ can be calculated.

## 3.3 System Features

The main features of CAPS are that the navigation signals and navigation messages are generated and uplinked directly to the satellite from the ground control station, before being downlinked and broadcasted to users via the communication satellite transponder [7].CAPS does not require the onboard atomic clock and use a higher stability atomic clock group in the ground station [3]. Therefore, the complicated onboard technique can be greatly simplified, and the system can be easily and flexibly built at a low cost. In addition, the constellation of CAPS can be flexibly selected by renting the transponders on the communication satellites [8]. Furthermore, CAPS uses C-band frequencies for navigation. Therefore, the error caused by the ionosphere and multipath effect is smaller, and the precision of carrier-phase measurement is higher than that of the GNSS. These features hold the promise of achieving PVT service with a relatively high accuracy.

## 4 Test and Analysis

Timing and positioning test had been carried out in NTSC in August 2017, with receiver used 4-Hz sampling frequency.

### *4.1 Test Campaign*

The basic structure of CAPS receiver is almost like the GPS receiver. The main difference is timing and positioning algorithm and the radio frequency. For no anti-multipath antenna, the antenna was installed on a 40-m high tower to reduce the multipath effects. Figure 4 shows the basic frame of the test.

The single-satellite timing test was performed by observing a GEO satellite. The timing error data was collected by the counter SR620. The 1PPS signal from the atomic clock in the ground station is used as the opening signal of the counter, and the output 1PPS signal of the receiver was used as the closing signal. The output of the counter is the timing error of the CAPS system.

The constellation type of 4 GEO satellites cannot provide 3D positioning because the GEO satellites are all located in orbit over the equator [4]. Therefore, when IGSO satellite is available, positioning tests can be performed by using 3 GEO satellites and 1 IGSO satellite.

### *4.2 Test Results and Analysis*

The communication satellite APSTAR 7 was selected to perform the single-satellite timing test. The timing results based on the code measurement and the carrier measurement are illustrated in Fig. 5. The standard deviation of the single-satellite timing errors based on the code measurement are 0.3028 ns, while the result base on carrier

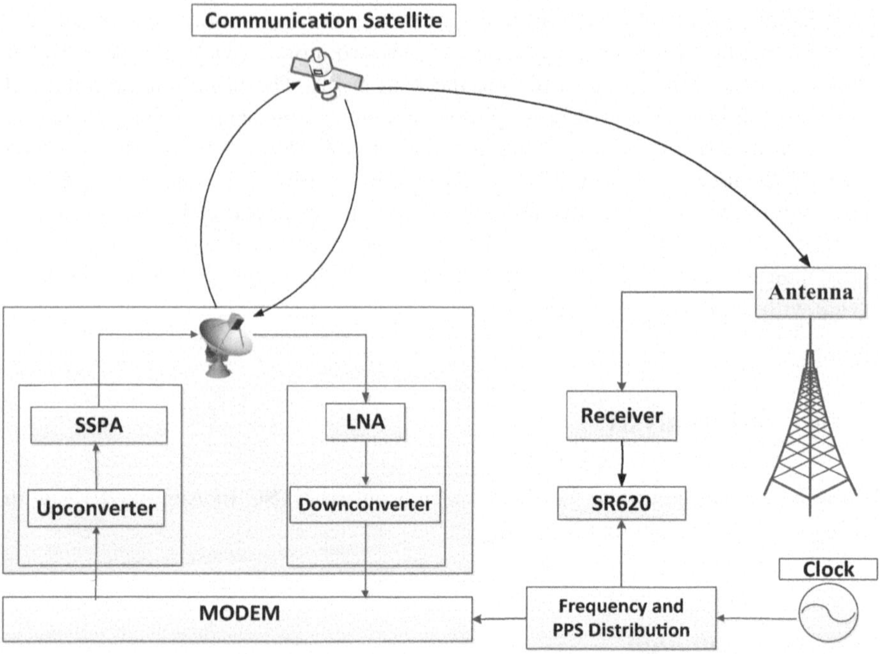


**Fig. 4** Frame of the test

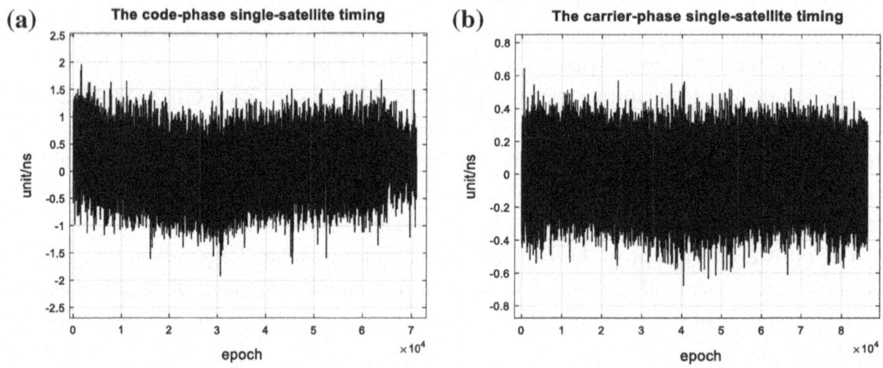


**Fig. 5** Single-satellite timing error of the CAPS based on the code-phase measurement (**a**) and the carrier-phase measurement (**b**)

measurement is 0.1566 ns. It can be seen that the timing errors based on the carrier measurement is smaller than the code measurement.

The positioning errors in the east, north, and vertical components and synthetic ENU for the static positioning based on the code measurement are illustrated in Fig. 6, while the corresponding results for the carrier-phase measurement are shown

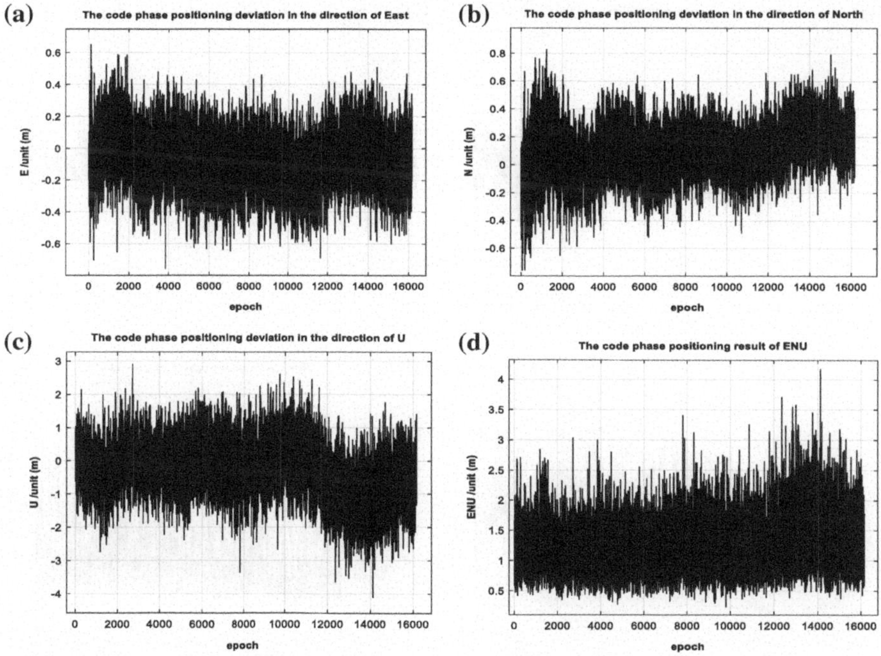


**Fig. 6** Positioning error in the east (**a**), north (**b**), and vertical (**c**) components and synthetic ENU (**d**) for the CAPS static positioning based on the code measurement

in Fig. 7, respectively. It can be seen that the positioning result in the east-west component is much better, both for the code-phase and for the carrier-phase measurements. This is mainly because of the distribution of the current CAPS constellation. It can be seen that there is a slow fluctuation in the positioning results based on the code measurement from Fig. 6. This is caused by the multipath error. Although the multipath effect can be reduced by placing the antenna position, it is still difficult to eliminate.

The root mean square (RMS) of those positioning errors results are shown in Table 2. The accuracies of the positioning based on the code measurement are about 0.1876 m, 0.2128 m, 0.8749 m, and 1.3391 m in the east-west, north-south, and vertical components and synthetic, respectively, while the corresponding results based on carrier measurement are about 0.0867 m, 0.1462 m, 0.1150 m, and 0.2058 m. The standard deviations of the positioning results are shown in Table 3.

## 5 Conclusion

CAPS is a regional navigation system based on commercial communication satellite transfer mode which is proposed and built by Chinese Academy of Sciences. It operates on C-band, and it is a technical experiment and validation platform. According

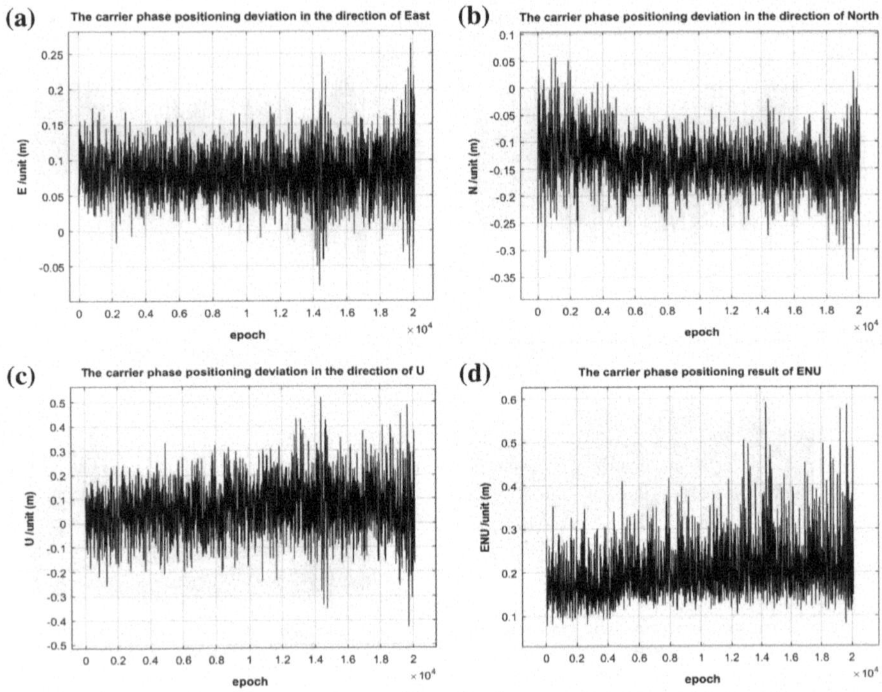


**Fig. 7** Positioning error in the east (**a**), north (**b**), and vertical (**c**) components and synthetic ENU (**d**) for the CAPS static positioning based on the carrier measurement

**Table 2** RMS of the positioning errors

| RMS/m | East-West | North-South | Vertical | Synthetic |
|---|---|---|---|---|
| Code measurement | 0.1876 | 0.2128 | 0.8749 | 1.3391 |
| Carrier measurement | 0.0867 | 0.1462 | 0.1150 | 0.2058 |

**Table 3** Standard deviation of the positioning errors

| STD/m | East-West | North-South | Vertical | Synthetic |
|---|---|---|---|---|
| Code measurement | 0.1741 | 0.1848 | 0.8317 | 0.4452 |
| Carrier measurement | 0.0306 | 0.0440 | 0.0971 | 0.0565 |

to the principle, verification tests, and the analysis, the following conclusions can be drawn:

(1) Based on the architecture and the principle of the CAPS, this novel system has obvious advantages include: (a) higher precision because of C-band; (b) no need of onboard atomic clock; (c) short construction period; and (d) low investment.

(2) The precision of CAPS single-satellite timing is about 0.3028 ns and 0.1566 ns based on the code measurement and the carrier measurement, respectively.
(3) The accuracy of CAPS static positioning based on the code measurement and the carrier measurement is about 1.34 m and 0.2 m, respectively. The static positioning precision of CAPS is higher than GPS. The positioning error in the east-west component is smaller because of the distribution of the constellation.

CAPS also has a number of disadvantages. For example, the PDOP of CAPS is not good as that of GNSS. The navigation signal is retransferred by satellite transponder which will destroy the continuity of carrier. The CAPS is still in development, and some of the disadvantages may be overcome. In addition, higher positioning accuracy can be achieved by improving the orbital accuracy or by using precise ephemeris for data post-processing.

**Acknowledgements** We thank the students and the staff in the Key Laboratory of Precision Navigation and Timing Technology at NTSC for their technical and tests assistance.

## References

1. Ai GX, Shi HL, Wu HT, Yan YH, Bian YJ, Hu YH et al (2008) A positioning system based on communication satellites and the Chinese area positioning system (caps). Chin J Astron Astrophys 8(6):611–630
2. Kaplan E, Hegarty C (2012) Understanding GPS. Principles and applications. Artech House 2006
3. Jiaojiao Z, Zishen L, Jian G, Liang W, Ningbo W, Kai Z et al (2018) The first result of relative positioning and velocity estimation based on caps. Sensors 18(5):1528
4. Li BH, Andrew D (2010) China's regional navigation satellite system-CAPS. Inside GNSS, pp 59–63
5. The Virtual satellite atomic clock method in CAPS. Patent No. 201610012134.0.2006, 5 December 2007
6. Xiaohui D, Huli S, Lirong Z, Cheng L (2012) Study of a forwarding approach of satellite time service. Astron Res Technol 75(1):69–78
7. Ai GX, Shi HL, Wu HT, Li ZG, Guo J (2009) The principle of the positioning system based on communication satellites. Sci China Ser G (Phys Mech Astron) 52(3):472–488
8. Fen C, Xuhai Y, Mudan S, Zhigang L, Liang C, Weichao L et al (2014) Evaluation of c-band precise orbit determination of geostationary earth orbit satellites based on the Chinese area positioning system. J Navig 67(02):343–351

# Data-Augmented Design of Turbulence Modeling

**Yizhi Zhang and Weipeng Li**

**Abstract** As for the high Reynolds flow, aerodynamic design is mostly based on Reynolds-Averaged Navier–Stokes equations (RANS). As introducing the ensemble average hypothesis, the accuracy of RANS equation is widely doubted in predicting the transition and flow separation, for example, the laminar separation bubble or stalls under high angle of attack. This article takes an airfoil as an example and conducts a research on the data-augmented turbulence modeling design. Based on the high-fidelity prior data from experiment, a spatially varying term which will act as a multiplier of the viscous production term in Spalart–Allmaras model equation can be constructed using primal N-S flow and adjoint flow. In order to handle the issue of the extreme high dimension of this optimization problem (which is close to the number of grids), an adjoint method is used to solve the derivatives efficiently. The posterior result states that using a data-augmented turbulence modeling could predict the flow characteristics more accuracy, which can let the prediction of aerodynamic parameters like lift and drag more precise.

**Keywords** Data augmented · Adjoint method · Turbulence model

## 1 Introduction

Computational fluid dynamics is widely used in physics and engineering problems such as aerodynamics and aerospace industry, weather simulation and environmental engineering. There are some of the computational methods such as direct numerical simulation (DNS) and large eddy simulation (LES) which can provide relatively accurate solution. However, their calculation cost is expensive and require a long period of time consuming. From the consideration of affordability, Reynolds-Averaged Navier–Stokes (RANS) equation is still the majority method in industrial

Y. Zhang · W. Li (✉)
School of Aeronautics and Astronautics, Shanghai Jiao Tong University, Shanghai, China
e-mail: liweipeng@sjtu.edu.cn

Y. Zhang
e-mail: Itszyz919@sina.com



Z. Jing (ed.), *Proceedings of the International Conference on Aerospace System Science and Engineering 2019*, Lecture Notes in Electrical Engineering 622,
https://doi.org/10.1007/978-981-15-1773-0_27

and academic flow solvers. Through the introducing of ensembled average assumptions, explicit expression [1] and partial differential equation [2] are introduced to solve the Reynolds stress. The turbulent models, such as Spalart–Allmaras and Menter k-omega SST, model the spatiotemporal multiscale turbulent structures and significantly improve the efficiency of calculation.

Meanwhile, the ability of RANS model to perform a high-fidelity computation is restricted due to the introduction of the spatiotemporal assumption and the constrains of model parameters. The restriction can be stated from two aspects, the first is that some of the fluid which included complex effects such as flow separation and large adverse pressure gradient cannot be accurately modeled [3]. The second point is that the non-dimensional parameters are setting based on a small set of canonical problems and are of less generality when solving flows under different condition or geometrics. In addition, during the application, the selection of turbulence models requires personal judgment which relies on designers' engineering experiences.

As for the basic research of turbulent fluid, Prandtl-con-Karman's log theories [4] describe the local approximate solution on logarithmic region, but failed to provide a whole prediction of velocity and kinetic energy profile. Schaefer et al. [5, 6] conduct a sensitivity analysis on the S-A model closure coefficients. It was stated that the von Kaman constant ($\kappa$) and the turbulent Prandtl number ($\sigma$) play an important role on the changing of quantities uncertainty while the primary sources of uncertainty come from the formation of the model.

Early in the twentieth century, Parneix [7] used DNS datasets to improve the terms in turbulence model such as the second-moment closure. They modified the equation to improve the model accuracy. Most of the previous work for model discrepancy improvement use data to calibrate the parameters in existing models such as S-A model. Recently, data-augmented methods are used to address the model discrepancies. Xiao et al. [8] computed the distribution of disturbance in the anisotropy Reynolds stress tensor based on DNS datasets. The disturbance to the anisotropic tensor is calculated with ensurance of realizability of the resulting perturbed stresses. The perturbations are reconstructed as a function of local flow variables.

Duraisamy [9] built a FIML approach that combined the flow inverse and machine learning. A specific objective function was used to quantificationally describe the model discrepancy, and a set of correction terms are embedded into the turbulence model to improve the model accuracy. The mapping relationship between flow field characteristics and correction terms is established by using neural network method, which can be used to improve the accuracy of turbulence model under different flow conditions.

Ling et al. [10] developed a machine learning classifier based on DNS and LES results to identify the turbulence in RANS models. They studied the flow field regions with significant deviations or irrational assumptions and compared the training effects on certainty of different machine learning algorithms, such as support vector machine, decision tree and random forest. The results show that in the flow with different conditions, and these classifiers can also identify the flow characteristics.

Wang et al. [11] state that the large deviation of Reynolds stress is the main reason that limits the prediction accuracy of the RANS model. It is important to identify these differences for improving the RANS model. Based on the averaged flow field characteristics of DNS data, a random forest is established to reconstruct the Reynolds stress tensor. The reconstruction is based on the fully developed turbulence in square pipe and the flow field with large flow separation. After learning, the RANS calculation can obtain more accurate results than the base turbulence model in the flow field with different geometric conditions.

In this work, we managed a more comprehensive approach toward the discrepancy improvement of RANS calculation on an airfoil RAE2822. Flow characteristics information is used to improve the accuracy of numerical calculation of RANS. The next section introduces the numerical approach, and the third part follows the method of constructing a data-augmented turbulence model which is based on the discrete adjoint method. The fourth section gives the result and analysis toward the discrepancy and correction.

## 2 Numerical Method

There are several widely used RANS turbulence models, such as the one-equation model Spalart–Allmaras, two-equation model k-omega SST and some other sophisticated models. The construction of turbulence models is based on dimensional analysis and provides the closure terms of high order fluctuations in N-S equations. These terms are based on the dimensionless empirical parameters.

The flow solver is settled on a cell-centered finite volume formulation of the compressible RANS equations on structured grids. The inviscid fluxes are discretized under the three-order MUSCL scheme combined with Steger-Warming and LUSGS time-stepping method is adopted.

A C-type mesh with 305 points in the wrap-around direction and 65 points in the wall-normal direction is applied. The total cell number is approximately 24,000. The first layer $y^+$ is set below unity and a grid verification test is carried out to make sure the calculation accuracy. In the boundary condition, the flow variables at the far field are set in the freestream condition and the eddy viscosity use the fully turbulent value at $\nu_{t,\infty}/\nu_\infty = 3$. Boussinesq assumptions are used to close the RANS equation.

The freestream flow conditions correspond to a Mach number of 0.729 and Reynolds number of 6.5 million based on the chord length of 1.0 m with an 2.31° angle of attack. The static pressure was computed based on the specified Reynolds number and Mach number and an assumed value of static temperature. The mesh is shown in Fig. 1.

The Spalart–Allmaras [12] is chosen as the turbulence model in this paper. It can be written as

$$\frac{D\hat{\nu}}{Dt} = P(\hat{\nu}, U) - D(\hat{\nu}, U) + T(\hat{\nu}, U)$$

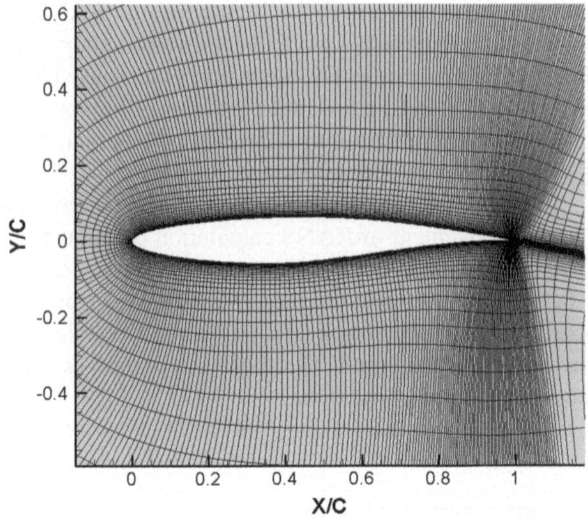


**Fig. 1** Mesh of airfoil Rae2822

In which, $U$ represents the Reynolds averaged flow variables and $P(\hat{\nu}, U)$, $D(\hat{\nu}, U)$ and $T(\hat{\nu}, U)$ are the production, destruction and transport terms respectively. They have the formation as,

$$P(\hat{\nu}, U) = C_{b1}(1 - f_{t2})\hat{S}\hat{\nu}$$

$$\hat{S} = \Omega + \frac{\hat{\nu}}{\kappa^2 d^2} f_{v2}$$

$$D(\hat{\nu}, U) = c_{w1} f_w \left(\frac{\hat{\nu}}{d}\right)^2$$

$$T(\hat{\nu}, U) = \frac{1}{\sigma}\left[\nabla \cdot \left((\nu + \hat{\nu})\nabla\hat{\nu}\right) + C_{b2}(\nabla\hat{\nu})^2\right]$$

Turbulence eddy viscosity can be gained through:

$$\nu_t = f_{v1}\hat{\nu}$$

To check the RANS solver, we perform a numerical calculation toward a channel flow under two different Reynolds number. Figure 2 shows the velocity profile of RANS calculation and DNS database [13]. Table 1 shows the comparison of solution from RANS and DNS, the relative error of velocity and friction coefficient is reasonable which is smaller than 2%.

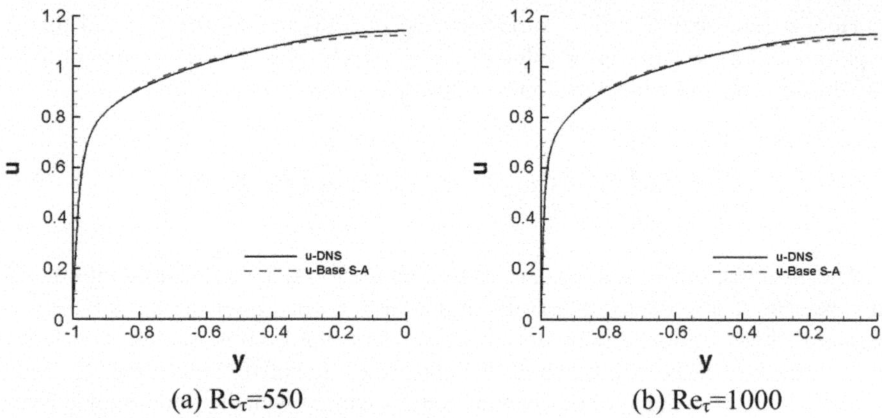


**Fig. 2** Comparison of u profile between DNS and S-A turbulence model

**Table 1** Verification of RANS solver by the relative error of friction coefficient

| $C_f$ $(\times 10^{-3})$ | DNS | S-A-base | Relative error (%) |
|---|---|---|---|
| $Re_\tau = 550$ | 5.8909349 | 5.9877831 | 1.64 |
| $Re_\tau = 1000$ | 5.0954817 | 5.1198286 | 0.48 |

# 3 Data-Augmented Modeling Based on Adjoint Method

The coefficients in the turbulence models are decided based on small set of prior flows, so it is hard to construct connection with those important physical characteristic parameters such as local Reynolds number and Mach number, which are changing with different flow conditions or geometry profiles. The major discrepancy comes from the turbulence model formation, rather than the option of coefficient. Thus, the conventional methods that adjusting the coefficient have limited effect toward the accuracy of calculation. In this work, a spatially varying factor $\beta(x)$ is added into the turbulent model equation as a multiply of production term. $x$ refers to each grid point in the domain. The modified model version can be written as,

$$\frac{D\hat{\nu}}{Dt} = \beta(x) \cdot P\left(\hat{\nu}, U\right) - D\left(\hat{\nu}, U\right) + T\left(\hat{\nu}, U\right) \tag{3.1}$$

The influence of factor $\beta(x)$ is global rather than for production term only, and it can be seen as adding a correction term $\delta(x)$ as $\delta(x) = (\beta(x) - 1) \cdot P\left(\hat{\nu}, U\right)$. Through the introducing of this factor, modified RANS equation can gain a higher accuracy solution. The objective function is stated as:

$$J = \min\left\{\sum_{j=1}^{N_c}\left(d_{j,\exp} - d_j\right)^2 + \sum_{j=1}^{N_{\text{cell}}}\left(\beta - \beta_{\text{prior}}\right)^2\right\} \tag{3.2}$$

Here $d$ represents the flow variables, and $d_{j,\exp}$ is the data from wind tunnel experiment [14]. $\beta$ has a prior value of unity. We choose pressure coefficient for $d$ and thus the objective function can be stated as,

$$J = \min\left\{ \sum_{j=1}^{N_{\text{surface}}} \left(Cp_{j,\exp} - Cp_j\right)^2 + \sum_{j=1}^{N_{\text{cell}}} \left(\beta - \beta_{\text{prior}}\right)^2 \right\} \tag{3.3}$$

The objective function is regularized by a regulation factor λ. It can be introduced to biases the $\beta$ solution to sit near the initial unity setting. Engineering judgment is required when setting the value of λ. A lower value of λ leads to over-fitting while a higher one may cause fitting failure. We choose a value of $10^{-8}$ to let the $\beta$ occupy a small order value in the objective function comparing with the flow variable part.

In order to minimize the objective function, we need to solve out the derivatives of objective function with respect to the design variables. The number of design variables is the same as the grid scale which is very large in an airfoil CFD solver and the conventional finite different method requires large computational resources. Due to the expansive computational cost, the adjoint method is applied to solve the gradient. The adjoint method is efficient and spends approximately only two times of the flow solver's time cost. In the process of solving, the amount of calculation is independent of the number of design variable and can greatly reduce the calculation time. The adjoint method is based on the control theory of partial differential equation system, and the flow control equation is set as constrains. The aerodynamic design problem is transformed into an optimization problem with specific constrains. In 1927, Jameson [15] first applied adjoint method to aerodynamic design. Since then, continuous adjoint method and discrete adjoint method have been developed. Considering the complexity of code, the discrete adjoint method [16] is adopted in this paper.

$$\left[\frac{\partial R}{\partial U}\right]^T \psi = \left[\frac{\partial J}{\partial U}\right]^T \tag{3.4}$$

$$\frac{dJ}{d\beta} = \frac{\partial J}{\partial \beta} - \psi^T \frac{\partial R}{\partial \beta} \tag{3.5}$$

Through constructing and solving of adjoint equation in Eq. (3.4), the derivative of the objective function with respect to the design variable $\beta(x)$ can then be obtained from Eq. (3.5). Then, the steepest descent method with fixed step is used to settle the optimization problem. The iteration process is terminated when the objective function attains a steady value. The final set of $\beta(x)$, which is the optimization solution can be embedded into the turbulence model to obtain more accurate calculation results. The coefficient matrix on the left side of the adjoint equation $\left[\frac{\partial R}{\partial U}\right]^T$ is a large sparse matrix and this sparse linear system of algebraic equations is solved using FGMRES routine, an iterative solver in Math Kernel Library which use the generalized minimal residual method (GMRES) [17].

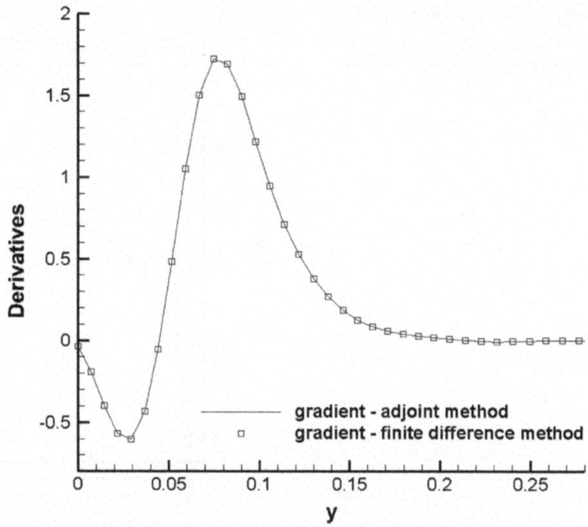


**Fig. 3** Verification of adjoint method with the comparison of finite difference method

Accuracy of the adjoint method is verified through a flat plate case. In this case, the coefficient $\beta(x)$ is added into the solver as the way we state. The finite difference method is used by introducing small disturbance to the design variables. The comparison of adjoint method and finite difference method can be seen in Fig. 3, and the relative error is quite reasonable.

## 4 Result and Analysis

The value of $\lambda$ in objective function is estimated by both the experimental error and the error in base solver and set at $10^{-8}$. The result of surface pressure coefficient is shown in Fig. 4. The modified S-A model has a higher accuracy in objective function. The u velocity field and pressure fields are shown in Figs. 5 and 6. The location of shock wave, which is approximately recognized by the flattening of the pressure curve, is closer to the experimental data. The lift coefficient, shown in Table 2, is improved as well that the absolute relative error decreases from 4.07 to 0.427%.

Figure 7 shows the field of correction factor $\beta(x)$, we could see that the regions with significant changes are concentrated near the surface of airfoil and the wake region. The region near the upper and lower airfoil surface is directly related to the pressure coefficient and also dominates the correction field. In the comparison of pressure coefficient, values at $X/C = 0.55$ see a notable change and this is consistent with the peak value in the distribution of correction term.

**Fig. 4** Surface pressure for the Rae2822 airfoil at Re = 6,500,000 and Ma = 0.729

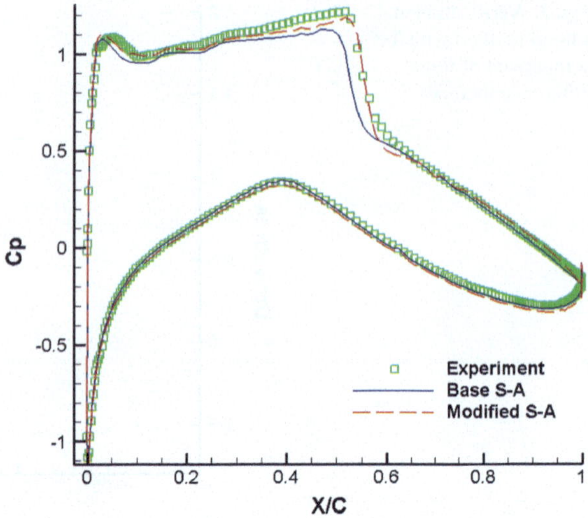


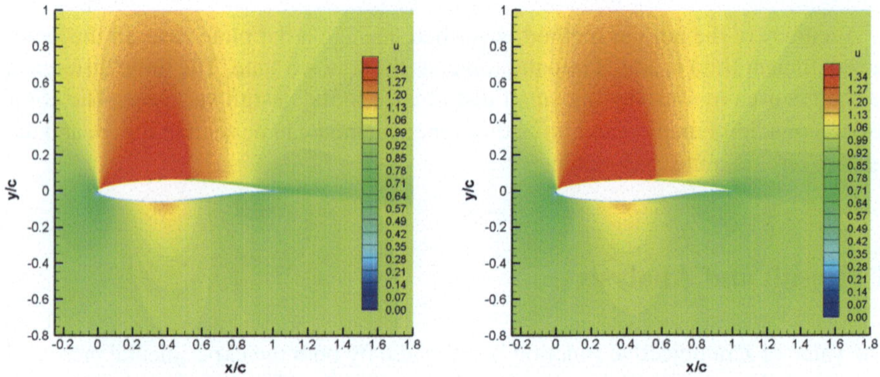


**Fig. 5** Comparison of velocity fields of airfoil

# 5 Conclusion

The present study proposed a generalized data-augmented turbulence model based on the discrete adjoint formulation which can eliminate the error of turbulence model by adding a correction factor to the production term in S-A model. After solving the adjoint equation and optimize the distribution of correction factor, the discrepancy of origin RANS solution and high-fidelity experiment result is corrected. In the process, elements of regulation factor $\lambda$ and step length affect the correction effect and are designed carefully. Appropriate value of regulation factor affects degree of fitting, and step length is related to the convergence speed and steadiness. In the result, the

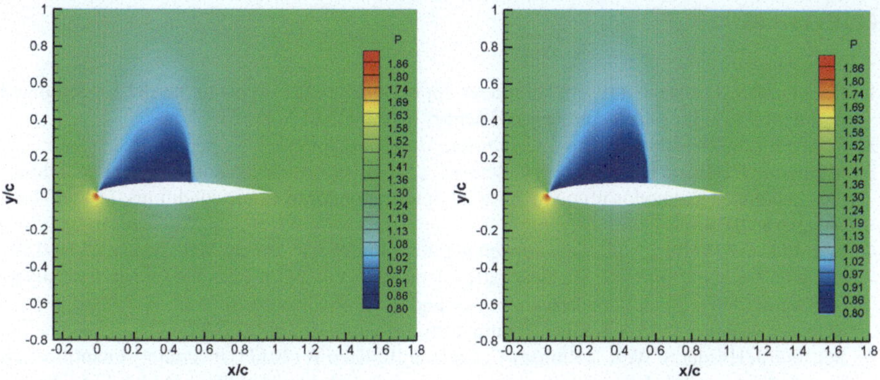


(a)Base S-A (b)Modified S-A

**Fig. 6** Comparison of pressure fields of airfoil

**Table 2** Comparison of lift coefficient

| | Experiment | Base S-A | Modified S-A |
|---|---|---|---|
| $C_L$ | 0.7309 | 0.70111274 | 0.727782 |
| Relative error | / | −4.07% | −0.427% |

**Fig. 7** Field of correction factor $\beta$

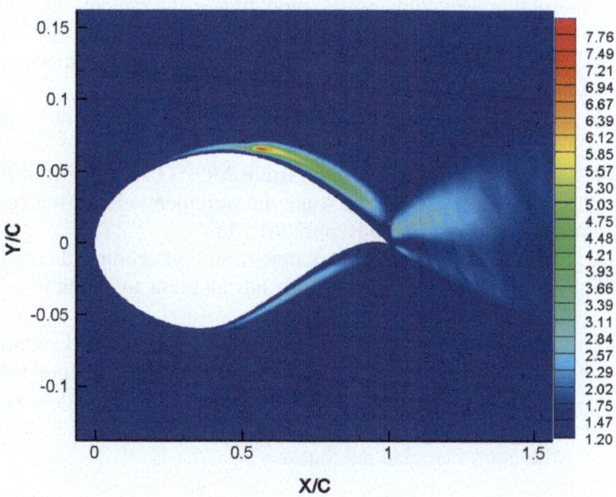


accuracy of lift coefficient and pressure coefficient distribution is improved, and the distribution of corrective factor is shown.

**Acknowledgements** Supported by project (National "973" program (2014CB744802) and National Natural Science Foundation (11772194)).

## References


1. Baldwin BS, Lomax H (1978) Thin-layer approximation and algebraic model for separated turbulent flows. In: AIAA Aerospace Sciences Meeting AIAA, pp 78–257
2. Pope SB (2000) Turbulent flows. Cambridge University Press
3. Mollicone J, Battista F, Gualtieri P, Casciola C (2018) Turbulence dynamics in separated flows: the generalised Kolmogorov equation for inhomogeneous anisotropic conditions. J Fluid Mech 841:1012–1039
4. Prandtl L (1925) Berichtüber die entstehung der turbulenz. Z Angew Math Mech 5:136–139
5. Schaefer J, Cary A, Mani M, Spalart P (2017) Uncertainty quantification and sensitivity analysis of SA turbulence model coefficients in two and three dimensions. In: 55th AIAA Aerospace Sciences Meeting, AIAA SciTech Forum, https://doi.org/10.2514/6.2017-1710
6. Schaefer J, Hosder S, West T, Rumsey C, Carlson JR, Kleb W (2017) Uncertainty quantification of turbulence model closure coefficients for transonic wall-bounded flows. AIAA J 55(1):195–213. https://doi.org/10.2514/1.J054902
7. Parneix S, Laurence D, Durbin P (1998) A procedure for using DNS databases. J fluids eng 120(1):40–47
8. Xiao H, Wu JL, Wang JX, Sun R, Roy CJ (2015) Quantifying and reducing model-form uncertainties in Reynolds-averaged Navier-Stokes simulations: a data-driven, physics-informed Bayesian approach. J Comput Phys 324:115
9. Singh AP, Duraisamy K (2016) Using field inversion to quantify functional errors in turbulence closures. Phys Fluids 28:045110
10. Ling J, Jones R, Templeton J (2016) Machine learning strategies for systems with invariance properties. J Comput Phys 318:22–35
11. Wang JX, Wu JL, Xiao H (2017) Physics informed machine learning approach for reconstructing Reynolds stress modeling discrepancies based on DNS data. Phys Rev Fluids 2:1–22
12. Spalart PR, Allmaras SR (1994) A one-equation turbulence model for aerodynamic flows. Rech Aerospatiale 1:5–21
13. Lee M, Moser RD (2015) Direct numerical simulation of turbulent channel flow up to Re_tau = 5200. J Fluid Mech 774:395–415
14. Cook PH, McDonald MA, Firmin MCP (1979) Aerofoil RAE 2822—pressure distributions, and boundary layer and wake measurements. Experimental data base for computer program assessment, AGARD Report AR 138
15. Jameson A (1988) Aerodynamic design via control theory. J Sci Comput 3(3):233–260
16. Giles M, Pierce N (2000) An introduction to the adjoint approach to design. Flow Turbul Combust 65(3–4):393–415
17. Saad Y, Schultz M (1986) GMRES: A generalized minimal residual algorithm for solving nonsymmetric linear systems. SIAM J Sci Stat Comput 7(3):856–869
18. Slotnich J, Khodadoust A, Alonso J et al (2014) CFD vision 2030 study: a path to revolutionary computational aerosciences, NASA/CR-2014-218178 [R]. Mchenry County Natural Hazards Mitigation Plan, United States

# Spacecraft Anomaly Detection via Transformer Reconstruction Error

**Hengyu Meng, Yuxuan Zhang, Yuanxiang Li and Honghua Zhao**

**Abstract** Anomaly detection refers to the problem of finding patterns in data that do not conform to expected behavior. Nowadays anomaly detection method deployed to production is based on reconstruction error generated by LSTM sequence modeling. Recently, the remarkable improvement achieved by BERT model in language translation demonstrated that the self-attention-based transformer is superior to LSTM models, due to its ignoring distance. In this paper, we continue the research on transformer and propose a transformer-based architecture, masked time series modeling, applying transformer in data stream, which has two novel components (1) the attention mechanism used for updating timestep in parallel and (2) the mask strategy used to detect the anomaly in advanced time. We compared the performances of our method with state-of-the-art AD methods on challenging public NASA telemetry dataset. The experiment results demonstrated our method saves about 80% time cost because of parallel computing compared with LSTM methods and achieves 0.78 F1 point-based score, moreover achieving a better score on range-based indicators.

**Keywords** Spacecraft · Anomaly detection · Transformer · Mask · Deep learning · Attention mechanism

---

H. Meng · Y. Zhang · Y. Li (✉)
School of Aeronautics and Astronautics, Shanghai Jiao Tong University, Shanghai, China
e-mail: yuanxli@sjtu.edu.cn

H. Meng
e-mail: lulvhp@sjtu.edu.cn

Y. Zhang
e-mail: yuxuanzhang@sjtu.edu.cn

H. Zhao
Eastern Airlines Technic Co, China Eastern Airlines, Shanghai, China
e-mail: hhzhao@ceair.com

 
Z. Jing (ed.), *Proceedings of the International Conference on Aerospace System Science and Engineering 2019*, Lecture Notes in Electrical Engineering 622,
https://doi.org/10.1007/978-981-15-1773-0_28

# 1 Introduction

Anomaly detection (AD) is the process of identifying non-conforming items, events, or behaviors [1, 2]. Efficient detection of anomalies can be useful in many fields. Examples include quantitative transaction, threat detection for cyber-attacks [3, 4], or safety analysis for self-driving cars [5]. Many real-world anomalies can be detected due to promotion of various industrial sensors. Especially for in-orbit spacecraft, failure to detect hazards could cause serious or even irreparable damage since spacecrafts are expensive and complex system. In the absence of remedial measures, anomaly detection is important and necessary to warn operation engineer of anomalies.

Current anomaly detection methods for spacecraft telemetry primarily can be grouped into two types: expert-based and reconstruction-based. Expert-based: Experts are required to define appropriate ranges for various monitoring indicators or to provide clear definitions of anomalies. Clustering or density-based approaches has been implemented for a small number of spacecrafts [6, 7]. However, as systems become more complex, these approaches are expensive and cannot handle increasing and complex anomalies.

Reconstruction-based anomaly detection is the most popular one and has been deployed into spacecraft [8, 9]. The main ideas of reconstruction-based anomaly detection methods are as follows: (1) What the "normal" sequence should look like, which means reconstructing sequence via RNN models trained by normal sequences. (2) Use the same model to reconstruct the sequences with anomalies and compare the reconstructed sequence with the input. (3) Set the error function and threshold. Where an abnormality occurs in the entire sequence, the reconstruction may be unideal, so classification or clustering can segment the anomalies [7, 10].

However, there exist two problems on reconstruction-based anomaly detection: (1) LSTM depends on uni-directional sequential propagation, which makes the computation of LSTM low-effective, and (2) one-direction propagation will delay the anomaly detection time due to the sparse unexpected data in the early observed anomaly subsequences.

**Contributions**: In this paper, we propose a masked time series modeling method based on transformer, as shown in Fig. 1, which has two novel components (1) the attention mechanism used for updating timestep in parallel, (2) the mask strategy used to detect the anomaly in advanced time. In this way, the reconstruction in the front of anomalies is able to be affected by the anomaly data, resulting in anomaly early detection. Once model reconstruction is generated, we apply a dynamic thresholding approach for evaluating reconstruction error. Experiments show that due to the characteristic of transformer encoder, the modeling greatly reduces time consumption without significant drop in accuracy compared with LSTM reconstruction. And using bidirectional data makes model to capture anomalies better for range-based precision and recall indicator. This work is tested on NASA spacecraft datasets, but can be applied to other anomaly detections task.

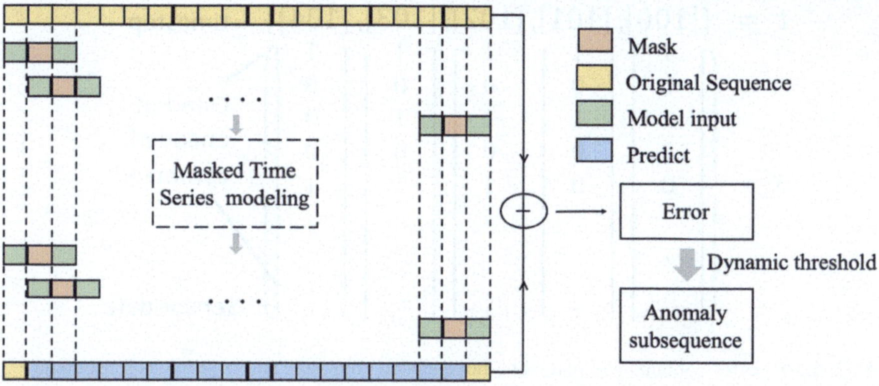


**Fig. 1** Framework of masked time series modeling. We intercept subsequences from the original sequence. The front and back parts of the subsequence are used as input to the model, and the middle part of the subsequence is output of reconstruct model. Detect anomalies based on the error between reconstructed sequence and the original

The structure of the article is as follows: Sect. 2 introduces the related work, the anomaly detection based on LSTM reconstruction and the transformer encoder generally used for NLP tasks. Section 3 presents our method, showing the inputs using contextual information, and gives the model reconstruction process. In Sect. 4, experiments were carried out on the NASA spacecraft dataset. The article compared the relationship between model consumption time and detection accuracy, and compared it with the previous methods on point-based and range-based indicators. Section 5 summarizes the full paper, puts forward the defects and shortcomings, and looks forward to the future work.

## 2 Related Work

Quite part of anomalies is caused by improper operations, especially in complex and stressful work of aerospace. Therefore, when it comes to abnormality detection, not only the sensor signal but also the operator's operation command should be considered. Since the one-hot coding representation (shown in Fig. 2) of the operation command in the time dimension is discrete, which is similar to the discrete word vector in natural language processing (NLP), methods proven effective in NLP such as transformer can be considered.

**Anomaly Detection Based on LSTM Reconstruction**

Figure 3 visually shows the anomaly detection based on LSTM reconstruction [11]. The researchers trained the reconstructed model using normal sequences. When a model reconstructs a sequence with anomalies (blue lines in the first column), there is a difference between the input and output (green lines in the second column)

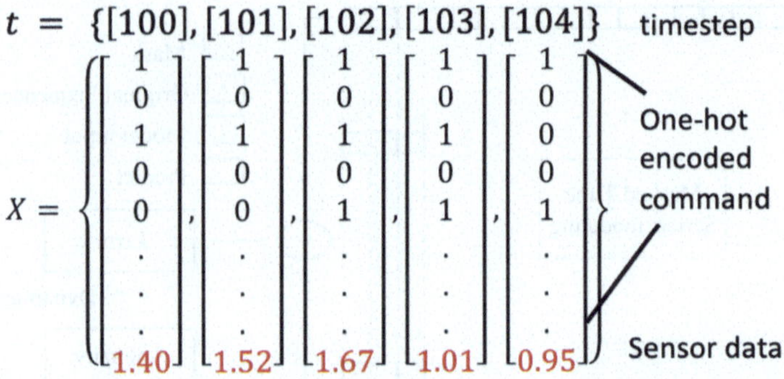


**Fig. 2** Each vector contains two parts: one-hot encoded command and continuous sensor data

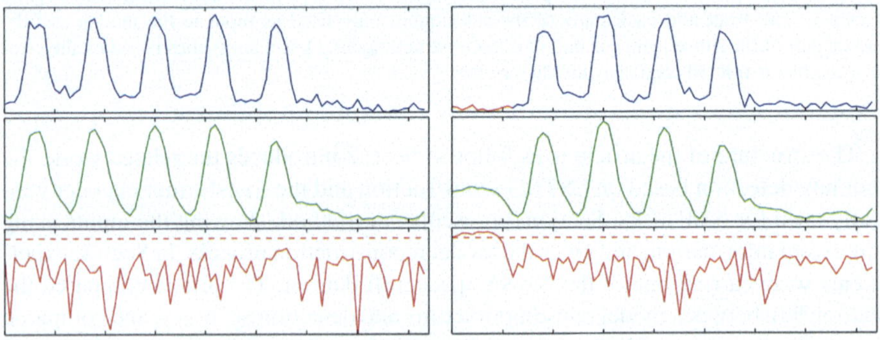

**Fig. 3** Examples of anomalies via reconstruction

sequences. Anomalous subsequences can be segmented applying a simple threshold segmentation or probability distribution model on reconstructed error (red line in the third column). For some anomaly detection studies, the size limit of the dataset is an inevitable topic. For other reconstruction tasks, such as GAN-based image generation, at least thousands of images are often required. However, some anomaly detection datasets often have only tens of hundreds of data, some specific anomalies even only have a little labeled data. And as the sampling density increases, length of the sequence also increases, making it increasingly difficult to generate the entire sequence.

Some anomaly detection researchers use time series local reconstruction to train models. In details, researchers intercept fixed-length subsequences from the time series data, using the front part as the input of the time series model and the tail part as the prediction object. This process is shown in Fig. 4.

Currently, LSTM (Long short-term memory) reconstruction error-based system instead of costly expert system has successfully identified several confirmed anomalies since deployed to the Soil Moisture Active Passive satellite (SMAP) and the Mars Science Laboratory rover (MSL), Curiosity.

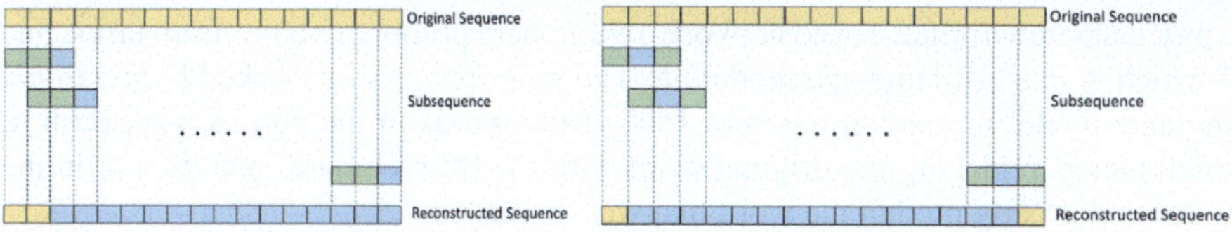


**Fig. 4** Fixed-length (e.g., 1000) sequence can intercept into 800 subsequences, each of which has a length of 200. Then, the model can use the previous 180 timesteps as input and the tail 20 as output

## Transformer Encoder

Self-attention is a special attention mechanism. In self-attention, query is equal to key equal to value.

Transformer, a self-attention mechanism that learns contextual relations between words (or sub-words) in a text, was proposed by [12]. As shown in Fig. 5, transformer includes two separate mechanisms—an encoder that reads the text input and a decoder that produces a prediction for the task. Since transformer-based BERT model has dominated effect in language modeling task, which shows transformer's strong ability to extract features.

As opposed to directional models, which read the text input sequentially (left-to-right or right-to-left) [11], the transformer encoder reads the entire sequence of words at once. Therefore, it can be considered bidirectional, though it would be more accurate to say that it is non-directional. This characteristic allows the model to learn the meaning of a word based on its surroundings (left and right of the word).

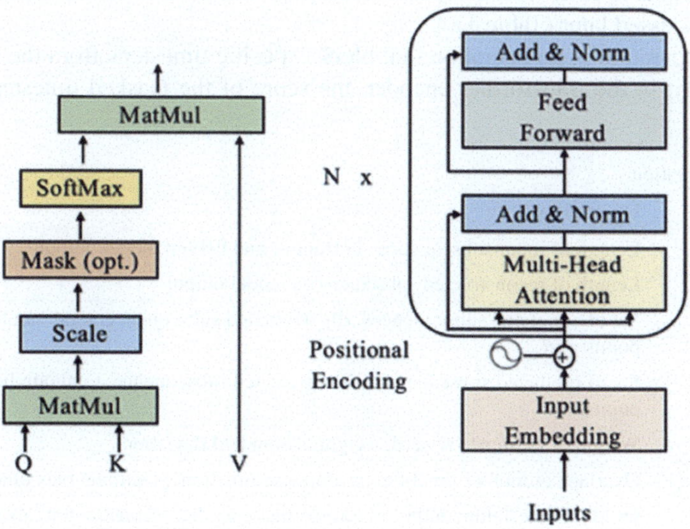


**Fig. 5** Dot-product attention (left) and transformer encoder (right)

For deep transformer-based network, researchers propose two pre-train tasks, one of which is masked language modeling. The main process of masked bidirectional language model is: randomly select 15% of the words in the corpus, and mask it, which means replacing the original word with the [Mask] mask, and then train the model to correctly predict the word that was discarded.

## 3 Method: Masked Time Series Modeling

In order to solve the shortcomings of the LSTM network, transformer encoder is applied to the spacecraft time series data modeling. Inspired by the masked language model, we propose a method using masked state prediction to reconstruct the time series.

### Model Input and Mask

Since LSTM can only propagate in one direction, the reconstruction process is actually more similar to prediction: predicting current data from historical data and not using future data. In fact, from other time series tasks, the comprehensive use of information in both directions will greatly enhance the capabilities of the model.

The input of masked time series model is intercepted from sequences like Fig. 4. The subsequence of this model input includes past and future information compared to the input of the LSTM model. In other words, in LSTM-based reconstruction, $L_{i0}$ is the length of whole input, and $L_{i1}$ is 0. In our method, the two are greater than 0(usually equal to each other). So, there exist problems about online detection which will be discussed later (Table 1).

Mask operation is an operation that blocks specific timesteps from the network. Specifically, in the transformer encoder, the score of the masked timesteps in the

**Table 1** Notation

| Notation | Description |
|---|---|
| $L_{i0}, L_{i1}$ | Length of input subsequences in front of and behind model output |
| $L_o$ | Length of reconstructed subsequences/model output |
| $R, R_i, N_r$ | Set of annotated subsequences, $i$th annotated subsequence, total number of points in $R$ |
| $P, P_j, N_p$ | Set of predicted subsequences, $i$th predicted subsequence, total number of points in $P$ |
| $\alpha, z$ | Weight of existence reward, weight of standard deviation |
| $\gamma(), \omega(), \delta()$ | Overlap cardinality function, overlap size function, positional bias function |
| $w_i, O_i, w_i'$ | Original timestamp, output of transformer encoder, reconstructed timestamp |
| $e, e_i, \varepsilon, \varepsilon_i$ | Error sequence and its $i$th timestamp; Threshold sequence and its $i$th timestamps |

relation matrix is set to infinity so that it can only accept the reconstruction via the remaining timesteps.

This article discusses the location and length for the mask operation. In fact, there are three variables that determine the mask, which are the length of time series before the mask, the length of time series after the mask, and the length of the mask itself. When the post length is 0, the model degenerates to be the same as LSTM, and only uses uni-directional information for prediction. Different mask lengths also have an impact on accuracy as shown in Sect. IV.

**Reconstruction Model Structure**

The model structure of masked time series modeling is shown in Fig. 6. The input to the model is the normalized raw signal $w_i$. After masking the middle part which will be reconstructed, the position encoding layer is added to avoid self-attention mechanism ignores the position information. Next is the transformer encoder. After obtaining the features $O_i$ extracted by the transformer encoder, the fully connected layer is used to obtain the final prediction of the masked state $w_i'$.

The model itself is similar to the masked language modeling. Here in this task, the input and output are continuous vectors. So we use regression instead of classification in the top layers. Moreover, researchers use multi-head mechanism to map word

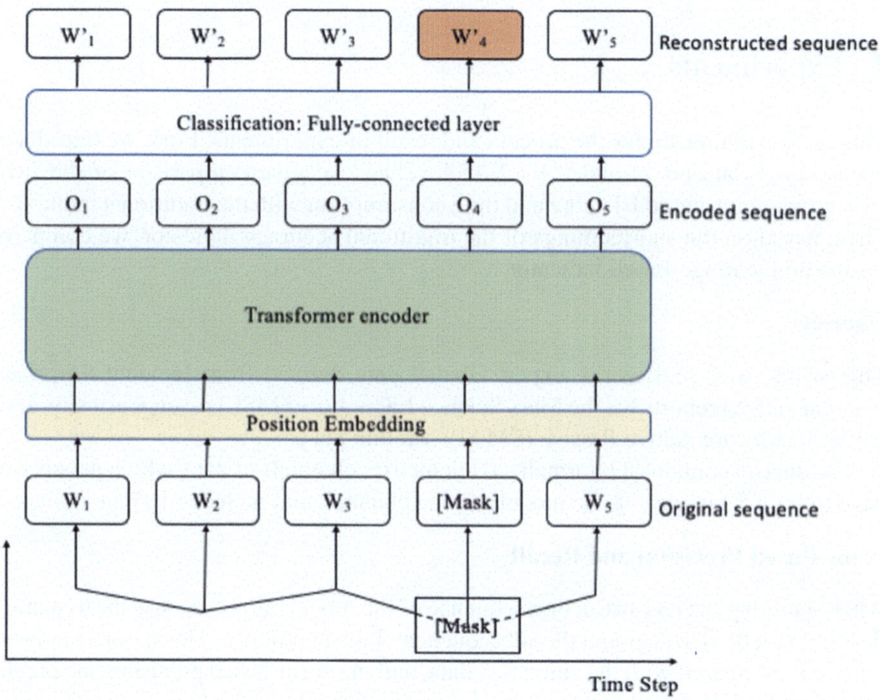


**Fig. 6** Process of masked time series reconstruction

vectors to different subspaces for parallel computation. But in the dataset we use, the signal has only 25 or 55 dimensions, so using too many heads makes no difference. The experiments also verify this.

**Anomaly Detection via Non-dynamic Threshold** [11]

After obtaining the reconstructed sequence $w_i'$, we can calculate the reconstruction error $e_i$ compared with the original sequence. Then, we segment anomalies via dynamic threshold $\boldsymbol{\varepsilon}$ which is defined as: $\varepsilon = \mu(e_i) + z\sigma(e_i)$

where $\varepsilon$ is determined by:

$$\varepsilon_i = \text{argmax}(\varepsilon) = \frac{\Delta\mu(e_i)/\mu(e_i) + \Delta\sigma(e_i)/\sigma(e_i)}{|e_a| + |P_j|^2}$$

Such that ($\mu$ is expectation and $\sigma$ is standard deviation):

$$\Delta\mu(e_i) = \mu(e_i) - \mu(\{e_i \in e_i | e_i < \varepsilon\})$$
$$\Delta\sigma(e_i) = \sigma(e_i) - \sigma(\{e_i \in e_i | e_i < \varepsilon\})$$
$$e_a = \{e_i \in e_i | e_i < \varepsilon\} \quad P_j = \text{sequence of } e_a$$

## 4 Experiments

This section demonstrates the process and result of experiments. First, we introduce open-source datasets provided by NASA. Then, we give comparison of the two indicators: point-based F1 score and time consumption with the traditional methods. Then, based on the shortcomings of the traditional accuracy indicator, we compare results on the range-based indicator.

**Datasets**

This article uses real world, expert labeled data derived from Incident Surprise, Anomaly (ISA) reports for the Mars Science Laboratory (MSL) rover, Curiosity, and the Soil Moisture Active Passive (SMAP) satellite [11].

The dataset contains 82 normalized telemetries channels of data, which means we have to train 82 models—one model for one channel, and includes 105 anomalies.

**Point-Based Precision and Recall**

After obtaining the reconstruction sequence of the model output, we use the dynamic threshold method to segment the subsequence of the anomalies. The anomaly subsequence is compared with the annotated data, and the point-based precision and recall result are defined as follows:

- True positive is defined as for any $R_i$ in $R$, there exists $P_j$ that the intersection of $R_i$ and $P_j$ is not none. In other words, $R_i$ overlaps with $P_j$. If there exist several $P_j$ overlapping with $R_i$, true positive set only count once.
- False negative means when there is no $P_j$ having an intersection with $R_i$, this $R_i$ is ignored by our model so FN(false negative) + 1.
  False positive means if a $P_j$ don't overlap with any $R_i$, we count it as FP.

The effect of different hyperparameters on the experimental results is shown in Table 2.

We also explore the balance between accuracy and time performance and whether multi-head mechanism would affect results. Figure 7 shows our method that consumes less time during training. At the same time, for the point-based precision and recall indicators, our method reaches the almost same precision and recall as the state of the art. And multi-head mechanism makes no sense in our task since the dimensions of time series are not complicated enough.

**Range-Based Precision and Recall**

Tae-Young K [13] presents a new mathematical model called range-based precision and recall metrics to evaluate the accuracy of time series classification algorithms. They expand the well-known precision and recall metrics to measure ranges. Figure 8 visualizes motivation of range-based.

$$\text{Recall}(R_i, P) = \alpha \times \text{ExistenceReward}(R_i, P) + (1 - \alpha) \\ \times \text{OverlapReward}(R_i, P)$$

The definition of existence reward is the approximately same as defined in point-based while overlap reward makes the differences. Overlap reward depends on three functions $\gamma()$, $\omega()$, $\delta()$, each of which captures the cardinality, overlap range size, and position bias of overlap

**Table 2** Experiments results comparing different hyper parameters

| Serial number | 0 | 1 | 2 | 3 | 4 | 5 |
|---|---|---|---|---|---|---|
| Stack of transformer | 6 | 6 | 8 | 12 | 12 | LSTM-based |
| Length of front input | 100 | 200 | 100 | 100 | 100 | 200 |
| Length of post input | 100 | 0 | 100 | 100 | 100 | 0 |
| Length of output | 5 | 5 | 5 | 5 | 1 | 1 |
| Precision | 64.9% | 62.0% | 68.7% | 78.3% | 85.4% | 87.5% |
| Recall | 62.9% | 59.0% | 64.8% | 68.6% | 72.4% | 80% |
| Time consumption(min) | 114.3 | 118.8 | 129.3 | 138.5 | 150.7 | 757.3 |

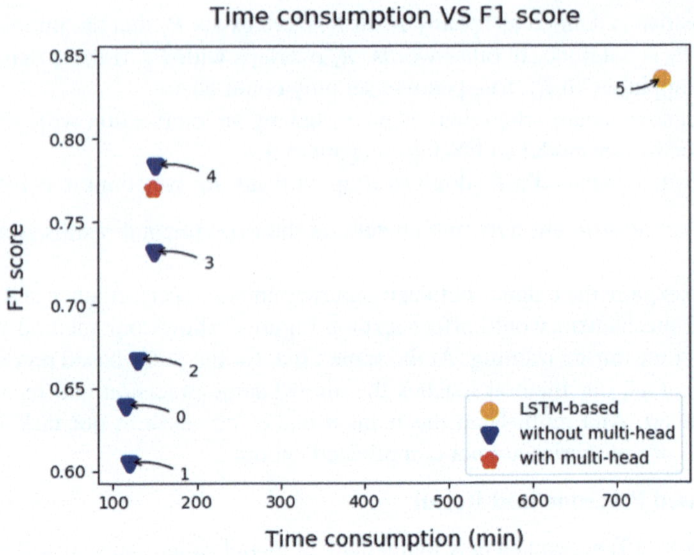


**Fig. 7** Time consumption vs accuracy. The annotation numbers near the scatters are the experimental serial number. And the red one shows transformer with multi-head do not work better

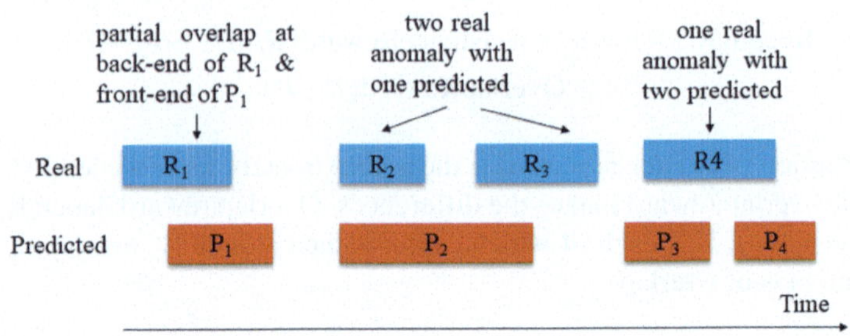


**Fig. 8** Three cases are all counted as true positive; however, the first case detects anomalies delayed from the true value, the second case two anomalies are identified as one, and the third case counts one as two. Physical meanings are different

$$\text{OverlapReward}(R_i, P) = \left[\sum_{j=1}^{N_p} \omega\big(R_i, R_i \cap P_j, P_j\big)\right] \times \text{CardinalityFactor}(R_i, P)$$

$$\text{CardinalityFactor}(R_i, P) = \begin{cases} 1 & \text{if } R_i \text{ overlap with at most one } P_j \\ \gamma(R_i, P) & \text{otherwise} \end{cases}$$

In this experiment,$\alpha$ is set to 0.5 because we want to pay equal attention to existence reward and overlap reward. $\gamma(.)$ is set to the reciprocal of the number of overlaps. $\delta(.)$ is set to front-end bias since we want to detect anomalies as early as possible.

**Table 3** Results evaluated on range-based indicator

| Method | Our method | LSTM-based |
|---|---|---|
| Precision | 0.64 | 0.62 |
| Recall | 0.50 | 0.49 |

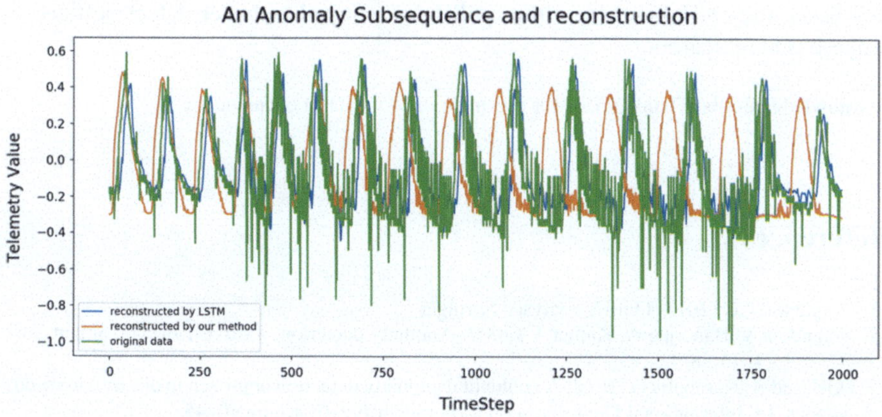


**Fig. 9** Comparison of reconstruction results. This is one of anomaly subsequences

And range-based precision only consists overlap reward. The parameters are set to the same in range-based recall.

The comparison results between LSTM-based and our method is listed in Table 3.

Even for point-range recall, our method is slightly worse than the baseline. But on range-based recall, our approach is slightly better. Figure 9 is a comparison of the reconstructed sequence of one of the channels. As shown in the figure, the sequence reconstructed by our method has smaller gradients and less fluctuations due to the use of bidirectional information. On the one hand, the sequence reconstructed by our method reflects the abnormality of the input signal more quickly; on the other hand, the smaller the fluctuation means that it is not easy to be recognized as two anomalies.

**Summary**: According to the experimental results, our method has greatly reduced the consumption time while the accuracy has not dropped significantly. At the same time, our approach has slightly improved the range-based indicator due to the use of contextual information.

## 5 Conclusion

LSTM-based anomaly detection method has been deployed into production now. This paper proposes a reconstruction algorithm for time series data anomaly detection—masked time series modeling. This algorithm is based on the transformer model and is significantly faster than RNN. At the same time, the advantages of using both front and back information make it better highlight the abnormal point. The experimental results verify our vision.

In the future, we will continue to explore how contextual information reacts to anomalies. In particular, visualize the relationship matrix in the self-attention mechanism to explore the relationship of anomalous points in the full sequence [14]. On this basis, the end-to-end algorithm will be developed to better detect outliers on range-based indicators.

**Acknowledgements** Thank NASA for making a so cool dataset open-source.

## References

1. Aggarwal CC (2013) Outlier analysis. Springer
2. Chandola V, Banerjee A, Kumar V (2009) Anomaly detection: a survey. ACM Comput Surv 41(3):15:1–15:58
3. AlEroud A, Karabatis G (2012) A contextual anomaly detection approach to discover zero-day attacks. In: International Conference on CyberSecurity (ICCS), pp 40–45
4. Chandola V, Mithal V, Kumar V (2008) Comparative evaluation of anomaly detection techniques for sequence data. In: IEEE International Conference on Data Mining (ICDM), pp 743–748
5. Anava O, Hazan E, Zeevi A (2015) Online time series prediction with missing data. In: International Conference on Machine Learning (ICML), pp 2191–2199
6. Zong B, Song Q, Min MR, Cheng W, Lumezanu C, Cho D, Chen H (2018) Deep autoencoding gaussian mixture model for unsupervised anomaly detection
7. Sölch M, Bayer J, Ludersdorfer M (2016) Variational inference for on-line anomaly detection in high-dimensional time series. arXiv preprint arXiv:1602.07109
8. Hundman K, Constantinou V, Laporte C, Colwell I, Soderstrom I (2018, July) Detecting spacecraft anomalies using LSTMs and nonparametric dynamic thresholding. In: Proceedings of the 24th ACM SIGKDD international conference on knowledge discovery & data mining, pp 387–395, ACM
9. Li FT, Chana D (2017) Multi-level anomaly detection in industrial control systems via package signatures and LSTM networks. In: 2017 47th annual IEEE/IFIP international conference on Dependable Systems and Networks (DSN), Denver, CO, pp 261–272
10. Mingyi Z, Kejiang Y, Yang W, Cheng-Zhong X (2018) A deep learning approach for network anomaly detection based on AMF-LSTM. In: 15th IFIP international conference on Network and Parallel Computing (NPC'2018), pp 137–141
11. Malhotra P, Vig L, Shroff G, Agarwal P (2015) Long short term memory networks for anomaly detection in time series. In: Proceedings of Presses universitaires de Louvain
12. Vaswani A, Shazeer N, Parmar N, Uszkoreit J, Jones L, Gomez A et al (2017) Attention is all you need. In Adv Neural Inf Process Syst 5998–6008
13. Tae-Young K, Cho S-B (2018) Web traffic anomaly detection using C-LSTM neural networks. Expert Syst Appl 106:66–76. Tatbul N, Lee TJ, Zdonik S, Alam M, Gottschlich J (2018) Precision and recall for time series. Neural Inf Process Syst (NIPS) 2018
14. Qi L, Rudy K, Chao C (2017) Unsupervised detection of contextual anomaly in remotely sensed data. Remote Sens Environ 202:75–87

# Altitude Ignition by Reducing IDT at Low Pressure with Additive

**Zhijia Chen, Xiaobin Huang and Hong Liu**

**Abstract** Altitude ignition is essential for the design of aero-engines, which is critical to aircraft flight safety. Various factors have been studied to stimulate the success of ignition. However, the study on the ignition at low pressure (less than 1 atm) is rare. At the altitude of 10,000 m, the typical pressure of combustor inlet lowers to 0.26 atm. Such low pressure leads to a significant increase in the ignition delay time (IDT) and may result in flameout and unsuccessful relight. Not only the range of flight envelope is limited, but also it is a hidden danger during flight. This paper is dedicated to investigate an additive to enhance the low-pressure ignition by reducing the IDT of the fuel. Particularly, methoxydiethylborane (MDEB) was chosen as the additive with *n*-decane as the surrogate fuel. The modified fuel with different concentrations all performed well in a hot surface experiment. The results show that the IDT is decreased from 1235 to 70 ms at 0.9 atm with 20% MDEB even with a lower ignition temperature of 623 K with respect to 923 K. The results on other pressures also validate the efficacy of the MDEB additive for reducing the IDT or even enhance the ignition success. Two kinetic parameters A and B were introduced to explain the reaction mechanism among different conditions. This work shows a novel paradigm to reduce the IDT at low pressure by adjusting the fuel combustion characteristic. Moreover, the IDT with different initial pressures can also be derived by the proposed predictor with boundary condition.

**Keywords** Low pressure · Fuel additive · *n*-decane · Ignition delay time · Correlation

Z. Chen · X. Huang (✉)
School of Aeronautics and Astronautics, Shanghai Jiao Tong University, Minhang, Shanghai, People's Republic of China
e-mail: xbhuang@sjtu.edu.cn

Z. Chen
e-mail: czj2018@sjtu.edu.cn

H. Liu
School of Aeronautics and Astronautics, Shanghai Jiao Tong University, 800 Dongchuan Road, Minhang 200240, Shanghai, People's Republic of China



Z. Jing (ed.), *Proceedings of the International Conference on Aerospace System Science and Engineering 2019*, Lecture Notes in Electrical Engineering 622,
https://doi.org/10.1007/978-981-15-1773-0_29

## 1 Introduction

Aero-engine plays a vital role in the performance of flight, as it is the power plant for almost all aircrafts. Thus, several basic requirements of combustor are listed although the relative importance varies between different engine types, e.g., reliable ignition and short ignition delay, not only on the ground but also at high altitude [1]. At high altitude, fuel may be unignited due to the narrowed ignition range [2], and the flight safety of aircraft will not be guaranteed [3]. The characteristic "time" is generally used to describe the limit of ignition, which associates with multiple physical and chemical processes such as evaporation, mixing and chemical reaction [4]. A countermeasure against high altitude ignition is to reduce the characteristic "time."

There are a lot of published researches on fuel preparation and operating conditions. Tomoaki et al. [5] found that the evaporation rate of droplet is decreased with the increasing of pressure, and nature convection will accelerate droplet evaporation more effective. Experiments were also conducted on different samples at atmospheric pressure to investigate the evaporation characteristics, which were used to verify the practicable of PME as an alternative fuel [6]. A new trapped vortex annular combustor was developed, in which the concept of the trapped vortex combustion is used in aero-engines design to improve the high altitude ignition [7]. Fuel will be introduced into the cavities, which will enhance the mixing of fuel and air to reduce the mixing time. These studies set out with the same aim of discovering the influencing factors of physical processes and minimizing the required characteristic times.

Among the combustion properties of fuel, ignition delay time makes a contribution to total ignition delay and is examined carefully to evaluating the combustor design [8]. Zhukov et al. [9] measured the IDT for mixture of Jet-A with air at pressures of 10 and 20 atm. The experimental data are fitted to an expression where the IDT has the quasi-Arrhenius dependence on pressure, temperature and equivalence ratio. Zhang et al. [10] studied the IDT of gas-phase RP-3/air mixture in a heated shock tube at a wide pressure range of 1–20 atm. And these data are in good agreement with those obtained from Jet-A. Mao et al. [11] carried out similar studies in a heated rapid compression machine and a heated shock tube, which provides further insight into the effect of pressure on IDT of RP-3. The IDT of *n*-decane which is used as jet fuel surrogates [12] was investigated both at a low-pressure range of 1.82–20 atm [13] and at a high-pressure range from 10 to 80 atm [14]. In these detailed studies, it conclusively shows that IDT is strongly associated with pressure. The value of IDT increased with the decline of pressure within the test situation.

Fuel additives offer an effective way of enhancing the ignition characteristic. It has been extensively reported that metal nanoparticles can be implemented to reduce the IDT due to the availability of high reactive surface area [15]. But the stability aspect of nanofluids and the limited experimental findings hindered the usage in commercial applications. In order to increase the sensitivity of JP-10 in air with high temperature and high pressure, some additives have been studied, which are methylated PCU alkene dimer, nitronorbornane, dinitronorbornane and ethyl–hexyl nitrate. But the

**Table 1** Properties of two reagents

| | Physical state | Density, g/cm3 @20 °C | Molecular weight, g/mol | Flash point, °C |
|---|---|---|---|---|
| *n*-Decane | Liquid | 0.735 | 142.28 | 46 |
| MDEB | Liquid | 0.868 | 99.97 | −6 |

results showed that the IDT of JP-10 did not lower obviously [16]. While some aluminum alkyls show a significant capability in reducing the IDT of JP-10 at an initial pressure approximate to 30 atm in a constant volume combustion bomb [17]. This rather surprising finding might provide a feasible method in reducing IDT of fuel by using metal alkyls as fuel additives under an extreme condition. However, rare studies have attempted to investigate the effect of additive acting in low pressure.

This study set out to explore the effect of fuel additive on IDT and proposed a predictor of IDT with different initial pressure. *n*-decane and MDEB were chosen as the surrogate fuel and fuel additive, respectively. The data obtained from these experiments show a well performance of this method at low pressure. A further exploration of the effect of low pressure on IDT is provided.

## 2 Experimental Methods

### *2.1 Modification of Fuel*

First, in this study, *n*-decane (CAS No: 124-18-5) with a purity of 99% was procured from Aladdin Industrial Corporation and MDEB (CAS No: 7397-46-8) with a purity of 97% was procured from Infinity Scientific (Beijing) Co. Ltd. The basic physical and chemical properties of these two reagents are listed in Table 1.

The formula of modified fuel was as follows: the volume ratios of *n*-decane to MDEB were 9:1, 8:2 and 7:3. The pure *n*-decane and modified fuel with different MDEB concentrations will be tested both in the atmosphere and in low pressure.

### *2.2 Experimental Setup*

Figure 1 shows an overview of the experiment device. It is observed that a hot surface is placed on the stainless chamber. The hot surface is made of copper and a rectangular dimension of 150 mm × 150 mm × 2 mm. The temperature of the hot surface is measured by a K-type thermocouple which is located below the hot surface and has an accuracy of ±2 K. The instantaneous temperature data are transmitted to the temperature controller. A PID pattern is used to keep the constancy of temperature. The temperature of hot surface may range from ambient temperature to 950 K. A vacuum

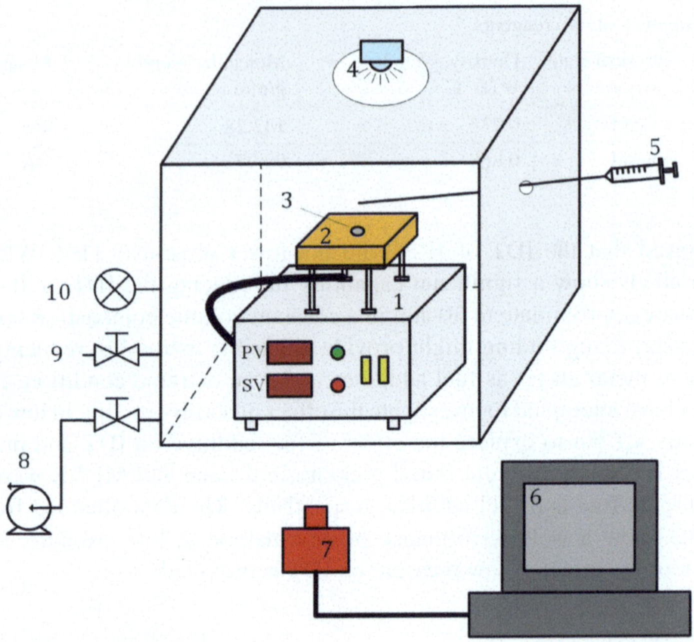


**Fig. 1** Scheme of the experimental device. (1) Temperature controller, (2) hot surface, (3) fuel drop, (4) LED light, (5) injector, (6) computer, sealed chamber, (7) high-speed camera, (8) vacuum pump, (9) gas inlet/outlet pipes and (10) pressure gauge

pump is used to regulate the pressure inside the stainless chamber. The pressure varies from 1 to 0.1 atm which is measured by a pressure gauge with an accuracy of 0.02 atm. In the center of the surface, a hemispherical groove, with a radius of 5 mm, was built for inhibiting the falling droplets from sliding off. The needle tube enters the chamber through an opening. The droplet is allowed to fall at a height of 15 mm. The ignition process is recorded by a high-speed camera with a sample rate of 500 pps. A typical process is followed during these experiments. The droplet is generated by a medical syringe and falls onto the groove. After a short delay, the droplet may be ignited and the whole process is recorded.

## 3 Results and Discussions

### 3.1 Derivation of the Arrhenius Expressions for IDT

Ignition will occur when the rate of chemical reaction suddenly increases sharply. While ignition phenomenon is of great relevant to the chemical process which contains many interwoven reactions among varies reactant. It is meaningful to build a simple model to describe the ignition behavior. By assuming that an equivalent aggregate reaction will substitute for complex combustion process, the gross parameters have a general form [18]:

$$\frac{\mathrm{d}(x)}{\mathrm{d}t} = \emptyset_1(P,\ T,\ t), \emptyset_2(F,\ \theta,\ \text{etc}) \tag{1}$$

where $(x)$ in mol/L, $t$ in millisecond, $P$ in atm and $T$ in K are concentration of pertinent reaction products, reaction time, pressure and temperature, respectively. $F$ is a proportional coefficient that represents the fuel–air ratio. $\Theta$ means some properties of different chemical compositions. And $\Phi_{12}\Phi_2$ are empirical functions.

For the ignition occurs in the gas phase, it is practicable to have the following expression [19]:

$$\frac{\mathrm{d}(x)}{\mathrm{d}t} = K_1 P^n \emptyset_2(F,\ \theta,\ \text{etc}) \tag{2}$$

where $K_1$ is a rate constant related to temperature, which can be determined either in a computational method or in an experimental method. The power $n$ varies from 0.5 to 2.5 in the literature, which represents the reaction order. Particularly, $n$ is a constant without unit. The unit of $K_1$ depends on the value of $n$ and has a general formula [20], $(\text{mol/L})^{1-n} \cdot \text{s}^{-1}$.

Then, the temperature is considered. Temperature influences the rates of chemical reactions almost always in terms of what is known as the Arrhenius equation [21]. In this equation, the rate constant $K_1$ can be calculated by multiplying the front factor $A$ and the exponential term. So, Eq. (2) can be written as:

$$\frac{\mathrm{d}(x)}{\mathrm{d}t} = A\mathrm{e}^{\left(-\frac{E}{RT}\right)} P^n \emptyset_2(F,\ \theta,\ \text{etc}) \tag{3}$$

where the ideal gas law constant $R$ is 8.314 $\text{J} \cdot \text{mol}^{-1} \cdot \text{K}^{-1}$, and $E$ is the activation energy in $\text{J} \cdot \text{mol}^{-1}$. Often $A$ and $E$ can be treated as temperature independent. Thus, $\frac{E}{R}$ can usually be rewritten as $B$.

When pressure and temperature are constant, kinds of species and equivalent ratio are also determined before the reaction. Time $t$ at which ignition occurs is obtained by a straightforward integration of Eq. (3) from the start of the reaction to ignition. It becomes [22]:

$$t = AP^{-n} \exp \frac{B}{T} \tag{4}$$

Equation (4) is the most common method of correlating ignition delay to temperature and pressure. Once the empirical constant $A$, $n$ and $B$ are determined, the IDT at a specific temperature and pressure can be predicted by this expression.

## ...ffect of Additive on Ignition Delay Time

...e, IDT is defined as the time interval from the droplet contacts with the surface of ...e hot plate to the occurrence of fire. To assess whether the selected additive works well in this system, the IDT of pure *n*-decane and modified fuel with different MDEB concentrations were measured. The ignition data measured at 1 atm are presented in Fig. 2. Particularly for pure *n*-decane, its IDT are 1094 and 1235 ms corresponding to the pressure of 1 atm and 0.9 atm at 873 K, respectively. And the pure *n*-decane was not ignited when the pressure reduced to 0.8 atm.

Since $P$ is a constant equal to 1 atm in this case, Eq (4) can be rewritten as:

$$t = A \exp \frac{B}{T} \tag{5}$$

The correlation between IDT and temperature is established in Fig. 2. Table 2 presents the value of parameters and correlation coefficients $R^2$ of these expressions. Although it shows a good fit to the Arrhenius expression in MDEB concentrates of 20 and 30%, the relevance coefficient is only 0.731 when the concentrate of MDEB is 10%.

It is noticed that the IDT increased when the temperature changed from 573 to 623 K. It might attribute to the NTC (negative temperature coefficient) phenomenon [23] which is related to the formation of alkylperoxy: $R + O_2 = ROO$. In this condition, a two-stage ignition must be employed, instead of the single-stage ignition

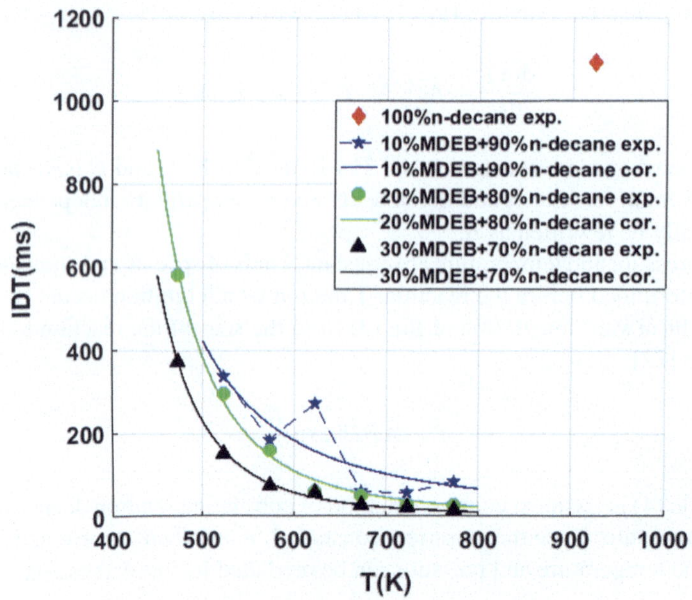


**Fig. 2** Ignition delay time versus temperature with different concentrations of MDEB

**Table 2** Correlation parameters for mixtures using $t = A\exp\left(\frac{B}{T}\right)$ at 1 atm

| | $A$ | $B$ | $R^2$ |
|---|---|---|---|
| 10%[a] | 3.0010 | 2471 | 0.7310 |
| 20% | 0.2212 | 3730 | 0.9943 |
| 30% | 0.0559 | 4160 | 0.9971 |

[a]concentration of MDEB

as we assumed before. Considering this situation and the convenience of operation, the modified fuel with an MDEB concentration of 20% was chosen for further study.

According to the experimental data, the expression of ignition delay time over a certain range of operating conditions is fitted by Eq. (4). With a correlation coefficient ($R^2$) of 0.99, the IDT can be calculated with the formula:

$$t = 0.0583P^{-1.522}\exp\frac{4408}{T} \tag{6}$$

where the power $n$ of $P$ equals to 1.522, which is consistent with the empirical range of 0.5 to 2.5 mentioned in the literature [20]. And this predictor may play an important role in verifying combustion reaction mechanism and chemical kinetics modeling. Furthermore, Fig. 3 depicts IDT as a function of temperature and pressure in a three-dimensional view. It may be helpful when the data match the engine conditions.

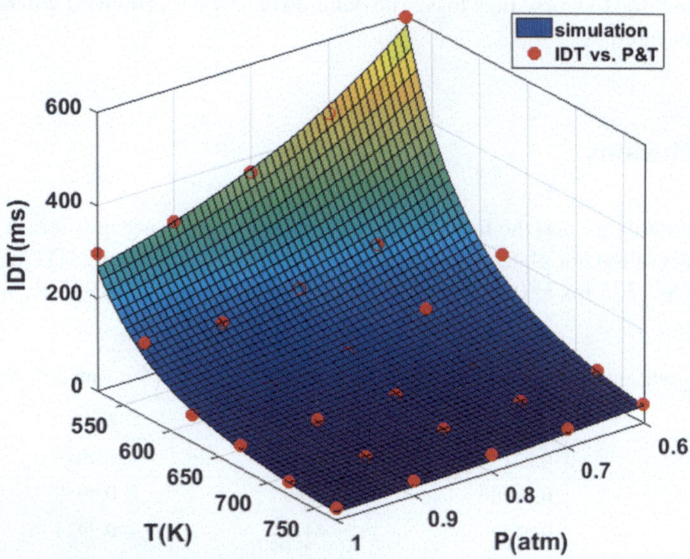


**Fig. 3** Three-dimensional view of the ignition delay time surface for 20% MDEB

order to get a better understanding of how the pressure affects the IDT, a series xperiment data of the IDT versus temperature at several different pressures were ıployed. Although a modified fuel with an MDEB concentration of 20% can be gnited at a lower pressure, only the case of 0.6 atm or more was studied due to the difficulty in realization of experimental operation. According to different pressure classifications, these data were also fitted to the form of Eq. (4), respectively. It should be emphasized that the value of n here is 1.522, which is calculated before. The parameter results from the fitted data are shown in Table 3. In these classes, all values show a good fit with correlation coefficients ($R^2$) higher than 0.99.

The correlation between IDT and pressure is interesting because the curves representing 1 and 0.9 atm are so close that the effect of pressure on the IDT in this range cannot be easily distinguished from Fig. 4. Turning now to Table 3, a tendency of kinetic parameters *A* and *B* to be associated with pressure was found. In fact, the physical meaning represented by *A* is the collision frequency and *B* is a parameter positively related to activation energy. The parameter *A* decreases as the pressure reduces, which may be due to the lower concentration of species at low pressures resulting in low collision frequencies. The parameter *B* increases with decreasing pressure, which indicates that the global activation energy increases when the pressure decreases.

Moreover, there is a significant difference in the case of 0.8 atm. It does not meet the trend mentioned above. Since this difference has not been reported in previous experimental studies, it is probably due to a coupling of gas-phase species accumulation and collision frequency. Although the low pressure causes a decrease in the frequency of collision of fuel with oxygen, it contributes to a thicker fuel profile in the gas phase due to worse heat loss. The detailed reaction mechanism will be studied in future research.

## 4 Conclusions

This study has examined the impact of pressure on low-pressure ignition. According to the results of the hot plate test, the pure *n*-decane will have a long IDT which may be as high as 1235 ms at 0.9 atm when the setting temperature is 923 K, while fuel

**Table 3** Correlation parameters for mixtures using $t = AP^{-1.522}\exp(\frac{B}{T})$ with 20% MDEB

| | *A* | *B* | $R^2$ |
|---|---|---|---|
| 1 atm | 0.2212 | 3730 | 0.9943 |
| 0.9 atm | 0.1828 | 3769 | 0.9958 |
| 0.8 atm | 0.0559 | 4419 | 0.9974 |
| 0.7 atm | 0.0818 | 4213 | 0.9960 |
| 0.6 atm | 0.0553 | 4447 | 0.9841 |

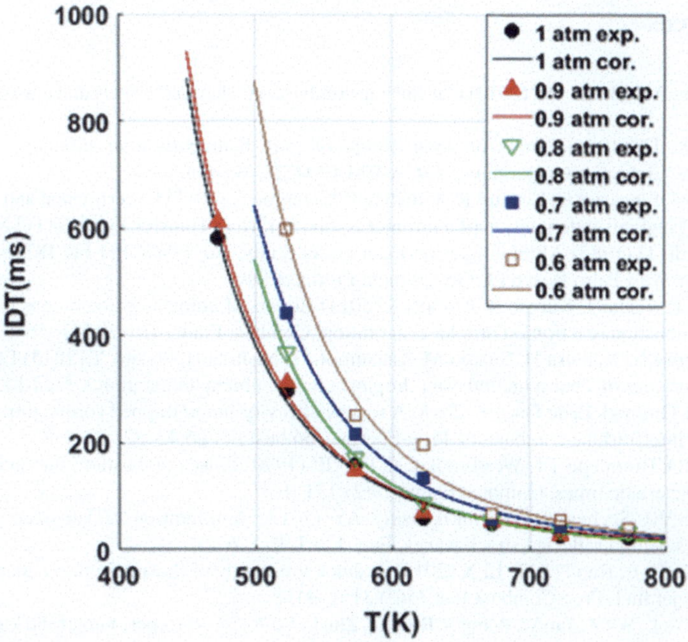


**Fig. 4** Ignition delay time versus temperature with different pressure

modified by MDEB shows a good ability in reducing ignition delay. Experiments have been carried out to prove that the ignition delay time will be more effectively reduced as the concentration of the additive increases. With 20% MDEB, all recorded IDTs are less than 700 ms with a wide pressure ranged from 0.6 to 1 atm and a lower temperature ranged from 473 to 773 K. These results suggest that MDEB is an effective additive in *n*-decane.

In order to find out the variation law of IDT and mechanism of effects of on ignition, a correlation expression was derived. As expected, IDT is positively correlated with temperature and pressure. Then, the change of IDT is attributed to the kinetic constants *A* and *B*. Since there are still some unverified assumptions and the expression of the ignition delay time is successfully constructed, a further study examining the detailed reaction kinetics would be necessary and practicable.

Overall this study confirmed that it is a feasible method for reducing IDT at low pressure with MDEB as the fuel additive. It provides a viable idea for improving the ignition performance at high altitude. It is believed that this method will contribute to widening the flight envelope through careful design.

**Acknowledgements** The authors gratefully appreciate the financial supports from the National Natural Science Foundation of China (No. 91441205).

## erences


1. Lefebvre AH, Ballal DR (2010) Gas turbine combustion: alternative fuels and emissions, CRC press
2. DiPiazza JT (1951) Limits of flammability of pure hydrocarbon-air mixtures at reduced pressures and room temperature. NACA-RM-E51C28, Washington
3. Rosa NG, Linassier G, Lecourt R, Villedieu P, Lavergne G (2011) Experimental and numerical study of high-altitude ignition of a turbojet combustor. Heat Transfer Eng 32(11–12):949–956
4. Andreadis D (1985) Altitude Ignition/Lean Decel Study. No. P/W/GPD-FR-18710. Pratt and Whitney West Palm Beach FL Government Products Div
5. Kitano T, Nishio J, Kurose R, Komori S (2014) Effects of ambient pressure, gas temperature and combustion reaction on droplet evaporation. Combust Flame 161(2):551–564
6. Hashimoto N, Nomura H, Suzuki M, Matsumoto T, Nishida H, Ozawa Y (2015) Evaporation characteristics of a palm methyl ester droplet at high ambient temperatures. Fuel 143:202–210
7. Zhao D, Gutmark E, de Goey P (2018) A review of cavity-based trapped vortex, ultra-compact, high-g, inter-turbine combustors. Prog Energy Combust Sci 66:42–82
8. Keller JO, Bramlette TT, Westbrook CK, Dec JE (1990) Pulse combustion: the quantification of characteristic times. Combust Flame 79(2):151–161
9. Zhukov VP, Sechenov VA, Starikovskiy AY (2014) Autoignition of kerosene (Jet-A)/air mixtures behind reflected shock waves. Fuel 126:169–176
10. Zhang C, Li B, Rao F, Li P, Li X (2015) A shock tube study of the autoignition characteristics of RP-3 jet fuel. Proc Combust Inst 35(3):3151–3158
11. Mao Y, Yu L, Wu Z, Tao W, Wang S, Ruan C, Zhu L, Lu X (2019) Experimental and kinetic modeling study of ignition characteristics of RP-3 kerosene over low-to-high temperature ranges in a heated rapid compression machine and a heated shock tube. Combust Flame 203:157–169
12. Pitz WJ, Mueller CJ (2011) Recent progress in the development of diesel surrogate fuels. Prog Energy Combust Sci 37(3):330–350
13. Olchanski E, Burcat A (2006) Decane oxidation in a shock tube. Int J Chem Kinet 38(12):703–713
14. Zhukov VP, Sechenov VA, Starikovskii AY (2008) Autoignition of *n*-decane at high pressure. Combust Flame 153(1–2):130–136
15. Saxena V, Kumar N, Saxena VK (2017) A comprehensive review on combustion and stability aspects of metal nanoparticles and its additive effect on diesel and biodiesel fuelled CI engine. Renew Sustain Energy Rev 70:563–588
16. Mikolaitis David W, Segal Corin, Chandy Abhilash (2003) Ignition delay for jet propellant 10/air and jet propellant 10/high-energy density fuel/air mixtures. J Propul Power 19(4):601–606
17. Ryan TW III, Schwab ST, Harlowe W (1995) Aluminum alkyl derivatives-Ignition and combustion enhancers for supersonic combustors. J Propul Power 11(1):124–129
18. Livengood JC, Wu PC (1955) Correlation of auto ignition phenomena in internal combustion engines and rapid compression machines. In: Symposium (international) on combustion 5(1): 347–356
19. Lefebvre AH, Freeman WG, Cowell LH (1986) Spontaneous ignition delay characteristics of hydrocarbon fuel-air mixtures. NASA-CR-175064, Indiana
20. Levenspiel O (1999) Chemical reaction engineering. Ind Eng Chem Res 38:4140–4143
21. Laidler KJ (1984) The development of the Arrhenius equation. J Chem Educ 61(6):494
22. Douaud AM, Eyzat P (1978) Four-octane-number method for predicting the anti-knock behavior of fuels and engines. SAE Trans 1978:294–308
23. Fu X, Aggarwal SK (2015) Two-stage ignition and NTC phenomenon in diesel engines. Fuel 144:188–196

# Author Index




Z. Jing (ed.), *Proceedings of the International Conference on Aerospace System Science and Engineering 2019*, Lecture Notes in Electrical Engineering 622,
https://doi.org/10.1007/978-981-15-1773-0